AF335192

GRAVITY AND THE LUNG

LUNG BIOLOGY IN HEALTH AND DISEASE

Executive Editor

Claude Lenfant
Director, National Heart, Lung and Blood Institute
National Institutes of Health
Bethesda, Maryland

ADDITIONAL VOLUMES IN PREPARATION

The opinions expressed in these volumes do not necessarily represent the views of the National Institutes of Health.

GRAVITY AND THE LUNG
LESSONS FROM MICROGRAVITY

Edited by

G. Kim Prisk

University of California, San Diego
La Jolla, California

Manuel Paiva

Université Libre de Bruxelles
Brussels, Belgium

John B. West

University of California, San Diego
La Jolla, California

MARCEL DEKKER, INC.   NEW YORK · BASEL

ISBN: 0-8247-0570-X

This book is printed on acid-free paper.

Headquarters
Marcel Dekker, Inc.
270 Madison Avenue, New York, NY 10016
tel: 212-696-9000; fax: 212-685-4540

Eastern Hemisphere Distribution
Marcel Dekker AG
Hutgasse 4, Postfach 812, CH-4001 Basel, Switzerland
tel: 41-61-261-8482; fax: 41-61-261-8896

World Wide Web
http://www.dekker.com

The publisher offers discounts on this book when ordered in bulk quantities. For more information, write to Special Sales/Professional Marketing at the headquarters address above.

Current printing (last digit):
10 9 8 7 6 5 4 3 2 1

PRINTED IN THE UNITED STATES OF AMERICA

INTRODUCTION

Weightlessness certainly is not a new phenomenon. Indeed, it was about 400 million years ago that some vertebrates transitioned from an aquatic to a terrestrial environment. An aquatic environment is a de facto weightless environment, since the surrounding water provides support to its inhabitants. In contrast, the terrestrial vertebrate must support itself against the forces of gravity. Adaptation to gravitational forces, in turn, leads to specific "structural" and functional changes.

However, all this occurred over millions of years and, of course, no one knew about it until the seventeenth century, when Galileo Galilei recognized gravity as a physical entity. That century has been termed the "Age of the Scientific Revolution," and during that time speculation was replaced by experimentation. Among all the giants of the times, Galileo ranks as possibly the greatest; he developed the mathematical basis of the laws of motion in the terrestrial environment.

Today we live in another age of scientific revolution. Although we can safely pinpoint its beginning—the middle of the twentieth century—there is no end in sight, nor can we define the limits of this revolution. Just think, from space to molecules and genes!

The International Space Station is being built and is becoming operational, and space may also become the "recreational" area of the future (at least for some). But we have to recognize that the full potential of the Space Station rests on how physiological systems built to operate under forces of gravity can function in the absence of these forces.

This is the subject of this monograph. The editors of *Gravity and the Lung: Lessons from Microgravity*, Drs. G. Kim Prisk, Manuel Paiva, and John B. West, and the contributors to this volume are all experts in their own fields and bring with them years of experience and research.

Over the years, and by way of its many contributors, the Lung Biology in Health and Disease series has focused on lung physiology and pathology in the most common, not to say usual, environment. The series has also explored known responses to one extreme environment, that is, the undersea world (Vol.

132). The current volume presents some aspects of human response to another extreme!

As the Executive Editor of this series of monographs, I am as proud to introduce this volume as I am grateful to the editors and the authors for having given me the opportunity to do do.

Claude Lenfant, M.D.
Bethesda, Maryland

PREFACE

Although humans have been traveling in space for 40 years (since the first flight of Yuri Gagarin in 1961), for the first 20 or so years only a few scientific studies were performed in space, and those studies were limited. Given the difficulties of placing and sustaining humans in that distant and hostile environment, this is hardly surprising. However, the first flight of the European-built Spacelab module (a laboratory placed in the cargo bay of the space shuttle) in 1983 provided the opportunity for well-designed and properly controlled experiments. The pressurized Spacelab module flew 14 times during the period from 1983 to 1998. Of those 14 flights, 11 carried a significant proportion of physiological experiments, and 3 of those were dedicated to life sciences research. Many of those flights carried payload specialists—professional scientists who were recruited because of their highly specialized knowledge and skills. Spacelab was decommissioned in 1998, following the Neurolab mission, which carried the most complex series of life sciences experiments performed in space. In the future, the International Space Station will provide the opportunity for more elaborate studies in microgravity.

Knowledge of environmental physiology comes from understanding how the organism reacts to changes in the environment. For example, thermoregulatory processes might be studied by raising or lowering ambient temperature. However, for subjects here on earth, gravity is a constant force that cannot be eliminated. We know that gravity has a large effect on the behavior between the top and bottom of the lung in an upright subject. There are substantial vertical gradients in alveolar size and ventilation, in perfusion, and in gas exchange. Therefore, we would expect to see large differences in the behavior of the lung in microgravity. In addition, because the lung presents such a large surface area to the environment (~ 50–$100 \ m^2$), it is vulnerable to polluted atmospheres such as might occur in the closed environment of a spacecraft. Yet none of these factors are unique to the microgravity environment. People routinely change the gravitational influence on their lungs every time they lie down, and many terrestrial dwellers are exposed to both indoor and outdoor atmospheric pollution. Thus, this volume focuses on using measurements made in the absence of gravity (or, in some cases, at high gravity levels) to understand how the lung is affected by gravity here on earth. The title *Gravity and the Lung: Lessons from Microgravity* reflects this approach.

We had the good fortune to be closely involved with several experiments on Spacelab, and its demise provides an ideal time to summarize those findings in the field of respiratory physiology before the transition of the Space Station. We are grateful to Claude Lenfant, Executive Editor, who agreed to place this volume in the Lung Biology in Health and Disease series.

It is always challenging to perform first-class experiments in a difficult environment. Performing studies in space presents numerous difficulties, but the result is exhilarating. It has been a privilege to participate.

G. Kim Prisk
Manuel Paiva
John B. West

CONTRIBUTORS

Jay C. Buckey, Jr., M.D.   Research Associate Professor, Department of Medicine, Dartmouth Medical School, Hanover, New Hampshire

Bruce D. Butler, Ph.D.   Professor and Vice Chairman of Research, Department of Anesthesiology, The University of Texas–Houston Medical School, Houston, Texas

Chantal Darquenne, Ph.D.   Assistant Research Physiologist, Department of Medicine, University of California, San Diego, La Jolla, California

Marc Estenne, M.D., Ph.D.   Professor of Medicine, Chest Service, Erasme University Hospital, Brussels, Belgium

David H. Glaister, O.St.J., Ph.D., M.B., B.S., F.F.O.M.   Group Captain (retired), Royal Air Force, Talywern Consultancy Ltd., Powys, United Kingdom

Robb W. Glenny, M.D.   Associate Professor, Department of Medicine and Department of Physiology and Biophysics, University of Washington, Seattle, Washington

Michael P. Hlastala, Ph.D.   Professor, Department of Physiology and Biophysics and Department of Medicine, University of Washington, Seattle, Washington

Dag Linnarsson, M.D., Ph.D.   Professor, Department of Physiology and Pharmacology, Karolinska Institutet, Stockholm, Sweden

Giuseppe Miserocchi, M.D.   Professor of Physiology and Chairman, Department of Experimental and Environmental Medicine, Università Milano-Bicocca, Monza, Italy

Daniela Negrini   Associate Professor, Department of Medicine, Surgery and Dentistry, Università Degli Studi di Milano, Milan, Italy

William T. Norfleet, M.D.   Principal Coordinating Scientist for Space Medicine, National Aeronautics and Space Administration, Houston, Texas

Manuel Paiva, Ph.D.   Professor, Biomedical Physics Laboratory, Université Libre de Bruxelles, Brussels, Belgium

G. Kim Prisk, Ph.D.   Professor, Department of Medicine, University of California, San Diego, La Jolla, California

H. Thomas Robertson, M.D.   Professor, Pulmonary and Critical Care Medicine, Department of Medicine and Department of Physiology and Biophysics, University of Washington, Seattle, Washington

John B. West, M.D., Ph.D.   Professor, Department of Medicine, University of California, San Diego, La Jolla, California

CONTENTS

Contents   xi

1

Historical Introduction

JOHN B. WEST

University of California, San Diego
La Jolla, California

I. Normal Gravity

A. Introduction

It could be argued that, of all the organs in the body, the lung is the most vulnerable to gravity, increased acceleration, and weightlessness. One of the reasons is that the blood in the pulmonary capillaries is separated from the air in the alveoli by an extremely thin blood–gas barrier over a vertical height of some 30 cm. Because blood has a much greater density than air, substantial pressure differences across the capillary walls therefore exist at different levels in the lung, and consequently there is a striking topographical inequality of blood flow. Another reason why gravity affects the lung is that the lung is very distensible and therefore it distorts under its own weight. Consequently, there are regional differences of alveolar expansion, mechanical stresses, intrapleural pressures, and ventilation. Finally, the fact that ventilation and blood flow do not match each other at different levels in the upright lung means that there are topographical differences of pulmonary gas exchange, and these can have important effects on overall gas exchange.

The normal gravitational field in which we live therefore causes marked differences of blood flow, ventilation, gas exchange, intrapleural pressure, alveolar expansion, and parenchymal stresses in the upright human lung. These will be the subject of this first section. Subsequently, we shall see that increased acceleration markedly exaggerates these topographical differences with correspondingly greater degrees of interference with normal lung function. Finally weightlessness, or microgravity, results in a more uniform distribution of these various aspects of lung function and, in some cases, improvements of gas exchange. However, recent measurements done in microgravity allow us to see that the lung has some intrinsic inequality of ventilation and blood flow.

B. Early Predictions

One of the first persons to recognize that gravity may have an important effect on pulmonary function was Johannes Orth (Fig. 1), a pathologist working in Göttingen in the 1880s. In 1887, he wrote a short treatise (1) entitled *Atiologisches und Anatomisches uber Lungenschwindsucht* (Etiological and Anatomical Considerations of Phthisis) in which he speculated on the cause of the apical localization of adult tuberculosis. He stated on page 20:

> First, anemia has to be considered. It occurs more readily at the apex than in other parts. Presumably the weight of the blood under normal conditions will not be of very great importance due to the small difference in height between the hilum and the apex of the lung, especially as this only exists in the erect posture. However I believe that if the total quantity of blood is reduced and there is incomplete filling of the vessels in the smaller circulation, the apex will be particularly affected, especially if the heart's action is reduced and the blood pressure is low. That an existing anemia as such contributes to a disposition for tuberculosis is suggested by the fact (which cannot be disregarded) that a relative very high percentage of those individuals with a stenosis of the lung arteries die from tuberculosis.

Note that he argued that gravity would not have great importance under normal conditions, but he was clearly aware of the possible influence of the weight of the blood in determining its distribution. It is interesting that he emphasized that the effects of gravity would be seen more clearly when the ''blood pressure is low'' (he presumably meant the pressure in the pulmonary circulation). This is precisely what is seen in animal preparations where the pulmonary artery pressure can be controlled and reduced (2). Orth also argued that pulmonary stenosis (which he knew increased the risk of pulmonary tuberculosis) would reduce blood flow to the apex of the lung, and subsequent measurements in patients with the tetralogy of Fallot have confirmed this (3).

Some 60 years later, the first measurements of right ventricular pressures in humans were reported by Cournand et al. (4) using the new technique of cardiac catheterization. William Dock (5) became aware of these data and argued that the pulmonary artery pressure might not be sufficient to raise blood to the top of the lung (Fig. 2). He reasoned that the reduced blood flow would impair the

Figure 1   Johannes Orth (1847–1923), professor of pathology in Göttingen in the 1880s. He argued that the weight of the blood would affect its distribution in the lung, and was therefore perhaps the first person to recognize that gravity has an effect on pulmonary function. (From Ref. 59.)

defenses of the lung to infection by the tubercle bacillus. Dock also made an additional remarkable prediction. He knew that the incidence of adult pulmonary tuberculosis was slightly greater in the right lung compared with the left, and therefore argued that the blood flow to the right apex would be less than that on the left side. Subsequent measurements with radioactive carbon dioxide confirmed this prediction in normal subjects (3), the reason presumably being the slight inclination of the main pulmonary artery trunk to the left, which therefore preferentially distributes blood flow to the left lung. Dock also referred to the

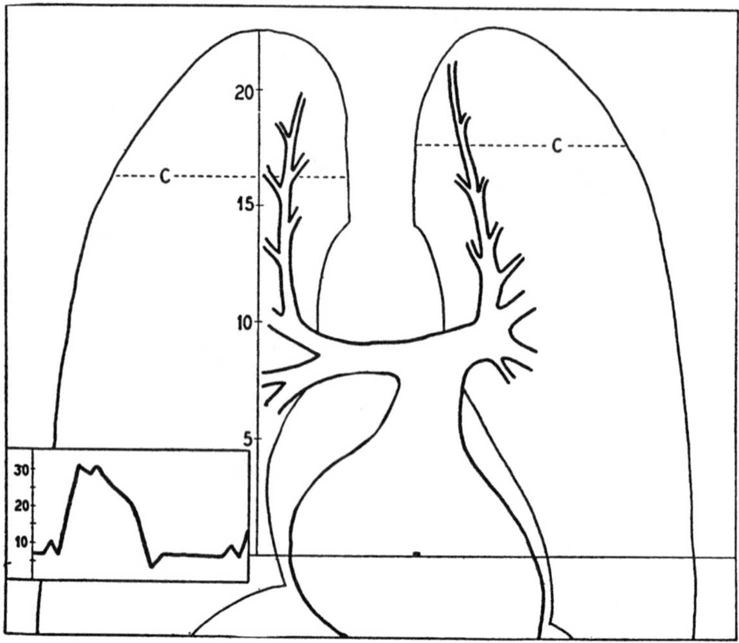

Figure 2 Diagram from Dock (5) in which he suggested that the pulmonary artery pressure would not be sufficient to raise blood to the top of the upright human lung. The two broken lines marked C indicate the extent to which the blood would rise based on the right ventricular pressure tracing shown in the inset (lower left). He used a reference point in the middle of the right ventricle indicated by the horizontal line. Note that he also thought that the blood flow to the apex of the right lung would be less than that of the left. (From Ref. 5.)

high incidence of pulmonary tuberculosis in patients with pulmonic stenosis (as did Orth) and added that patients with mitral stenosis who have an increased blood flow to the lung apices are known to have a low incidence of the disease.

A particularly colorful set of studies was carried out by Rothlin and Undritz (6) when they showed that while humans develop tuberculosis at the apex of the lung, and quadrupeds such as the rabbit and cow do the same in the dorsal (top) region of the lung, the bat, which spends much of its life inverted, tends to develop pulmonary tuberculosis at the lung bases (Fig. 3). Incidentally, we now know that the reason why adult tuberculosis has a predilection for the apex of the human lung is not so much that the blood flow is reduced there, but that the ventilation–perfusion ratio is high (largely because the blood flow is reduced) and this results in a higher alveolar P_{O_2}, which provides a better environment for growth of the tubercle bacillus.

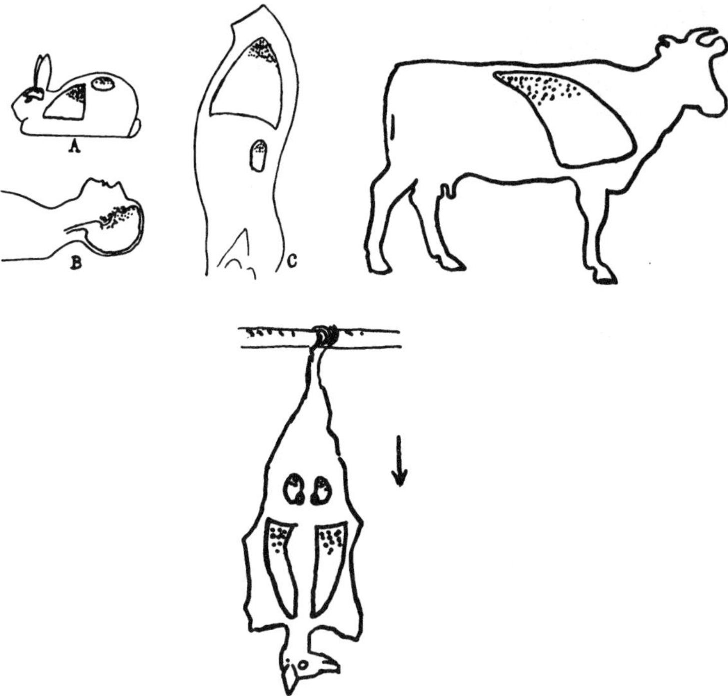

Figure 3 Composite of diagrams from the article by Rothlin and Undritz (6) where they pointed out that tuberculosis affects the upper parts of the lungs and kidneys, even in the bat, which spends most of its life upside down. They also claimed that the disease occurs in the upper part of the human brain. (From Ref. 6.)

C. Early Measurements by Sampling Gas in Different Lobes of the Lung

Evidence that gravity caused regional differences of blood flow and gas exchange was obtained by passing catheters into different regions of the lung and analyzing the gas that was withdrawn. Martin et al. (7), for example, passed fine catheters into individual lobes of the lungs of human subjects who were seated upright, and showed that the respiratory exchange ratio was higher in the upper lobes than in the lower. As already stated, this results from the higher ventilation–perfusion ratio in the upper regions of the lung, primarily because the blood flow there is so low. Mattson and Carlens (8) used a special bronchospirometry catheter that allowed them to separate the gas exhaled from the right upper lobe from that exhaled from the rest of the right lung, and they showed a very low oxygen uptake in the right upper lobe that increased when the subject lay supine. Rahn et al. (9) demonstrated a higher ventilation–perfusion ratio in the uppermost lobe of a dog's lung when the animal was tilted than would be predicted from the lung's ventilation. All these findings are consistent with the hypothesis that the upper lobes are relatively poorly perfused when the subject is upright.

D. Introduction of Radioactive Techniques

A disadvantage of sampling from individual lobes of the lung is that, because of their anatomy, the lobes span a relatively large vertical distance. More precise localization of topographical differences in the lung had to wait for the use of radioactive techniques. The first group to use radioactive gases to investigate pulmonary function was Knipping et al. (10) in Cologne who in 1957 employed radioactive ^{133}Xe to localize areas of defective ventilation in patients with lung disease. The patients breathed the radioactive gas from a spirometer circuit, and the gradual increase in counting rate over different regions of the lung was used as an index of the local ventilation. However, no systematic studies of topographical inequality of ventilation in normal subjects were carried out.

In 1955, the first cyclotron to be designed specifically for medical research purposes was dedicated at Hammersmith Hospital in London. It was built by the Medical Research Council and, in 1957, ^{15}O was produced by bombarding nitrogen with deuterons. The first report of the use of ^{15}O to study pulmonary function was made at the Second United International Conference on the Peaceful Uses of Atomic Energy on May 6, 1958 (11). In that report, it was shown that regional ventilation could be determined from the initial increase in counting rate following a single inspiration of ^{15}O-labeled O_2, and the subsequent fall in counting rate during breathholding (called the clearance rate) could be used as a measure of regional blood flow. A more extensive article was subsequently published in the *British Medical Journal* (12). In these early studies, attention was concentrated on patients who had regional lung disease, such as fibrosis or a cyst. There was an emphasis on the preparation of the ^{15}O and the optimal counting procedures, and it was pointed out that the radiation dose was very small in spite of the large amounts of radioactivity inhaled, because of the extremely short life of ^{15}O of only 2.07 min. In those early studies, no systematic measurements of regional blood flow and ventilation were made in the normal lung.

In 1959, the ^{15}O-labeled O_2 was converted into ^{15}O-labeled CO_2 by passing it over heated carbon at 450°C and then over hot copper oxide at 800 to 900°C. The first furnace converted all the oxygen to a mixture of CO_2 and CO, and the second furnace oxidized all the CO to CO_2. The advantage of ^{15}O-labeled CO_2 for physiological measurements was that the clearance rate was much faster than that of ^{15}O-labeled O_2, and it soon became apparent that there was a striking degree of blood flow inhomogeneity down the upright normal lung (13) (Fig. 4). Furthermore, in this early study not only was the marked topographical inequality

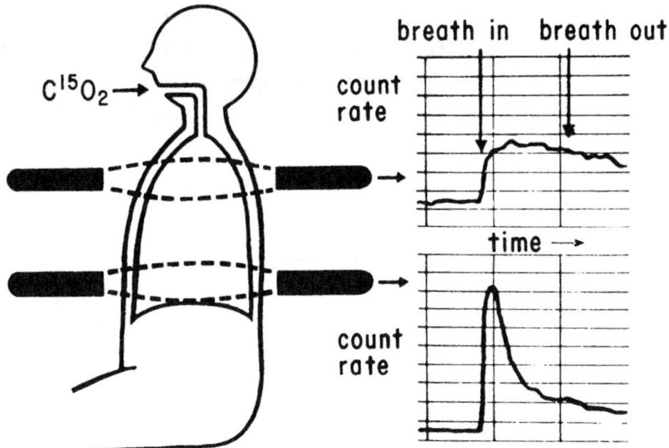

Figure 4 First demonstration of the lower blood flow at the apex of the upright human lung compared with the base. This was measured from the rate of removal of radioactive carbon dioxide following a single breath of this gas. (From Ref. 13.)

of blood flow demonstrated, but the difference in blood flow between the apex and base of the lung was shown to be essentially abolished in the supine position. In addition, measurements of regional ventilation from the initial rise in counting rate following a single inspiration of the gas were combined with the differences of blood flow to give the regional ventilation–perfusion ratio, and the effects on overall gas exchange were calculated. These topographical differences in gas exchange were subsequently described in more detail (14).

On the other side of the Atlantic, in Montreal, Ball et al. (15) refined Knipping et al.'s technique using ^{133}Xe and showed how this could be adapted to measure both regional ventilation and blood flow. The blood flow studies were made by dissolving the ^{133}Xe in saline and injecting it into a peripheral vein so that it was evolved into the alveolar gas when it reached the pulmonary capillaries. This group went on to make many important studies on regional pulmonary function, especially on the topographical distribution of ventilation (16).

Subsequently, a number of other radioactive techniques were used to study the effects of gravity on the lung, including macroaggregated albumin labeled with ^{131}I or ^{99m}Tc, and other radioactive gases such as ^{13}N and ^{81m}Kr.

E. Blood Flow

As indicated above, the first measurements of the topographical inequality of blood flow in the human lung were made using radioactive CO_2 (13). Figure 4 shows two typical tracings from this original study, and Figure 5 shows the pooled results. These data are from 16 normal volunteers who took six breaths each of the gas. The test inspiration was 1 L of radioactive CO_2 in air from functional residual capacity. Note the very striking decrease in blood flow from the bottom to top of the upright human lung with the apex almost unperfused. Measurements made with the subjects in the supine position showed that the difference between apex and base was essentially abolished, and also that moderate exercise in the upright position increased both the apical and basal blood flows.

Measurements with other techniques soon confirmed this pattern. For example, Bryan et al. (17) showed a large decrease in blood flow from base to apex of the upright lung that was abolished by the supine posture. Similar results were reported by other workers using radioactive xenon (18,19), and radioactive macroaggregated albumin (20). Measurements with radioactive xenon often showed a greater proportion of blood flow to the apex of the lung than was found with radioactive carbon dioxide, probably because the softer radiation of xenon does not permit such accurate localization, and also because of counts coming from blood in the chest wall. However, Anthonisen and Milic-Emili (19) reported that the extreme apex was actually unperfused in their series.

The cause of the uneven distribution of blood flow was extensively studied using an isolated lung preparation where it was possible to accurately control the pulmonary arterial, venous, alveolar, and intrapleural pressures. It was soon shown that with normal pulmonary vascular pressures, a distribution of blood flow essentially identical to that shown in Figure 5 could be obtained. Then, by

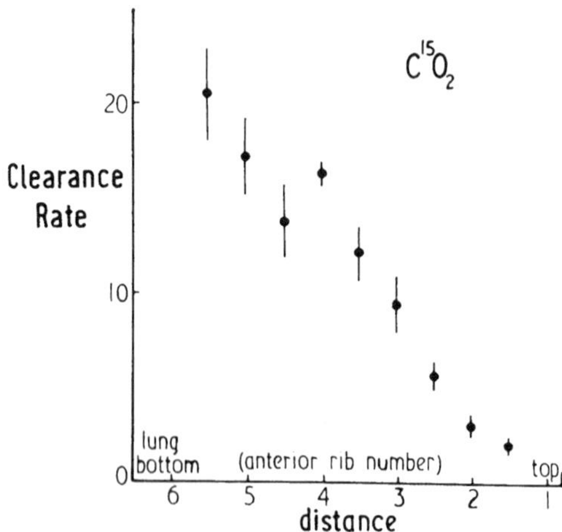

Figure 5 Distribution of blood flow in the upright human lung measured with radioactive carbon dioxide. (Modified from Ref. 13.)

manipulating the pressures, a three-zone model of the factors determining the topographical distribution of blood flow in the lung was developed (2). Subsequently, it was shown that this pattern was sensitive to lung volume, in that a zone of reduced blood flow developed at the bottom of the lung when lung volume was decreased. The explanation of this was a decrease in caliber of the extraalveolar blood vessels, and therefore an increase in their vascular resistance. This region was labeled zone 4 (21).

F. Intrapleural Pressure

One of the earliest suggestions that gravity may affect intrapleural pressure and regional expansion of the lung was made by Karl Wirz (Fig. 6). Although he

Figure 6 Karl Wirz (1896–1978), who was one of the first persons to suggest that the weight of the lung would cause regional differences in intrapleural pressure. (From Ref. 60.)

only spent two years as a junior assistant to the eminent physiologist Fritz Rohrer, Wirz made some remarkable contributions. Figure 7 shows his experimental setup for measuring regional differences of intrapleural pressure (22). He wrote:

> A further problem which has not yet been studied is whether, and how much, the weight of the lung itself influences the pleural pressure.
>
> If we consider that at each unit of the pleural surface not only the elastic recoil force of the lung but also a tension or pressure is exerted, corresponding to the weight of the column of lung tissue vertically below or above this surface unit, it is feasible to assume, at least when the lungs are heavy, that pressure differences can exist between the upper and lower parts of the thorax. If the weight forces are completely transmitted to the top of the lung, the negative pleural pressure at the top of the lung in man in a standing position can be estimated as being 1.5 to 2 cmH_2O higher than at the diaphragm, due to the weight of the lung. This difference would be much greater in large animals (horse, cow, elephant). The importance of the weight of the lung for the pleural pressure is evident from an interesting deviation in the structure of the respiratory organs in a very large animal, the Indian elephant, where, according to Boas (23) the leaves of the pleura are connected by a whitish elastic tissue. When the lung is adherent the transmission of the gravitational force exerted by the weight on the upper parts of the lung is greatly limited, since the lung cannot slide along the wall of the thorax and transmit the pulling force to the thoracic wall.
>
> If the lung can move, it is also probable that only a limited proportion of the gravitational force will influence the pleural pressure, since because of the fixation

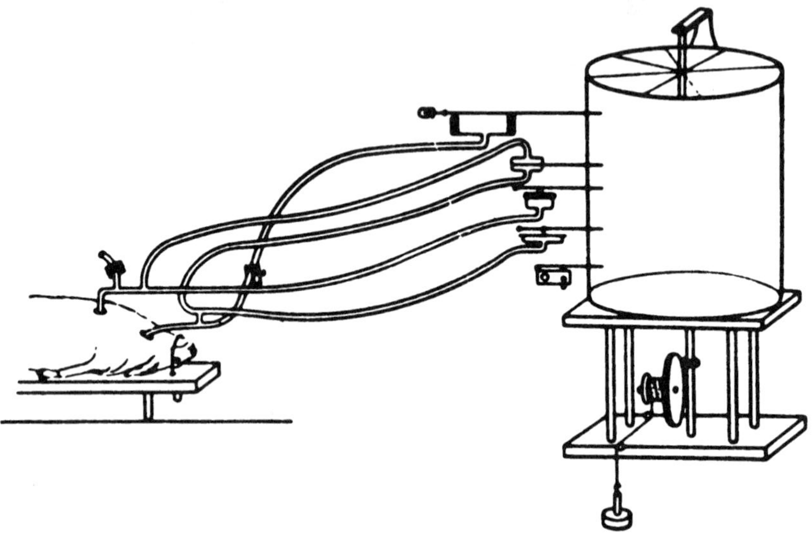

Figure 7   Method used by Wirz to record intrapleural pressures in different regions of the lung. (From Ref. 22.)

of the lung at the hilum, a large part of its weight is transmitted as a pulling force to the solidly constructed main bronchi and trachea.

Incidentally, the issue of whether the fused pleural surfaces in the elephant reduce the gravitational distortion of the lung has been revisited on several occasions, most recently by Brown et al. (24). However, it seems more likely that the obliteration of the pleural space in this animal is to protect the pleural blood vessels during snorkeling (25). The elephant is known to do this at a considerable depth by breathing through its raised trunk.

In a slightly later article, Rohrer (26) cited Wirz's calculations but went further to consider the effects of gravity on the diaphragm. He stated:

> Inside the lungs the gravitational forces have only minor importance. Whereas other tissues have approximately the same specific gravity as water, 1 ml lung parenchyma in an average state of expansion weighs only approximately 1/9 g. The pull exerted by the weight of a column of parenchyma of 1 cm² cross section, between the diaphragm and the lung apex, is approximately 1.5 to 2 g. It is probably that the gravitational forces have only a minor influence on the static pressure at the lung surface, since part of the weight of the lungs is borne by their attachment to the trachea.
>
> The bulk of the gravitational forces is derived from the abdominal contents. According to my measurements, the abdominal contents in cats constitute 10 to 13%, in rabbits up to 20%, of the body weight. In man we can estimate this at 7 to 10 kg, with variations according to the filling of the gastrointestinal tract, urinary bladder, and uterus.
>
> In an upright position the abdominal gravitational forces act in the same direction as the elastic pull of the diaphragm, downward, causing expansion of the thoracic contents and stretching of the abdominal wall. Part of the weight is borne in this position by the posterior wall of the abdominal cavity, via the mesenteric attachment.
>
> In an upside-down position (body straight, head downward), the tension of the abdominal wall, the weight of the abdominal contents, and the pull of the lungs work in the same direction, pushing the diaphragm passively toward the head, until its elastic counterpressure results in an equilibrium.
>
> In a horizontal position the effect is smaller, and the lower half of the diaphragm is influenced in a way opposite to that on the upper half.

Incidentally, although both Wirz and Rohrer allude to the support of the weight of the lung by the hilum (or trachea), there is some evidence from a study in a primate that this is minimal (27).

In 1933, Parodi (28) argued that the transpulmonary pressure decreases from the top to the bottom of the lung because of the weight of the lung. He argued that when the density of the lung is increased by disease, the transpulmonary pressure at the apex will be higher than normal. He also suggested that the higher transpulmonary pressure near the apex is partly explained by the large ratio of lung volume to the cross-sectional supporting area.

More recently, Milic-Emili et al. (29) and Bryan et al. (30) used esophageal balloons to show that there was an average vertical gradient of end-expiratory transpulmonary pressure of 0.2 to 0.25 cmH$_2$O. Subsequently, Proctor et al. (31) showed that the changes in pressure inside the esophagus depended on posture. Direct measurements of topographical differences in intrapleural pressure were made by Krueger et al. (32) and Turner (33), and they showed a vertical gradient of end-expiratory pleural pressure. A very extensive series of measurements of regional intrapleural pressure were made by Agostoni et al. (for a review, see Ref. 34).

G. Alveolar Expansion

Regional differences in intrapleural pressures, as described in the previous section, also imply regional differences of lung expansion. However, direct evidence of differences of alveolar size were not reported until the 1960s. One of the first demonstrations was by Glazier et al. (35) when they fixed dog lungs in situ by freezing, and alveolar size was measured by morphometric techniques (Fig. 8). It was found that the relative size of the alveoli depended on the posture of the animal and on where the sample was taken. For example, in the head-up animal, at functional residual capacity, the apical alveoli were about 4 times larger by volume than the basal ones, and most of the change in size occurred over the upper 10 cm of lung. No difference in size was found when the lungs were expanded using an alveolar pressure of 30 cmH$_2$O. In horizontal lungs, alveolar size was the same at the apex and base, but the most superior alveoli were larger than the dependent ones. When the animals were frozen during centrifugation with the acceleration vector from head to foot, the difference between the alveolar volume in apex and base increased to 11:1.

Further histological studies of regional alveolar size were made by Hogg and Nepszy (36). They froze exsanguinated dogs in the head-up position by surrounding them with solid carbon dioxide, and slices were then cut from the frozen lungs. The density of each slice was obtained by weighing and calculating its volume. From the density, the ratio of gas volume to tissue volume was computed, and regional lung volume as a percentage of regional total lung capacity was then derived. They reported that in dogs frozen at functional residual capacity, the expansion of the upper lung regions was 80 to 90% of total lung capacity, as against 28 to 40% of total lung capacity for the lower regions. The results were in good agreement with the regional differences in alveolar size found by Glazier et al. (35).

Additional studies were reported by D'Angelo (37) when he measured alveolar size in rabbits after rapidly freezing the lungs by applying liquid nitrogen to the chest wall. The relative volume of the alveoli was derived from the volume-to-surface ratio as in the study of Glazier et al. (35) described above. D'Angelo found that in head-up animals at functional residual capacity, the relative volume

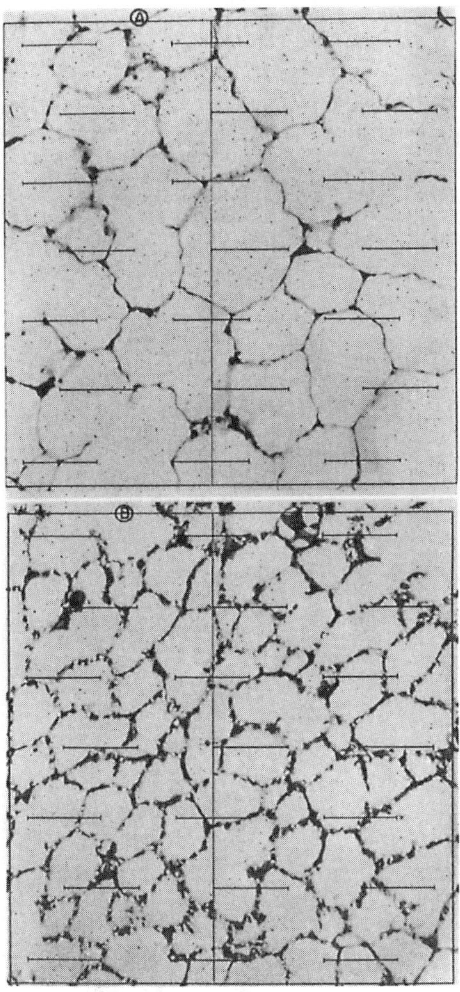

Figure 8 Histological appearance of lung from the apex (upper section) and 20 cm lower down (lower section) from the lung of a greyhound frozen in the vertical position. The grid used for determining alveolar size is superimposed on the fields. Each of the test lines has a length of 100 μm. (From Ref. 35.)

of the alveoli decreased markedly from apex to base, and the results were in good agreement with those found by Glazier et al. (35) and Hogg and Nepszy (36).

H. Mechanical Stress in Lung Parenchyma

Regional differences in lung expansion imply differences in the mechanical stresses in the alveolar walls. Orsos (38) considered this issue because he was intrigued by the apical localization of pigment on the pleural surface of patients with pulmonary tuberculosis. He reasoned that the lung behaved like a homogeneous elastic sheet which had a smaller cross-sectional area near the apex than the base. As a result, the forces stretching the lung in the vertical direction would cause more distortion in the tissue at the apex than the base (Fig. 9).

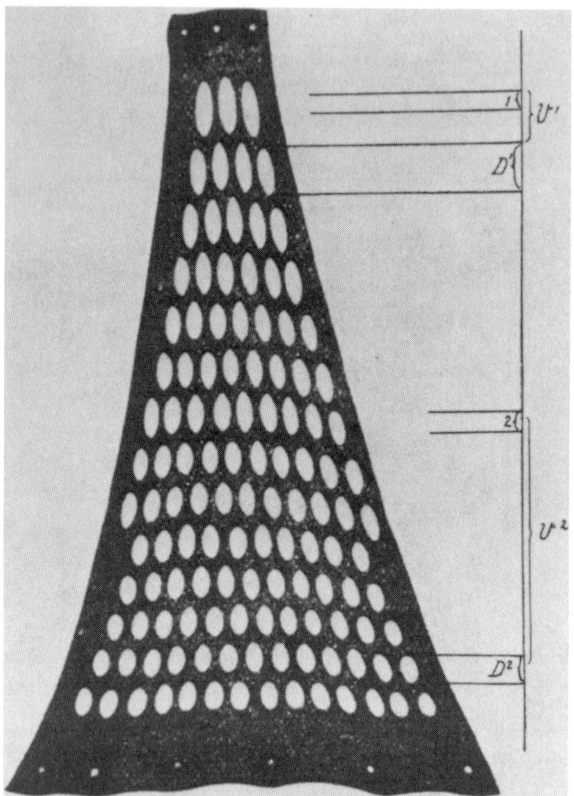

Figure 9   Early model of lung distortion suggested by Orsos. He argued that if the lung behaves as a uniform elastic sheet containing holes, the apical alveoli will be stretched more than the basal alveoli by vertical expansion of the lung because of the smaller cross-sectional area near the apex. (From Ref. 38.)

   A theoretical analysis of the distribution of stresses in the upright human lung was carried out using the technique of finite element analysis (39). The following question was posed: How does a lung-shaped homogeneous elastic body supported at its periphery distort under its own weight? Measurements of stress–strain behavior of isolated lung had shown that lung parenchyma is apparently isotropic in its elastic behavior (that is, that it behaves similarly in all directions) and also that the stress–strain properties are highly nonlinear. It was also assumed that the lung received essentially no support from the hilum. This conclusion was based on the radiographic appearance of the lung in a rhesus monkey compared with the shape of the lung when it was removed and supported in part by the hilum (27). It was also assumed that the pleural surfaces slide so easily over each other that the shear forces at equilibrium are negligibly small. This meant that the resultant stresses at the pleural surface were normal to the surface at equilibrium.

   The results of this analysis showed that both the stresses in the vertical direction and the lateral direction of the upright human lung were greater at the apex than the base (Fig. 10). The analysis also gave the relative volume of the

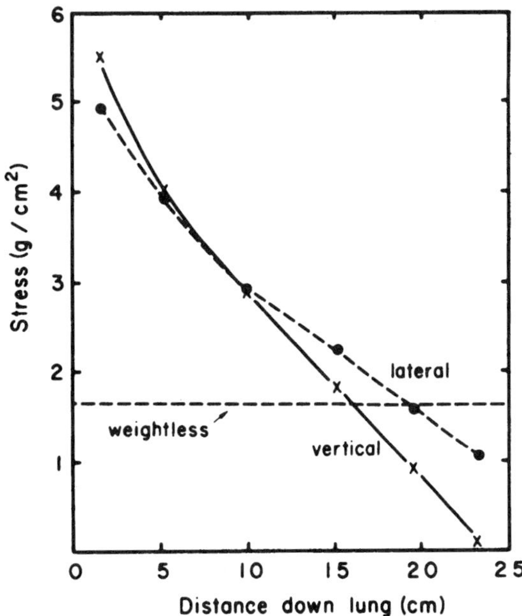

Figure 10   Distribution of expanding stresses down the upright lung at functional residual capacity as found in a theoretical study using finite element analysis. Note that the stresses in both the vertical and lateral directions are greatest at the apex. (From Ref. 39.)

lung units from apex to base and these were similar to those found in the experimental studies of Glazier et al. (35), Hogg and Nepszy (36), and D'Angelo (37).

I. Ventilation

As mentioned earlier, the original studies using radioactive CO_2 showed that ventilation per unit volume was less at the apex of the upright lung than at the base (13). However, the classical studies on topographical inequality of ventilation were made by Milic-Emili et al. (16) using ^{133}Xe, and their results are summarized in Figure 11. These are discussed in detail in Chapter 4 and only a brief summary is given here. The upright human lung is distorted by its own weight with the result that, at most lung volumes, the alveoli at the bottom of the lung are smaller than those at the top. However, because of the shape of the pressure–volume curve of lung tissue, the smaller basal alveoli expand more for a given inspired tidal volume than the larger apical alveoli. The result is that ventilation per unit volume is greater near the base of the lung than the apex. At total lung capacity, it is believed that all the alveoli become essentially identical in size.

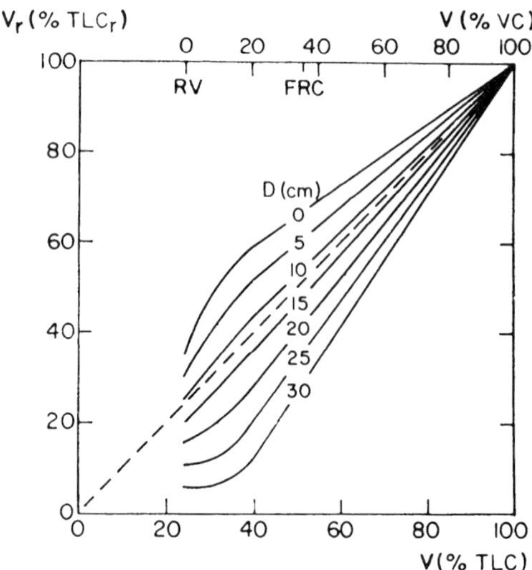

Figure 11 "Onion skin" diagram from Milic-Emili et al. (16) summarizing the regional expansion of the upright human lung. The vertical axis shows regional lung volume (Vr), expressed as percent total lung capacity. The horizontal axis shows overall lung volume (V), expressed as percent TLC below and as percent vital capacity above. The broken line (of identity) indicates the percentile degree of expansion of the regions if they were equal to that for the entire lung. Vertical distance (D) from top of the lung (in cm) is indicated on each curve. (From Ref. 16.)

At very low lung volumes there is closure of the small airways and the lung units assume their "trapped gas volume." The result is that when a small inspiration is made from residual volume, none of the gas enters the most dependent regions of the lung, and all of the inspired gas goes to the upper regions. However, after a certain lung volume has been exceeded, the most dependent regions begin to inspire gas, and from that volume up to total lung capacity, the base of the lung is better ventilated than the apex.

J.   Gas Exchange

Regional gas exchange is determined by the ratio of ventilation to blood flow. Since these two variables change at different rates down the upright human lung, there are topographical differences in the ventilation–perfusion ratio, and therefore in pulmonary gas exchange. These are considered in detail in Chapter 8 and therefore will not be dealt with further here.

II.   Increased Acceleration

A.   Early Predictions

The effects of increased acceleration during liftoff at the beginning of space flight were graphically imagined by the Russian visionary Konstantin Eduardovich Tsiolkovsky (1857–1935) at the end of the nineteenth century (Fig. 12). Here is his description (40, p. 99):

> The signal is given; the explosion, attended by a deafening noise, starts setting off. The rocket shakes and takes off. We have the sensation of terrible heaviness. My weight has increased tenfold. I am knocked down to the floor, severely injured and perhaps have even been killed—can there be any talk of observations? There are ways of standing up to this terrible weight, but only in a, so to say, compact form or being submerged in a liquid (we shall discuss it a little bit later).
>
> Even when submerged in liquid we will hardly be inclined to observe anything outside. Be all that as it may, the gravity in the rocket has apparently increased tenfold since takeoff. We would be informed of this by a spring balance or a load gauge, by the accelerated swinging of a pendulum (some 3 times faster), by the faster fall of bodies, by the diminished size of droplets (their diameter decreasing tenfold), by all things carried aboard the rocket becoming heavier, and many other phenomena.

B.   Early Centrifuges

Today, we normally associate increased acceleration with high-performance aircraft or the launch and reentry of space vehicles. However, examples of human centrifuges go back more than 200 years. For example, Erasmus Darwin (41, p. 159) described a colorful way of inducing sleep as follows:

Figure 12   Konstantin Eduardovich Tsiolkovsky (1857–1935), who was a mathematics teacher in provincial Russia but made remarkable predictions about space travel. (From Ref. 61.)

> Another way of procuring sleep mechanically was related to me by Mr. Brindley, the famous canal engineer, who was brought up to the business of a mill-wright; he told me, that he had more than once seen the experiment of a man extending himself across the large stone of a corn-mill, and that by gradually letting the stone whirl, the man fell asleep before the stone had gained its full velocity, and he supposed would have died without pain by the continuance or increase of the motion. In this case the centrifugal motion of the head and feet must accumulate the blood in both these extremities of the body, and thus compress the brain.

A few years later, a centrifuge of large dimensions (Fig. 13) was built in the psychiatric clinic of the Charity Hospital in Berlin, for the treatment of patients with mental disease. Its diameter was approximately 4 m (13 Rhenish feet!), and when rotated at 40 to 50 rpm, it could produce up to 5 G of acceleration at the periphery. Marked changes in respiration, heart rate, and blood distribution were apparently observed.

Figure 13 Early centrifuge used at the Charity Hospital in Berlin for the treatment of patients with mental disease. Note the people on the left who are turning a crank that is connected by a cable to the wheel driving the centrifuge. (From Ref. 44.)

In the late nineteenth century, the effects of acceleration on animals and humans were studied by Salathé in Marey's laboratory in France, and von Wenusch in Vienna. However, with the development of the Luftfahrtmedizinisches Forschungsinstitut in Berlin in the 1930s, the investigation of human and animal tolerances to centrifugal forces became much more scientific. This work was under the direction of von Diringshofen and resulted in a number of publications (42,43). The measurements on centrifuges were compared with others done using experimental aircraft. Figure 14 shows chest radiographs taken first in level flight and then with the subject exposed to 7.5 G_z, that is, with the acceleration vector from head to foot. Note the diminution of vascular lung markings at the apex, and the distortion of the mediastinum and the heart as a result of the increased G level. An interesting account of early centrifuge studies is in Gauer (44).

C. World War II

With the advent of World War II there was a large increase in the number of studies of increased acceleration, both using the centrifuge and high-performance

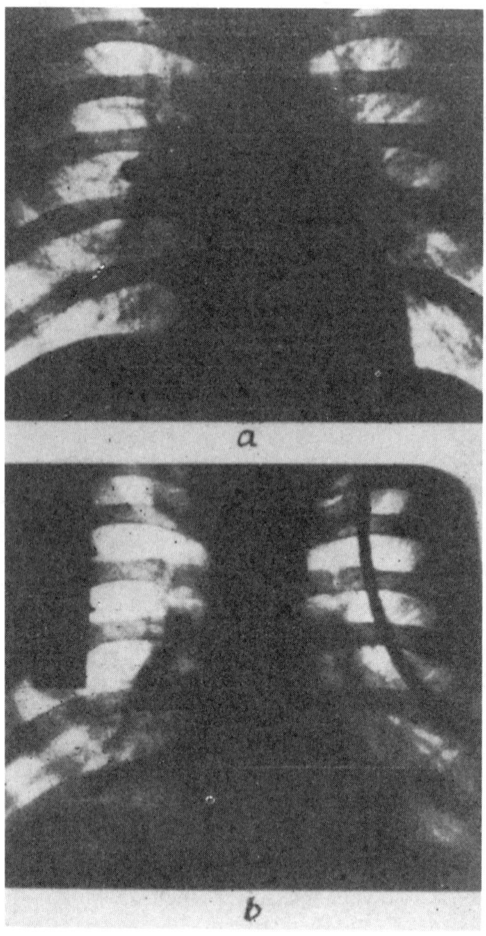

Figure 14 Chest radiographs taken first in level flight (a) and then with the subject exposed to 7.5 G_z, that is, with the acceleration vector from head to foot (b). (From Ref. 43.)

aircraft. The reason for this was that pilots were exposed to high accelerations during tight turns in dogfights, and it became important to understand how to avoid blackouts, which were caused by blood draining from the head. Among the centrifuges used for extensive physiological studies on the lung was that built at the Mayo Clinic and used extensively by Earl Wood and his colleagues. Developments after World War II are covered in Chapter 2.

III. Microgravity

A. Early Predictions

Reference was made earlier to the predictions of Konstantin Tsiolkovsky (see Fig. 12) on the effects of increased gravity during the liftoff of a spaceship. Here is his prediction of the sensations when the main engines shut down (40, p. 100). Recall that he was writing at the end of the nineteenth century and the first decade of the twentieth century.

The awful gravity that we experience will last until the explosion and the noise come to an end. Then, as dead silence sets in, the gravity pull will diminish instantaneously, just as it appeared. We are now out beyond the limits of the atmosphere at an altitude of 575 km. The gravity pull did not only diminish in force but vanished completely without a trace; we no longer even experience the terrestrial gravitation that we are accustomed to just as we are accustomed to the air, though it is not at all so necessary as the latter. The altitude of 575 km is very little, it is almost at the surface of the Earth, and the gravity should have diminished ever so slightly. And that actually is the case. But we are dealing with relative phenomena, and for them there is no gravity.

The force of terrestrial gravitation exerts its influence on both the rocket and bodies in it in the same way. For this reason there is no difference in the motion of the rocket and the bodies in it. They are carried along by the same stream, the same force, and, as far as the rocket is concerned, there is no gravity.

There are many things that convince us of this. All objects in the rocket, that were not attached, have left their places and are hanging in the rocket's air, out of contact with anything; and if they touch something, they do not exert any pressure on each other or on the support. We ourselves do not touch the floor and can have any position and be in any direction: we can stand on the floor, on the ceiling or on the wall; we can stand perpendicularly or have an inclined attitude; we float in the middle of the rocket like fish but without any effort whatsoever, and we do not come in contact with anything; no object exerts pressure on any other if they are not pressed together.

Water does not pour from a decanter, a pendulum does not swing and hangs to the side. An enormous mass hung from the hook of a spring balance does not make the spring taut—it always indicates zero. Lever scales are also useless: the balance beam takes up any position, quite irrespective of and indifferent to the equality or inequality of the weights in the pans. Gold cannot be sold by weighing its mass. Conventional ways of measuring mass cannot be employed here.

B. Early Parabolic Flights

Although normal gravity is always with us (like the poor), and increased acceleration is readily attained using a centrifuge (see Fig. 13), zero gravity is much more difficult to come by. The only exception is free fall, which is usually so brief that useful measurements cannot be made.

As early as World War I, a few seconds of microgravity could be produced in diving aircraft (45). With the development of higher performance aircraft during World War II, longer periods of microgravity could be obtained, and in 1950, the flight maneuver known as a Keplerian arc was proposed. This enabled periods of microgravity as long as 45 sec to be achieved (46,47). Later, rocket-launched capsules were developed and these allowed the study of small animals during periods of microgravity for many minutes during the free fall after the rocket had burned out (48).

A few measurements of the effects of microgravity on the human lung have been made by flying high-performance aircraft in a Keplerian arc. One of the earliest was on the effect of microgravity on the distribution of pulmonary blood flow using ^{133}I-labeled macroaggregated human serum albumin (49). The maneuver consisted of a 2.5-G pullup from straight and level flight, followed by 45 sec of weightlessness, and then a 3-G pullout. The tracer was injected 20 sec after entry into the microgravity portion of the flight, and a control study was done using the same initial pullup but this time flying the aircraft at 1 G instead of 0 G. It was found that there was a shift in blood flow toward the apex of both lungs of the seated subjects during microgravity, more marked on the left side than the right. This was similar to the distribution found in changing from the upright to the supine position.

In 1969, Foley and Tomashefski (50) measured forced expiratory volumes and flow rates on 12 normal subjects during Keplerian arcs in a KC-135 aircraft at Wright-Patterson Air Force Base, Ohio. The measurements were made with a waterless spirometer, and control studies were done during straight and level flight. There was a marked decrease in flow rate in microgravity but no reduction in forced vital capacity. The investigators attributed the changes to alterations in intrathoracic pressures.

Another set of experiments was carried out by Baumgarten et al. (51) using a Learjet aircraft that was flown in a roller-coaster profile. The main emphasis was on vestibular function but a few respiratory measurements were made. It was found that during hypergravity, there were increases in tidal volume and an increase in resting lung volume, whereas the opposite changes were seen in microgravity.

Additional studies were carried out by Michels et al. (52,53) in a NASA (National Aeronautics and Space Administration) Learjet flying in a parabolic profile, which allowed up to 27 sec of microgravity. In the first set of studies (52), the distribution of ventilation was investigated using a single-breath nitrogen washout, and the distribution blood flow was inferred from the pattern of expired oxygen and carbon dioxide. In both instances, measurements made during microgravity were compared with those at 1 G during straight and level flight, and 2 G during a controlled turn. The results showed a large reduction in topographical inequality of ventilation and blood flow in the lung, but there still appeared to be some uneven ventilation and blood flow at 0 G.

In the second set of studies (53), chest radiographs were taken in both microgravity and 1 G. There were no significant differences in the shapes of the ribcage and diaphragm, although there was a slight tendency for the lung to become shorter and wider at 0 G. The conclusion was that gravity produced topographical inequalities of ventilation in the upright lung by distorting the elastic lung tissue within the chest, rather than by altering the shape of the ribcage and diaphragm.

C. Sustained Microgravity

Most of the studies described in subsequent chapters were made possible by the development of space flight less than 50 years ago. However, as intimated already, there were remarkable visionaries who considered the possibilities of human space flight and its likely problems over a hundred years ago. One of the most important was Konstantin Tsiolkovsky (see Fig. 12) who, as a young, almost deaf, Russian mathematics teacher in a small provincial town, sketched a spacecraft design in 1883 (Fig. 15) and published his first article on space travel in

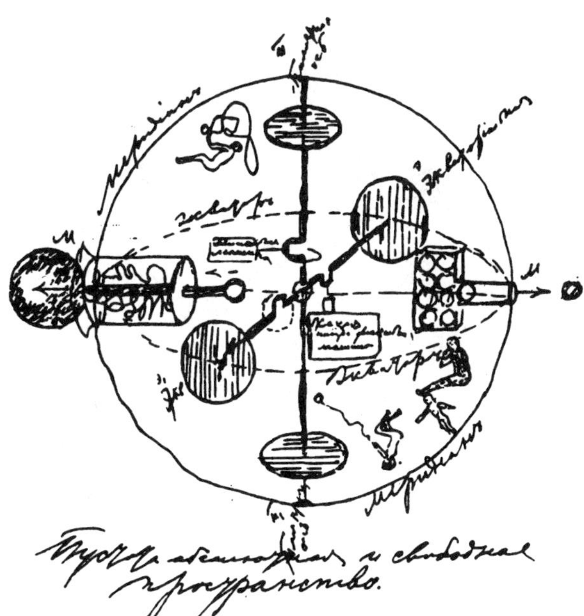

Figure 15 Sketch of a spacecraft made by Tsiolkovsky in 1883. (From Ref. 61.)

1895 (40). There is a dramatic memorial to him in Moscow showing the upward-sweeping trajectory of a spacecraft after liftoff. Tsiolkovsky derived the basic formula of rocket dynamics that relates the speed of rocket flight, jet exhaust velocity, propellant mass, and the mass of the rocket vehicle. He also recognized the importance of the orbital velocity of about 7900 m sec^{-1} which would allow a spacecraft to orbit the Earth, and also the escape velocity of about 11,200 m sec^{-1}, which is the velocity required for a spacecraft to escape the gravitational attraction of the Earth. He recognized that space flight would require liquid pro-pellants, because of their greater efficiency than solid propellants.

In 1926, the American, Robert Hutchings Goddard (1882–1945) (Fig. 16) launched the first liquid-propellant rocket, and in one of his papers he discussed the possibility of a rocket reaching the moon. However, his ideas were ridiculed in the *New York Times* and other quarters, and he largely retired from the public eye to work on his ideas in relative seclusion. Herman Julius Oberth (1894–

Figure 16   Robert Hutchings Goddard (1882–1945), who launched the first liquid-propellant rocket in 1926. (From Ref. 61.)

1989), in Germany, formulated many of the technical problems of space flight, but most of his ideas were dismissed as fantasies.

The development of the great booster rockets that were necessary to lift large masses into Earth orbit and beyond is a fascinating story, and only recently have some of the details come to light. A key person in the highly successful early Soviet space program was Sergei Pavlovich Korolev (1907–1966) (Fig. 17). His was a remarkable story that could only have taken place in the Soviet Union (54). Korolev carried out some highly successful experimental work on rockets in the 1930s, but then was incarcerated on trumped-up charges as an intellectual "enemy of the people" by Stalin. First he was sent to one of the most dreaded prisons in far eastern Siberia where he spent five months in the

Figure 17 Sergei Pavlovich Korolev (1907–1966), who was the chief designer of the early Soviet rocket systems. He had many firsts to his credit, including the launch of Yuri Gagarin. (From Ref. 61.)

winter digging in a surface gold mine and many of his fellow prisoners died. Later he was moved to a special gulag prison for scientists in Moscow where he did some engineering work under house arrest. This was the type of institution described by Aleksandr Solzhenitsyn in *The Gulag Archipelago* (55). He was in prison for a total of about seven years but was released after the war because the Soviets wanted to develop a missile program, and he was one of their most talented rocket engineers. Amazingly, he then continued to work at a breakneck speed until his death, and appeared to harbor no resentment toward the regime.

Initially, Korolev was asked to build a replica of the German A4 rocket that had been developed during World War II (see below), and although he argued that he could come up with a better design, he complied and produced a vehicle within two years. He then went on to mastermind the Soviet rocket and space systems. It is extraordinary that during almost all of his career, he was never referred to by name because of security reasons but only as the Chief Designer. In fact, even some of the cosmonauts who worked directly under him were apparently not aware of his last name. But his successes were spectacular. He was responsible for the first orbiting satellite, Sputnik-1, which was launched on October 4, 1957. This was made possible using the R-7 booster, which had five engines, each with four thrust chambers (Fig. 18) and generated 500 metric tons of thrust. The engine design was based on that of the A4. By contrast, the Atlas booster, which was the most powerful rocket in the United States at that time, had a thrust of only 200 metric tons. Sputnik-1 created a sensation everywhere, not least in the United States. Only a month after the launch of Sputnik-1, a second man-made satellite was placed in orbit, this time carrying the first living creature, the dog Laika. And, most dramatic of all, on April 12, 1961, the first human being, Yuri Alekseyevich Gagarin (1934–1968), was launched into orbit, and after a flight lasting 1 hour and 48 minutes, during which one Earth orbit was completed, he ejected from the spacecraft at an altitude of 7000 m to land safely by parachute.

These remarkable firsts were followed by many others. They included the first woman in space, Valentina Tereshkova, in June 1963; the first three man crew, on the vehicle Voskhod-1 in October 1964; the first extravehicular activity (space walk), by Aleksey Arkhipovich Leonov (1934–   ), on Voskhod-2 in March 1965; and the first man-made spacecrafts to orbit the sun and to reach Mars and Venus. Korolev was indeed a brilliant designer and an extraordinary man. Some aspects of the early development of the Soviet/Russian space medicine program were described by Gazenko (56).

The development of the U.S. manned space program followed a different route. As indicated earlier, Robert Goddard was an early rocket pioneer, but his innovative work was not generally supported. For example, during World War II, the U.S. military were not convinced of the value of rockets as delivery systems for weapons, and Goddard's skills were used for the development of rocket-

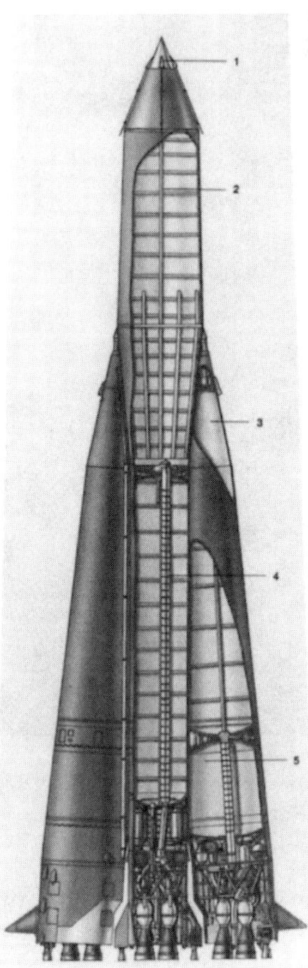

Figure 18 R-7 booster designed by Korolev and others, which put Sputnik-1 and many other firsts into Earth orbit. (From Ref. 61.)

assisted aircraft takeoffs. However, in Peenemunde, Germany, Wernher von Braun (1912–1977) (Fig. 19) and his coworkers developed the A4 rocket (Fig. 20) and its sophisticated guidance system. These workers immediately recognized the potential importance of what they had done for future space travel. At the time, the rocket was renamed the V-2 (for Vengeance-2) and many were launched against London with terrifying consequences.

After World War II, von Braun and other German rocketeers were brought to the United States and the A4 became the Redstone launcher. Other improvements in boosters followed, but the launch of Sputnik in 1957 caught the U.S.

Figure 19   Wernher von Braun (1912–1977), who developed the A4 rocket in Germany during World War II. (From Ref. 61.)

space program by surprise and it was clear that the Soviet boosters were considerably more powerful than those available in the U.S. program. This difference in the lifting power of the boosters had interesting effects on the two programs. For example, in the early U.S. manned space programs, including projects Mercury and Gemini, the astronauts breathed 100% oxygen at a pressure of 5 pounds per square inch, giving a cabin oxygen partial pressure of about 260 torr. This was in contrast to the Soviet program, where the human spacecraft from the beginning contained 21% oxygen–79% nitrogen, at 760 torr. One of the reasons for the differences in the cabin atmosphere is that a spacecraft containing a pressure of only one-third of the normal atmospheric pressure can be made with a smaller mass than one that has to withstand the full atmospheric pressure in the hard vacuum of space. In fact, the low cabin pressure with 100% oxygen was also used in the U.S. Apollo program, and there were tragic consequences when three astronauts died in a fire on the launch pad during a test. Subsequently, the atmosphere of the spacecraft during launch was changed from pure oxygen to 60% oxygen–40% nitrogen at normal atmospheric pressure, and the nitrogen was gradually replaced with oxygen after launch, as the cabin pressure was reduced.

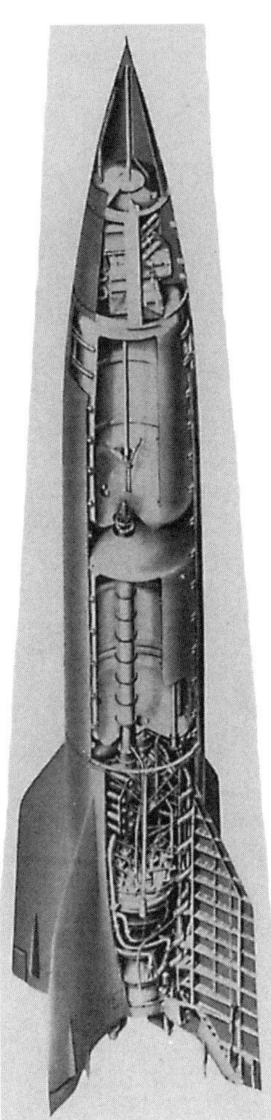

Figure 20 A4 rocket developed by Wernher von Braun and his coworkers. This was the first successful suborbital rocket. It was renamed the V-2 and many were launched against London. (From Ref. 62.)

When President Kennedy committed the United States to sending a man to the moon and return him safely to Earth within the decade of the 1960s, it was clear that a much larger booster would be required, and this resulted in the development of the Saturn V (Fig. 21). Its five main engines provided a total thrust of 3400 metric tons. The Soviets also planned to put human beings on the moon, and developed the enormous N-1 booster, but this proved to be unreliable,

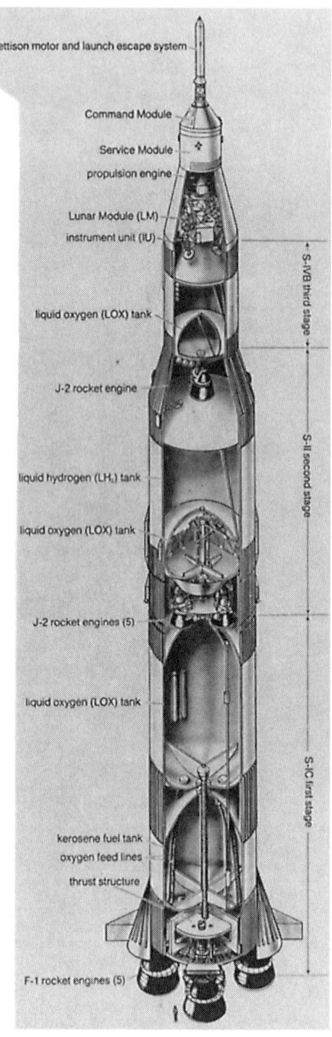

Figure 21   Saturn V booster, which started the astronauts on their historic voyage to the moon. (From Ref. 63.)

with one exploding during liftoff, and the program was abandoned. It goes without saying that the Apollo program, with the moon landings and exploration, constituted one of the great landmarks in the history of human exploration.

During the last 30 years, the U.S. and Soviet/Russian programs diverged in the sense that the United States concentrated on the Space Shuttle, while the

Soviets developed the Mir space station. The Shuttle has allowed frequent access to space, and the performance of highly sophisticated physiological studies using Spacelab. In fact, many of the studies described in this book were carried out on various Spacelab missions. The only exception to this pattern was the U.S. Skylab program, which took place in the early 1970s. Skylab was ingeniously built using the spent third stage of a Saturn V booster, and the crew of Skylab 4, which was launched in November 1973, spent a period of 84 days which, at the time, far eclipsed any previous duration record. Unfortunately, relatively few of the studies were reported in peer-reviewed journals.

Spacelab was a cylindrical pressurized laboratory, about 7 m in length and 4 m in diameter that was carried in the payload bay of the Shuttle (Figs. 22 and 23). Strictly, the term ''Spacelab'' refers to both the pressurized laboratory module and a variety of pallets, or laboratory test benches, that were located in the Shuttle bay behind the laboratory module. However, the studies described in this book were made in the cylindrical module. Both sides of the module were covered by standard instrument racks that were supplied with electrical power, cooling,

Figure 22   Artist's drawing of Spacelab in orbit. The pressurized cylindrical module contained instrument racks and was carried in the bay of the orbiter. It was entered from mid-deck through a tunnel. (Drawing by NASA.)

Figure 23 Spacelab SLS-1 during the flight. Note the instrument racks on both sides. The view is looking forward, and the entrance to the tunnel leading to the mid-deck of the orbiter can be seen. (Photograph by NASA.)

and vacuum. Thus, Spacelab provided a ''shirtsleeves'' research environment similar to that of a ground-based laboratory.

A feature of Spacelab was that for the first time in the U.S. program, non-career astronauts called Payload Specialists were part of the crew. In fact, most Spacelab flights had six or seven crew members. The commander and pilot were responsible for flying the mission and had no responsibility for the scientific project, although they often offered their services. Typically, there were two Mission Specialists who were career astronauts with a broad scientific training, and they took part in the scientific experiments as well as caring for the logistics of the flight. Finally, the two Payload Specialists were scientists who were chosen by the principal investigators of the experiments on board, and they took a substantial responsibility for carrying out the experiments.

The Spacelab missions were crammed with experiments, and crew time was at a premium. There was a very sophisticated data collection and transfer system that allowed the principal investigators and their colleagues to view the results of the experimental procedures in real time at the Johnson Space Center, in Houston. If there were problems, it was possible for the experimental team on the ground to communicate with the experimenters in the Spacelab. In all, there were 11 major Spacelab flights that concentrated on physiology and related life sciences (57). The last was Neurolab, which took place in April 1998. A feature of the Spacelab scientific program was the very close involvement of the university academic community, and a large number of peer-reviewed publications appeared in some of the top scientific journals. It was sad to see this extremely productive program come to a close, especially as it is hard to imagine that the International Space Station will provide such a sophisticated research environment in the foreseeable future.

An advantage of the Mir space station (Fig. 24) was that it allowed measurements of human responses to long-term space flight. Relatively few measurements could be made on Mir itself, at least in the early stages, because of the restricted space, power and equipment. However, extensive studies were done on cosmonauts before and after very long exposures to microgravity, including

Figure 24 Soviet/Russian Mir space station. This was launched in February, 1986, as the world's first permanently inhabited space station, and has now been in orbit for over 14 years. A considerable number of scientific studies have been carried out on Mir.

the physician Dr. Valeriy Polyakov, who spent 438 days continuously in space on Mir. It remains to be seen whether the upcoming International Space Station will combine the advantages of long-term exposure to space with sophisticated laboratory facilities.

IV. Epilogue

From time to time, it is natural for some of us to feel, like Wallace Fenn (58), that we were born 50 years too soon. When we look at the present advances in molecular biology, it is clear that medical science is rapidly moving into an extremely exciting era. On the other hand, it has been a privilege to be involved with the early human space program. I was lucky enough to spend a year at the NASA Ames Research Center in 1967–1968, and our group has been continuously supported by NASA since early 1969 (before Neil Armstrong set foot on the moon!). We have been able to make many stimulating studies of the effects of microgravity on the respiratory system and had the privilege of working closely with many of the astronauts. We have had our share.

References

1. Orth J. Atiologisches und Anatomisches uber Lungenschwindsucht. Berlin: August Hirschwald, 1887.
2. West JB, Dollery CT, Naimark A. Distribution of blood flow in isolated lung: Relation to vascular and alveolar pressures. J Appl Physiol 1964; 19:713–724.
3. Dollery CT, West JB, Wilcken DEL, Hugh-Jones P. A comparison of the pulmonary blood flow between left and right lungs in normal subjects and patients with congenital heart disease. Circulation 1961; 24:617–625.
4. Cournand A, Lauson HD, Bloomfield RA, Breed ES, Baldwin EdF. Recording of right heart pressures in man. Proc Soc Exp Biol Med 1944; 55:34–36.
5. Dock W. Apical localization of phthisis. Am Rev Tuberc 1946; 53:297–305.
6. Rothlin VE, Undritz E. Beitrag zur Lokalisationsregel der Tuberkulose. Schweiz Zeitschr Allgemeine Path Bakteriol 1952; 15:690–700.
7. Martin CJ, Cline F, Marshall H. Lobar alveolar gas concentrations: Effect of body position. J Clin Invest 1953; 32:617–621.
8. Mattson SB, Carlens E. Lobar ventilation and oxygen uptake in man: Influence of body position. J Thorac Surg 1955; 30:676–682.
9. Rahn H, Sadoul P, Farhi LE, Shapiro J. Distribution of ventilation and perfusion in the lobes of the dog's lung in the supine and erect position. J Appl Physiol 1956; 8:417–426.
10. Knipping HW, Bolt W, Valentin H, Venrath H, Endler P. Regionale Funktionsanalyse im der Kreislauf- und Lungen-Klinik mit Hilfe der Isotopenthorakographie und der selektiven Angiographie der Lungengefäße. München Med Wochenschr 1957; 99:46–47;42.

11. Dyson NA, Hugh-Jones P, Newbery GR, West JB. The preparation and use of oxygen-15 with particular reference to its value in the study of pulmonary malfunction. 2d U.N. International Conference on Peaceful Uses of Atomic Energy. Vol. 26. Geneva: United Nations, 1958:103.

12. Dyson NA, Hugh-Jones P, Newbery GR, Sinclair JD, West JB. Studies of regional lung function using radioactive oxygen. Br Med J 1960; 1:231–238.

13. West JB, Dollery CT. Distribution of blood flow and ventilation-perfusion ratio in the lung, measured with radioactive CO_2. J Appl Physiol 1960; 15:405–410.

14. West JB. Regional differences in gas exchange in the lung of erect man. J Appl Physiol 1962; 17:893–898.

15. Ball WC, Jr., Stewart PB, Newsham LGS, Bates DV. Regional pulmonary function studied with xenon-133. J Clin Invest 1962; 41:519–531.

16. Milic-Emili J, Henderson AM, Dolovich MB, Trop D, Kaneko K. Regional distribution of inspired gas in the lung. J Appl Physiol 1966; 21:749–759.

17. Bryan AC, Bentivolgio LG, Beerel F, MacLeish H, Zidulka A, Bates DV. Factors affecting regional distribution of ventilation and perfusion in the lung. J Appl Physiol 1964; 19:395–402.

18. Glazier JB, DeNardo GL. Pulmonary function studied with the xenon-133 scanning technique: Normal values and a postural study. Am Rev Respir Dis 1966; 94:188–194.

19. Anthonisen NR, Milic-Emili J. Distribution of pulmonary perfusion in erect man. J Appl Physiol 1966; 21:760–766.

20. Ueda H, Iio M, Kaihara S. Determination of regional pulmonary blood flow in various cardiopulmonary disorders. Jpn Heart J 1964; 5:431–444.

21. Hughes JM, Glazier JB, Maloney JE, West JB. Effect of lung volume on the distribution of pulmonary blood flow in man. Respir Physiol 1968; 4:58–72.

22. Wirz K. Das Verhalten des Druckes im Pleuraraum bei der Atmung und die Ursachen seiner Veränderlichkeit. Pflugers Arch 1923; 199:1–56.

23. Boas JEV. Fehlen der Pleurahöhlen beim indischen Elefanten. Morphol Jahrbuch 1906; 35:494–495.

24. Brown RE, Butler JP, Godleski JJ, Loring SH. The elephant's respiratory system: Adaptations to gravitational stress. Respir Physiol 1997; 109:177–194.

25. West JB. Snorkel breathing in the elephant (abstr). FASEB J 2000; 14(4): A610.

26. Rohrer F. Physiologie der Atembewegung. Handbuch Norm Pathol Physiol 1925; 2:70–127.

27. West JB. Physiological consequences of the apposition of blood and gas in the lung. In: de Reuck AVS, Porter R, eds. Ciba Foundation Symposium: Development of the Lung. London: J & A Churchill, 1967:176–195.

28. Parodi F. La Mécanique Pulmonaire. Paris: Masson, 1933.

29. Milic-Emili J, Mead J, Turner JM. Topography of esophageal pressure as a function of posture in man. J Appl Physiol 1964; 19:212–216.

30. Bryan AC, Milic-Emili J, Pengelly D. Effect of gravity on the distribution of pulmonary ventilation. J Appl Physiol 1966; 21:778–784.

31. Proctor DF, Caldini P, Permutt S. The pressure surrounding the lungs. Respir Physiol 1968; 5:130–144.

32. Krueger JJ, Bain T, Patterson JL Jr. Elevation gradient of intrathoracic pressure. J Appl Physiol 1961; 16:465–468.

33. Turner JM. Distribution of lung surface pressure as a function of posture in dogs. Physiologist 1962; 5:223.

34. Agostoni E. Mechanics of the pleural space. Physiol Rev 1972; 52:57–128.

35. Glazier JB, Hughes JM, Maloney JE, West JB. Vertical gradient of alveolar size in lungs of dogs frozen intact. J Appl Physiol 1967; 23:694–705.

36. Hogg JC, Nepszy S. Regional lung volume and pleural pressure gradient estimated from lung density in dogs. J Appl Physiol 1969; 27:198–203.

37. D'Angelo E. Local alveolar size and transpulmonary pressure in situ and in isolated lungs. Respir Physiol 1972; 14:251–266.

38. Orsos F. Die Pigmentverteilung der Pleura pulmonalis und ihre Beziehung zum Atmungsmechanismus und zur generellen mechanischen Disposition der Lungensptizen fur die Tuberkulose. Verh Deut Pathol Ges 1912; 15:136–149.

39. West JB, Matthews FL. Stresses, strains, surface pressures in the lung caused by its weight. J Appl Physiol 1972; 31:332–345.

40. Kosmodemiansky A. Konstantin Tsiolkovsky, 1857–1935. Moscow: General Editorial Board for Foreign Publications, Nauka Publishers, 1985.

41. Darwin E. Zoonomia: or, The Laws of Organic Life. London: Printed for J. Johnson, 1794.

42. von Diringshofen H. Röntgenaufnahmen und Durchleuchtengen des Herzens bei Fliekrafteinwirkungen im Flugzeug. Luftfahrtmedizin 1938; 2:281–286.

43. von Diringshofen H. Die Wirkung von Fliehkräften auf den Blutkrefslauf des im Flugzeug sitzenden Menschen. Luftfahrtmedizin 1942; 6:152–165.

44. Gauer O. The physiological effects of prolonged acceleration. German Aviation Medicine, World War II. Washington, DC: Department of the Air Force, 1950.

45. Ferry G. L'Aptitude à L'Aviation: Le Vol en Hauteur et le Mal des Aviateurs. Paris: Ballière, 1918.

46. Haber F, Haber H. Possible methods of producing the gravity free state for medical research. J Aviat Med (Aerosp Med) 1950; 21:395–400.

47. Roman JA, Ware RW, Adams RM, Warren BH, Kahn AR. School of Aerospace Medicine physiological studies in high performance aircraft. Aerosp Med 1962; 33:412–419.

48. Henry JP, Ballinger ER, Maher PJ, Simmons DG. Animal studies of the subgravity state during rocket flight. J Aviat Med (Aerosp Med) 1952; 23:421–432.

49. Stone HL, Warren BH, Wagner JrH. The distribution of pulmonary blood flow in human subjects during zero-G. Collected Papers Presented at the 22d Meeting of the AGARD Aerospace Medical Panel, Fuerstenfeldbruck Air Base, Germany, Sept 2–6, 1965. Vol. 2. Paris: Advisory Group for Aerospace Research and Development, NATO, 1965.

50. Foley MF, Tomashefski JF. Pulmonary function during zero-gravity maneuvers. Aerosp Med 1969; 40:655–657.

51. Baumgarten RJv, Baldrighi G, Vogel H, Thümler R. Physiological response to hyper- and hypogravity during rollercoaster flight. Aviat Space Environ Med 1980; 51:145–154.

52. Michels DB, West JB. Distribution of pulmonary ventilation and perfusion during short periods of weightlessness. J Appl Physiol 1978; 45:987–998.
53. Michels DB, Friedman PJ, West JB. Radiographic comparison of human lung shape during normal gravity and weightlessness. J Appl Physiol 1979; 47:851–857.
54. Harford J. Korolev: How One Man Masterminded the Soviet Drive to Beat America to the Moon. New York: Wiley, 1997.
55. Solzhenitsyn AI. The Gulag Archipelago, 1918–1956. New York: Harper & Row, 1974.
56. Gazenko OG. Milestones of space medicine development in Russia (establishment and evolution of the Institute of Biomedical Problems). J Grav Physiol 1997; 43: 1–4.
57. West JB. Historical perspectives: Physiology in microgravity. J Appl Physiol 2000; 89:379–384.
58. Fenn WO. Born fifty years too soon. Ann Rev Physiol 1962; 24:1–10.
59. Gottlieb BJ, Berg A. Das Antlitz des Germanischen Arztes im vier Jahrhunderten. Berlin: Rembrandt-Verlag, 1942.
60. Otis AB. History of respiratory mechanics. In: Fishman AP, Macklem PT, Mead J, eds. Handbook of Physiology. Section 3. The Respiratory System. Vol. III. Bethesda, MD: American Physiological Society, 1986.
61. Rauschenbach BV, Sokolskiy VN, Gurjian AA. Historical aspects of space exploration. In: Nicogossian AE, Mohler SR, Gazenko OG, Grigoryev AI, eds. Space Biology and Medicine. Vol. 1. Washington, DC: American Institute of Aeronautics and Astronautics, 1993:1–50.
62. von Braun W. Multi-stage rockets and artificial satellites. In: Marbarger JP, ed. Space Medicine: The Human Factor in Flights Beyond the Earth. Urbana, IL: University of Illinois Press, 1951:14–30.
63. Exploration. In: McHenry R, ed. The New Encyclopedia Britannica. Vol. 19 (Macropedia). Chicago: Encyclopedia Britannica, 1993:44–58.

2

Effects of Acceleration on the Lung

DAVID H. GLAISTER

Talywern Consultancy, Ltd.
Powys, United Kingdom

All the breathing spring (William Collins, 1721–1759)

By its own weight made steadfast and immovable (William Congreve, 1670–1729)

I. Introduction

The possibility that gravity affected the distribution of pulmonary blood flow was recognized many years ago as a factor in the apical distribution of pulmonary tuberculosis and one reason for its treatment by supine bed rest (1). Gravity, the force that gives a body its normal weight, will cause an unsupported body to fall at an acceleration of 9.81 m sec^{-2}, the gravitational constant, with which all land dwellers are so familiar that it is taken for granted. Increased levels of acceleration are less familiar, but are inevitably associated with change in speed (linear acceleration/deceleration), or change in direction when traveling at a uniform speed (centrifugal acceleration). Linear accelerations tend to be of modest magnitude and of short duration due to the great speeds generated, the exception being the launch forces in space flight, but considerable centrifugal forces are encountered in aerobatics, or even in car racing and fairgrounds.

From the equation:

$$a = v^2/r$$

(where a = acceleration in m/s^2, v = circumferential velocity in m/sec, and r = radius of turn in m), an aircraft flying at a speed of 500 knots (258 m/s) round a circular path with a radius of half a kilometer will expose its crew to a radial acceleration of 66.3 m/s^2, or 6.8 G, (the capital G denoting the ratio of the applied acceleration to the gravitational constant, small g). Furthermore, such a maneuver can be maintained for many seconds, or even minutes, long enough for increased hydrostatic pressure gradients within the cardiovascular system to lead to black-out (retinal arterial pressure falling below intraocular tension with cessation of retinal perfusion), or to loss of consciousness (cessation of cerebral perfusion). These effects were well researched during and following World War II as fighter aircraft became faster and more maneuverable. Since these effects can develop within a few seconds, and space flight demanded several minutes of acceleration exposure, NASA and the Soviet space program subsequently looked at the supine, or semisupine, posture as a means of protecting astronauts and cosmonauts on their way to orbiting the Earth.

Postural effects of gravity are simply defined according to body position, the force always acting perpendicular to the Earth's surface. Aircraft and space flight, on the other hand, can develop forces in any direction independent of the Earth's surface, so a standard acceleration nomenclature was devised based upon human body axes (2). In this system, forces along the body's long axis are termed + or $-G_z$, lateral forces are + or $-G_y$, and fore-and-aft forces are + or $-G_x$. Specific force vectors and terminology are given in Table 1. Note that the inertial force always acts in the opposite direction to the applied acceleration so that as a car accelerates forward, for example, the inertial reaction forces its driver back into his seat.

Because blackout and loss of consciousness comprised such effective acceleration exposure end points to aircrew performance, researchers in aviation physi-

Table 1 Acceleration and Inertial Force Terminology

Direction of acceleration	Direction of resultant inertial force	Standard terminology
Headward	Head to foot	$+G_z$
Footward	Foot to head	$-G_z$
Forward	Sternum to spine	$+G_x$
Backward	Spine to sternum	$-G_x$
To the right	Right to left	$+G_y$
To the left	Left to right	$-G_y$

ology had little incentive to investigate the effects of acceleration on the mechanics of breathing, despite the pioneering research of Otis et al. (3). Subsequently, the need for further research into the effects of $+G_x$ acceleration for both space flight and military aviation, together with the availability of radioisotope techniques pioneered by West and his colleagues at the Hammersmith Hospital, London, and single-breath techniques using inert insoluble tracer gases, led to a thorough investigation of the effects of acceleration on the lung. This chapter discusses research carried out in the main at the Royal Air Force (RAF) Institute of Aviation Medicine, Farnborough (now the Centre of Aviation Medicine, Royal Air Force, Henlow), and stimulated initially by the condition of acceleration atelectasis that developed in aircrew exposed to $+G_z$ while breathing oxygen (4,5).

II. Instrumentation

The centrifuge (Fig. 1) offers a unique environment for physiological research, but poses problems of instrumentation and experimentation. Most centrifuges, as with that at Farnborough, can carry only the experimental subject, so he must be trained to conduct any experimental procedures on himself during a test run. These may include breathing maneuvers, the turning of taps, intravenous injec-

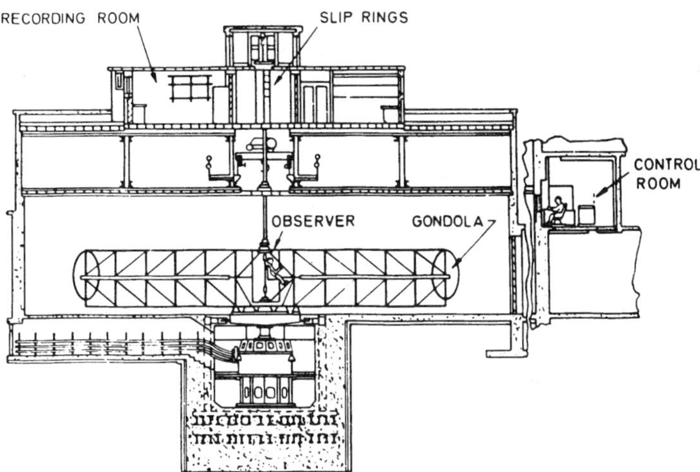

Figure 1 The layout of the man-carrying centrifuge at the RAF Institute of Aviation Medicine, Farnborough, United Kingdom. The centrifuge has a working radius of 30 ft (9.15 m) and was capable of 30 G at an onset rate of 1 G/s. The later addition of a 28.5-m long capillary line and rotary coupling allowed the use of a mass spectrometer in the recording room. (From Ref. 6.)

tions, arterial blood sampling, and so forth, often while blacked out, or nearly unconscious! Furthermore, any instrumentation may be affected by inertial forces and great care must be taken over the orientation of sensitive pressure transducers and to correct for hydrostatic pressure gradients in fluid-filled systems. Until the installation of a hypodermic gas sampling line and rotary coupling (7), gas analysis had to be by the Haldane sampling technique, or with then-available devices such as a nitrogen analyzer used for Aircrew mask leak detection. While the gas sample line gives good response times, the 10- to 12-sec delay required careful correction in the production of x–y plots, for example. The weight and space restrictions of the centrifuge gondola also necessitated the development of specialized instrumentation such as low-weight scintillation detectors and collimators (8,9) and the use of long leads connected to remote instrumentation through slip rings.

III. Pulmonary Ventilation

A. Mechanics of Breathing

Due to the increased weight of the anterior chest wall that has to be lifted against inertial forces in the supine posture, $+G_x$ causes a significant shift to the right in thoracic relaxation pressures (Fig. 2) and a consequent decrease in vital capacity (VC), this being halved at $+6\ G_x$ and reduced to equal tidal volume (V_T) by $+12\ G_x$, with breathing becoming impossible at higher levels of acceleration (10). The expiratory reserve volume (ERV) is also markedly reduced, falling close to 0 at even low levels of $+G_x$, though the residual volume (RV) appears to be unaffected by this or any other axis of acceleration. For simple mechanical reasons, other acceleration vectors have less effect on the divisions of lung volume, though the functional residual capacity (FRC) is highly sensitive to both acceleration axis and magnitude, just as it is highly sensitive to body posture under the effects of normal gravity.

B. Lung Compliance

Esophageal pressure measurement poses particular problems on the centrifuge with measurements highly sensitive to balloon placement and interpretation, especially when values become positive. $+G_z$ has little effect on lung compliance, but pressure measurements made during slow expirations from total lung capacity (TLC) to RV and smoothed to remove cardiogenic oscillations showed a progressive fall in compliance from 215 mL/cmH$_2$O at $-1\ G_x$ to 145 mL/cmH$_2$O at $-4\ G_x$ (11). However, this observation does not imply any increase in stiffness of individual lung units since the overall pressure–volume curve comprises the sum of a large number of individual curves distributed along a common pressure axis by the vertical gradient of transpulmonary pressure. This effect is discussed

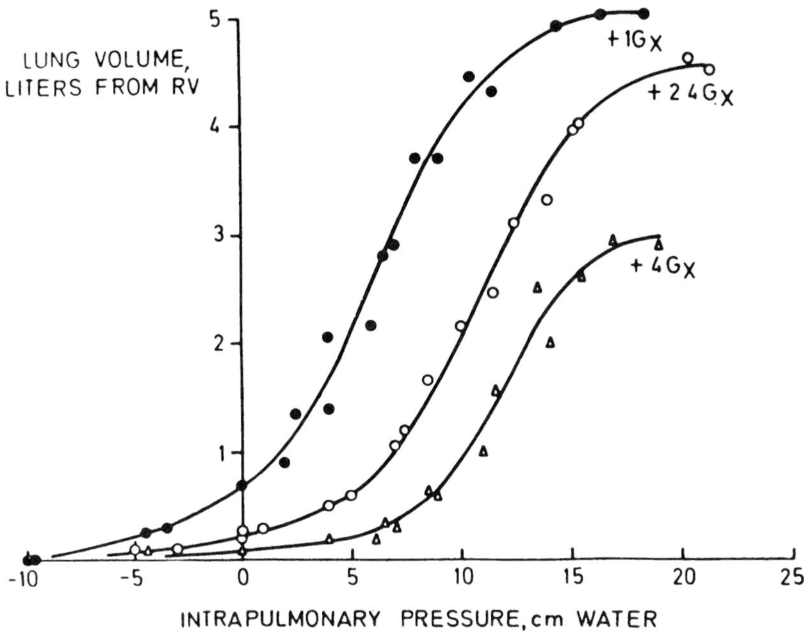

Figure 2 Effect of $+G_x$ acceleration on the relaxation pressure curve of the thorax. (From Ref. 6.)

in detail later (Secs. VI and VII). Direct measurement of the esophageal pressure gradient using a balloon with two bladders, one above the other, showed a linear increase in the recorded pressure difference with accelerations of up to $+4\ G_z$ (12), so that, by expanding the horizontal pressure axis, acceleration would be expected to make the lung's pressure–volume curve less steep.

C. Single-Breath Tests

Analysis of expired air made on the centrifuge at $+3\ G_z$ during a slow maximal expiration exhibited a terminal fall in CO_2 concentration (13). This increased in extent, and commenced earlier, with increasing $+G_z$ (Fig. 3). This fall was considered to indicate the preferential emptying of a poorly perfused lung region toward the end of expiration, or, with hindsight, the closure of well-perfused lower lung units when the closing volume (CV) was increased by acceleration. Similar terminal falls in CO_2 concentration can be seen with exposure to $+G_x$ acceleration (6) and with $\pm G_y$ (11), though in this latter case there is a double fall due to sequential emptying of the two lungs, which is discussed in detail later (Secs. VI and VII). Measurement of CV using a modification of Fowler's

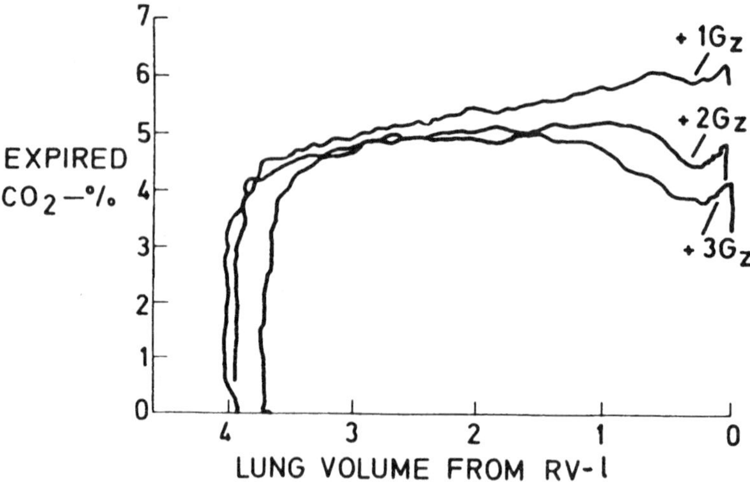

Figure 3   Expired CO_2 concentration plotted against expired volume for slow vital capacity expirations at $+1$ G$_z$, $+2$ G$_z$ and $+3$ G$_z$. (Redrawn from Ref. 13.)

single-breath technique confirmed a linear increase with $+G_z$ (Fig. 4), showing that closure is a feature of lung weight, rather than of regional differences in mechanical properties (14).

D.   Distribution of Pulmonary Ventilation

The acceleration-dependent gradient in transpulmonary pressure as demonstrated by measurement of esophageal pressure means that alveoli at differing vertical levels within the lung will, at any single overall lung volume, be at varying points on their pressure–volume curves and so have different regional volumes. This situation is illustrated in Fig. 5, which shows a typical pressure–volume curve with the volume change from RV to TLC taking a pressure change of some 30 cmH$_2$O. The lung is assumed to be 30 cm tall and to have a uniform specific gravity of 0.2. The inset diagram shows a lung at $+5$ G$_z$ with apical units expanded to near regional TLC (TLC$_r$) by a transpulmonary pressure of some 18 cmH$_2$O, while basal units are at their regional residual volume (RV$_r$) with a transpulmonary pressure of -5 cmH$_2$O. If the lung is then inflated by a 5-cmH$_2$O increase in transpulmonary pressure, the predicted regional volume changes will be as shown by the triangles. The inset graph shows the predicted regional volume change as a percentage of TLC plotted against transpulmonary pressure (or vertical position down the lung at $+5$ G$_z$). Ventilation increases with distance down the lung to peak at two-thirds distance before falling off to 0 at the base. A further

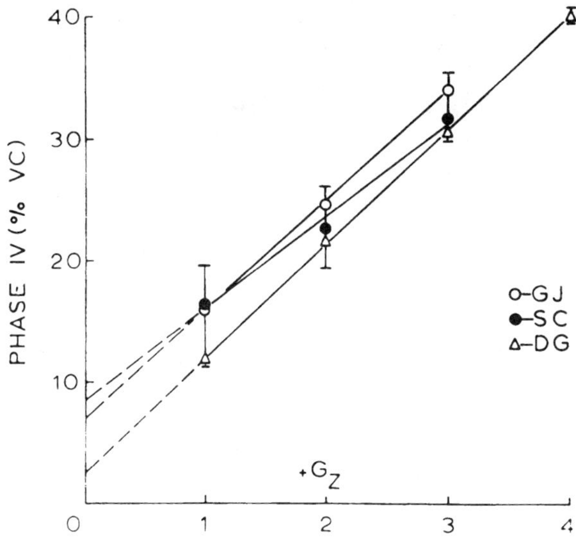

Figure 4   Effect of $+G_z$ on CV in three subjects, measured as the lung volume at which an upward inflexion (phase IV) appears on the washout of a tracer gas inspired as a bolus from RV. Bars indicate the overall range of measurements. (From Ref. 14.)

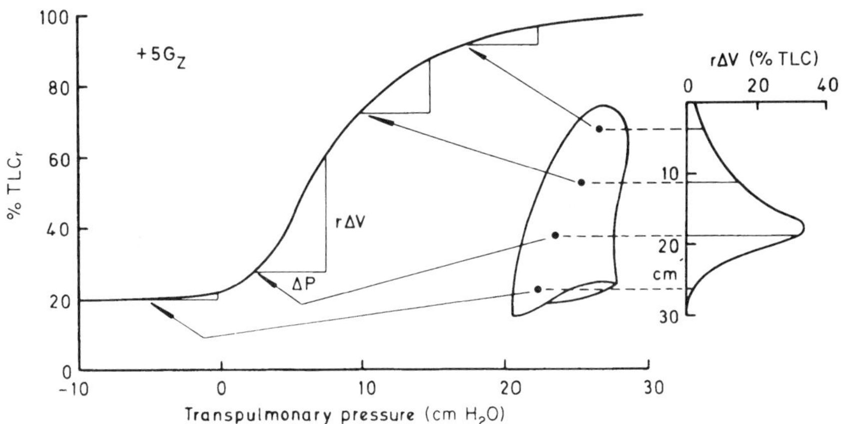

Figure 5   Transpulmonary pressure at four levels within a 30-cm tall lung indicated on the lung's pressure–volume curve for a lung with a specific gravity of 0.2, and at an acceleration of $+5$ G_z. The regional volume changes that would result from a uniform increase in inflation pressure of 5-cmH$_2$O are indicated by the triangles and plotted against distance down the lung in the diagram. (From Ref. 15.)

prediction of this simple model is that these units in the lower lung will have closed airways. These predictions are tested in the following two sections.

E. Xenon Studies of Ventilation

The regional ventilation of the lung was investigated on the centrifuge by the ^{133}Xe technique described by Ball et al. (16) using two focusing collimators to examine two 2.0-inch (~5-cm) diameter fields in the right lung, 1.0 inch (~2.5 cm) and 6.0 inches (~15 cm) below the clavicle (17). The rate of washin of ^{133}Xe was used as an index of ventilation and averaged 1.4:1 at +1 G_z and 2.1:1 at +3 G_z. Subsequently, the distribution of ventilation was measured using the same isotope and the lung-scanning method described by Dollery and Gillam (18). The results of lung scanning during exposure to +G_z were confounded by displacement of the diaphragm and the resulting poor definition of the lung base in consecutive scans carried out at different lung volumes, but this problem was eliminated in studies on supine subjects in whom the back of the chest was immobilized on a rigid formfitting couch and the anterior chest wall identified on each scan by use of a small radioactive marker. The collimators used in the scanning gave a vertical resolution of the order of 1.5 cm (9). Figure 6 shows the effect of +G_x acceleration on the distribution of ventilation per unit lung volume of 1.2 L (upper panel) and 2.2 L (lower panel) of air inspired from RV. While acceleration has no significant effect on the ventilation of the upper lung, it causes a dramatic falloff in the most dependent zone with, at +5 G_x, no ventilation for the lowermost 3 cm. The lower panel also clearly demonstrates the predicted increase in ventilation down the upper regions of the lung.

F. Acceleration Atelectasis

Acceleration atelectasis is a condition of absorptional collapse of the lung bases seen in fast-jet aircrew breathing 100% oxygen and pulling G, and exacerbated by the wearing of anti-G trousers (4,5). The acceleration-induced increase in CV, together with the decrease in FRC caused by the abdominal pressure exerted by the anti-G trousers, means that the aircrewman will be breathing with an end-inspiratory volume less than his CV. Airways in the dependent lung will remain closed and the trapped soluble gases will be absorbed. Symptoms comprise retrosternal pain on attempting to take a deep breath, an inspiratory catch, and tendency to cough. Basal lung collapse can be demonstrated by x-ray. The VC is markedly reduced (Fig. 7), and may remain so for many hours, but recovers rapidly following deep breaths or a fit of coughing. The addition to the breathing gas of a minimum of 40% nitrogen virtually eliminates the occurrence of atelectasis as its poor solubility keeps the nonventilated alveoli patent so that surface tension forces do not prevent the terminal bronchioles from reopening once the aircrewman returns to level flight.

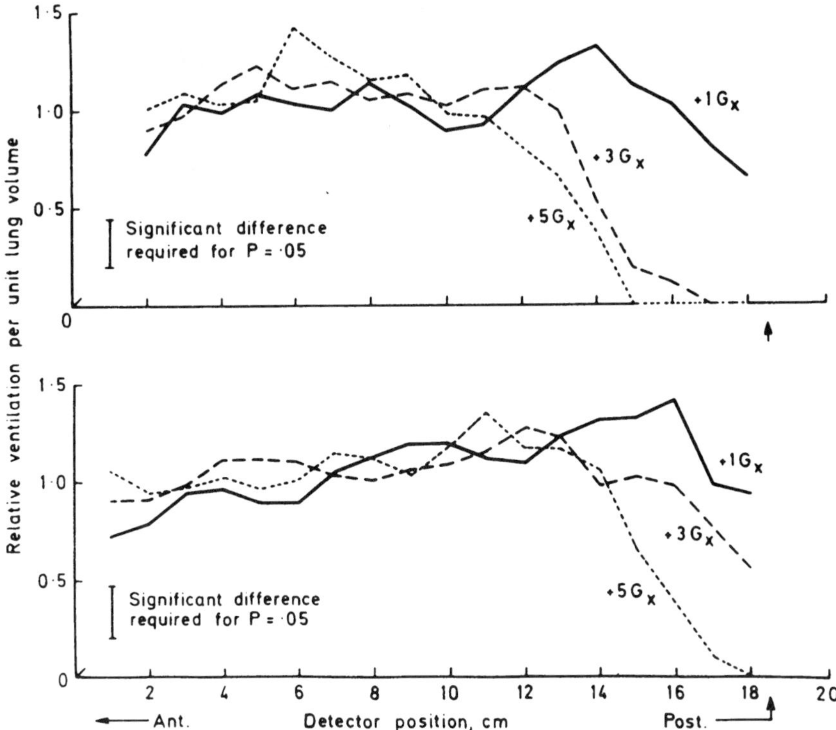

Figure 6 The effect of $+G_x$ on the regional distribution of inspired gas. The distribution of the first 1.2 L is shown in the upper panel, that of the next 1.0 L in the lower panel. Data are averaged from five subjects. (From Ref. 19.)

Studies in which the rate of washout of radioactivity from the lungs was recorded following a rapid intravenous injection of ^{133}Xe (17) showed that the washout from the lower lung at $+3\ G_z$, when anti-G trousers were worn, comprised two components, a fast one contributed by alveoli half cleared in 10 to 20 s, and a slow one taking more than 2 min to be half cleared. When the run was repeated following the breathing of 100% O_2 to induce atelectasis, the slow component disappeared and the VC was reduced by 56%. Furthermore, the placement of a scintillation detector over the abdomen to measure systemic ^{133}Xe showed a fourfold increase in radioactivity following the induction of atelectasis. It was concluded that, at $+3\ G_z$, the lower lung region contained two populations of alveoli, both containing gas so that the poorly soluble ^{133}Xe could diffuse into them from the pulmonary capillaries, but one was ventilated while the other was not. That the nonventilation was due to airway closure and gas trapping was

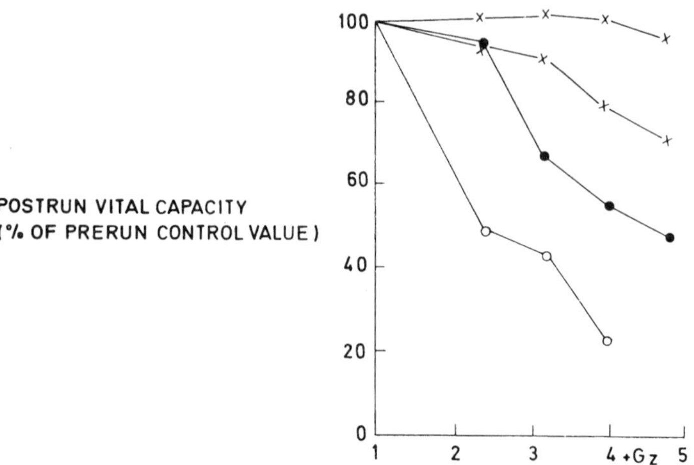

Figure 7 Reductions in VC as a percentage of the prerun control plotted against acceleration level for four subjects. Subjects wore inflated anti-G trousers and breathed 100% O_2 for 15 min before and during 1-min centrifuge runs. (From Ref. 6.)

proved by the fact that these alveoli became atelectatic when initially containing 100% O_2, and the resulting barrier to gas diffusion caused intravenously injected ^{133}Xe to pass on into the systemic circulation. Centrifuge studies have shown radiographically that the collapse always develops in the dependent lung, regardless of posture, and can be seen in the lung apices following exposure to $-G_z$ (5).

G. Isolated Lung Studies

The effect of a pleural pressure gradient was investigated on intact lungs dissected free from dogs and Macaque monkeys by immersion in a fluidized bed (20,21). This consisted of slightly expanded polystyrene beads through which was passed a uniform upward current of air that caused them to behave like a macromolecular fluid with a bulk density of about 0.2 g/ml. The lungs were continuously cycled between tracheal pressures of -5 cmH$_2$O and $+30$ cmH$_2$O with a cycle time of approximately 2 min and pressure–volume curves were recorded. In addition, boluses of ^{133}Xe gas were added to the circuit at minimal lung volume and the subsequent washout continuously analyzed for radioactivity, or the cycle was stopped after one full inflation and the lungs scanned to measure the resulting vertical distribution of radioactivity.

The upper portions of both the inflation and deflation limbs of the pressure–volume curves were accurately fitted by simple exponential functions of the form VC% $= a - be^{-kP}$ where a, b, and k are arbitrarily fitted constants. Constant k is an index of lung compliance and for any one lobe was unaffected by immersion in the fluidized bed. An example recorded from a dog lung is illustrated in Figure 8 with the best-fit exponentials shown as dashed curves. Points of departure of the recorded pressures from the extrapolated exponential fits are considered to represent the onset of airway closure on the deflation limb, P_c, and completion of airway opening on the inflation limb, P_o. The lung volume at which the point of departure occurred during deflation was the same as that of the onset of the phase IV in the simultaneously recorded bolus washout under all experimental conditions in both species investigated, strong evidence that it does indeed represent the onset of airway closure. The pressure at which P_c occurred was consistently positive (range $+2$ to $+7$ cmH$_2$O) and was raised by immersion of the lung in the fluidized bed. Similar correlations between P_c and the onset of phase IV were seen in human subjects with acceleration exposure (22) and with increasing age (23).

The major effect of immersing the lung in the fluidized bed, however, was on the distribution of the ''inspired'' gas. Figure 9 illustrates the distribution of ^{133}Xe in a Macaque monkey lung after a bolus of the gas had been given at minimal lung volume with the lung suspended in air (solid line) and then when immersed in the fluidized bed (dashed line). With the lung suspended in air, the inspirate was uniformly distributed, but in the presence of a pleural pressure

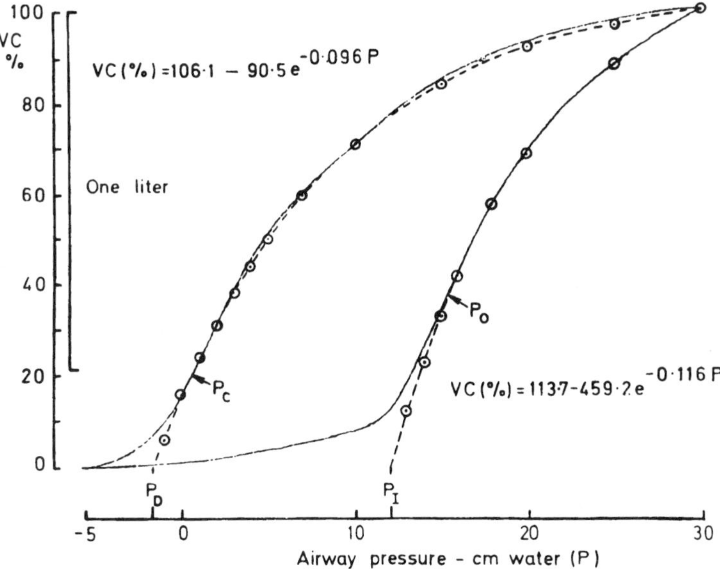

Figure 8 Pressure–volume curve for an isolated dog lung suspended in air. The open circles and dashed lines are the best-fit exponentials with constants as indicated. (From Ref. 20.)

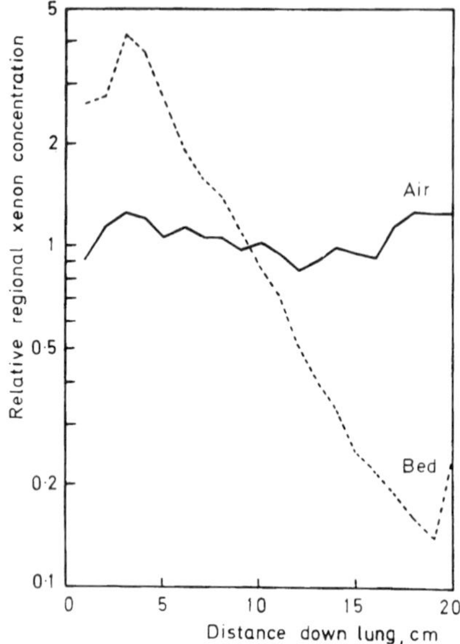

Figure 9 Vertical distribution of radioactivity in a monkey lung after the injection of a bolus of ^{133}Xe at minimal volume with the lung suspended in air (solid line) and immersed in a fluidized bed (dashed line). (From Ref. 21.)

gradient, the inspirate was initially distributed almost exclusively to the uppermost lung regions. A similar effect was noted in dog lungs, though the presence of regional differences in compliance meant that, even with the lungs suspended in air, inflation initially favored the upper lobe, and significant gradients in bolus distribution were also seen within individual lobes. Since the distribution of the inspired gas was grossly uniform in the monkey lung suspended in air, a primate with lung morphology similar to that of humans, it is reasonable to suppose that in humans, too, the lung's gross lobar elastic properties are uniform and that the uneven distributions of inspired gas seen in the human lung under normal gravity, as well as during exposure to increased acceleration, are the result of pleural pressure gradients.

IV. Pulmonary Perfusion

A. Distribution of Pulmonary Blood Flow

The three-zone lung model of West (24) can be used to predict the effect of acceleration on the distribution of pulmonary blood flow if the pulmonary arterial pressure is known at a given level. Pressures were measured in subjects seated on the centrifuge following pulmonary arterial catheterization (12), at accelerations of up to $+3$ G_z, and the findings are illustrated in Figure 10. The reference level for these measurements was that of the pulmonary trunk, but since hydrostatic forces will cause pressures to rise at lower levels, there will be a level of hydrostatic indifference at which the falls seen in Figure 10 are precisely balanced. This was estimated to lie some 5 cm below the hilum of the lung, or some 20 cm below the apex of a 30-cm tall lung.

Figure 11 shows the theoretical pressure–flow relationships in an upright lung at $+3$ G_z. Pulmonary arterial pressure is taken to be 15 cmH_2O at a level of hydrostatic indifference 5 cm below the lung hilum, with venous pressures a constant 10 cmH_2O less. Note that the use of a nonstandard pressure unit (cmH_2O

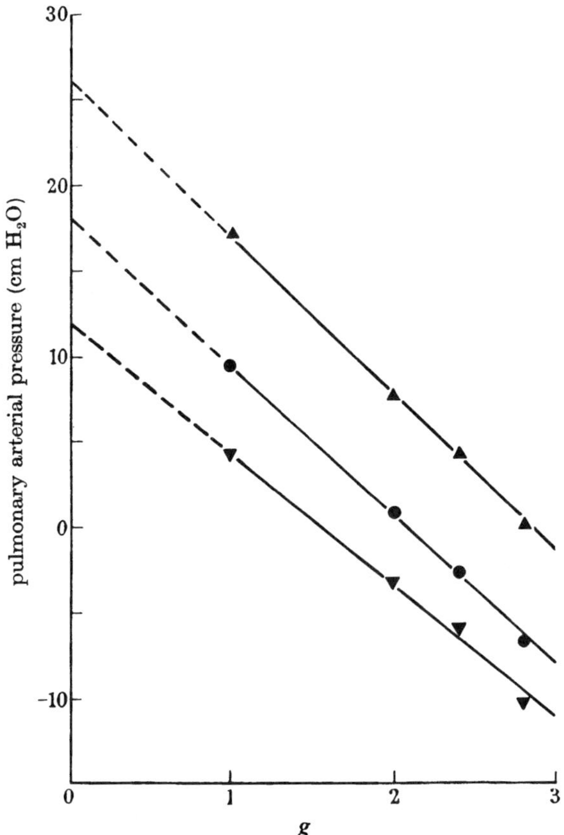

Figure 10 Effect of $+G_z$ acceleration on pulmonary arterial systolic (▲), mean (●), and diastolic (▼) pressures measured with the transducer at the level of the pulmonary trunk. Average data from four subjects. (From Ref. 12.)

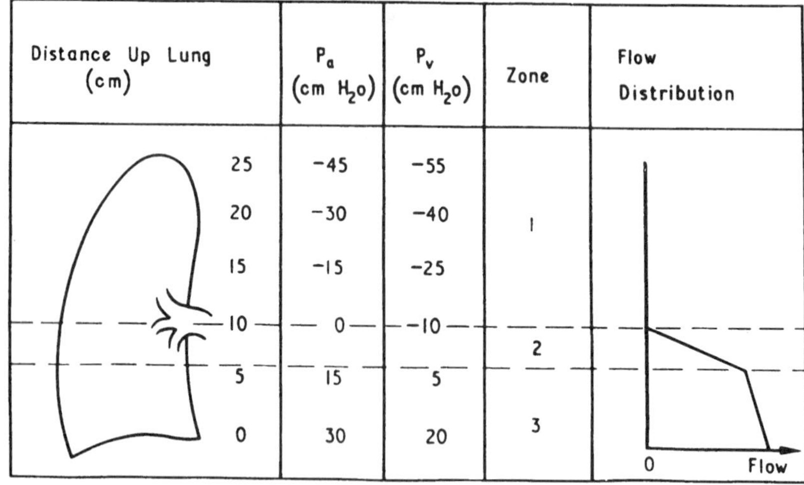

Figure 11 Theoretical pressure/blood flow relationships in an upright lung at $+3$ G_z, based on the three-zone model of West. (From Ref. 6.)

rather than mmHg) allows direct comparison with hydrostatic pressure effects, taking the specific gravity of blood to be unity. Above the level of the hilum, arterial pressure has fallen below alveolar pressure and there will be no blood flow (West's zone I). In zone II, arterial pressure is greater than, but venous pressure less than alveolar pressure and flow is proportional to the arterial–alveolar pressure difference (the waterfall or Starling resistor effect). Since the airways are open and alveolar pressure a constant, flow becomes simply proportional to arterial pressure. Thus, the hydrostatic increase in arterial pressure down zone II will cause a linear increase in flow that should be 3 times as steep at $+3$ G_z as under normal gravity. Below this level, both arterial and venous pressures are greater than alveolar pressure, the arteriovenous pressure difference remains constant and any increase in flow down zone III will be caused by passive distension of the capillary bed as hydrostatic pressures rise.

B. Xenon-133 Studies: $+G_z$

A late-emptying zone of unperfused lung was inferred from the expired CO_2 studies discussed previously (see Fig. 3), and the early ^{133}Xe studies using fixed scintillation counters showed an upper-to-lower ratio of blood flow per unit lung volume of 2.7:1 at $+1$ G_z and 5.9:1 at $+2$ G_z, while at $+3$ G_z blood flow was no longer detectable in the upper lung field centered 1.0 inch below the clavicle

(17). Bryan et al. (25) subsequently demonstrated a basal shift in perfusion using injections of macroaggregated albumin labeled with ^{131}I injected intravenously at up to +4 G$_z$, though the changes seen from +2 G$_z$ to +4 G$_z$ were small, and even at +4 G$_z$, flow extended well up the lung with no clear unperfused zone I. However, this method of measurement suffers from the inability to correct for the varying volume of lung seen down the scan, poor collimation of the energetic gamma radiation, and the fact that the scans were carried out later, away from the centrifuge. The development of flat-beam collimators and a scanning detector system that could be mounted in the centrifuge gondola, together with the use of ^{133}Xe, allowed precise measurement of the distribution of the pulmonary arterial blood flow (12). In each centrifuge run, an intravenous injection of ^{133}Xe dissolved in saline was made during breathholding and the lung scanned to determine its distribution following diffusion into the alveolar gas. A further scan was made at the same lung volume and level of acceleration after rebreathing ^{133}Xe to permit correction for the varying volume of lung seen down the scans. Typical scans obtained at +3 G$_z$ are illustrated in Figure 12 and clearly show that the

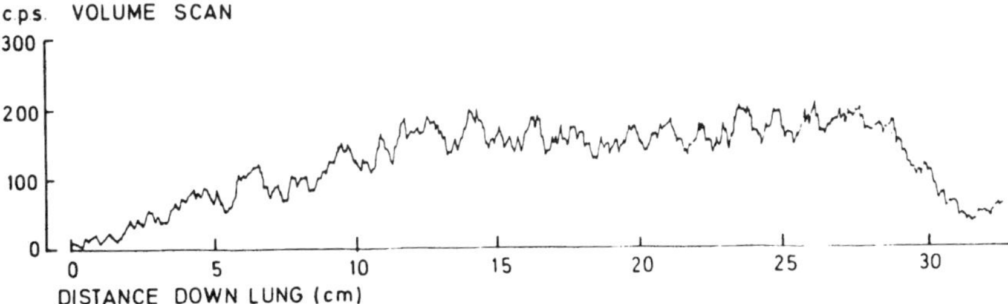

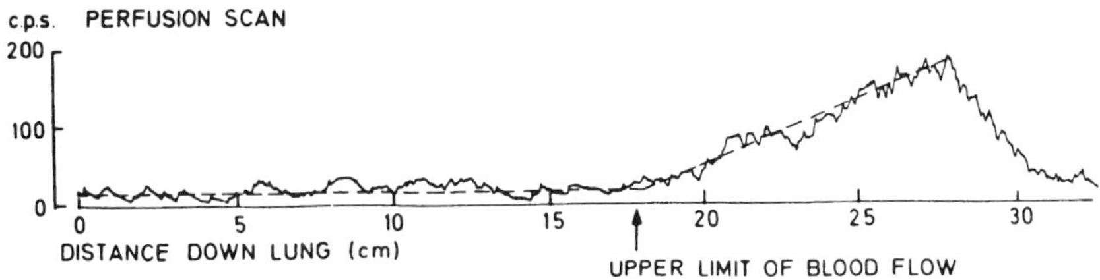

Figure 12 Distribution of radioactivity down the right lung at +3 G$_z$ following the intravenous injection of a saline solution of ^{133}Xe (lower, perfusion scan) and following rebreathing of the same isotope to give uniform distribution in the alveolar gas (upper, volume scan). (From Ref. 6.)

upper half of the lung is unperfused. When smoothed, these scans can be used to give blood flow per unit lung volume as illustrated in Figure 13. The slopes of the solid lines in this figure are in the ratios of 1:2:3, so that blood flow remains proportional to the hydrostatic pressure gradient, as predicted by the West model.

C. Blood Flow Distribution: $+G_z$

Hoppin et al. (26) used the ^{131}I-labeled macroaggregated albumin technique to measure the distribution of pulmonary blood flow in subjects riding on the Johns-

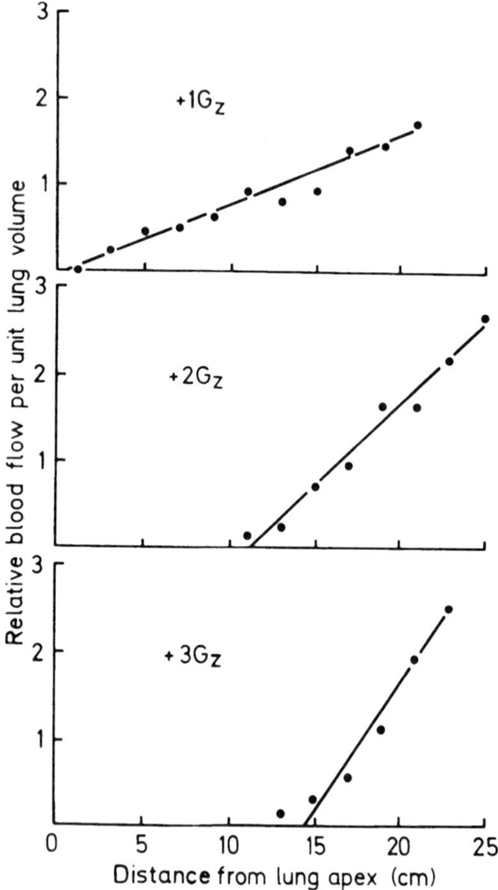

Figure 13 Effect of $+G_z$ acceleration on the vertical distribution of relative pulmonary blood flow per unit lung volume. (From Ref. 15.)

ville centrifuge at up to $+8\ G_z$, but found little change despite predicted arterial pressures of up to 90 torr. However, subsequent ^{133}Xe studies on the Farnborough centrifuge showed a highly significant increase in the gradient of blood flow per unit lung volume at $+5\ G_x$, with flow falling close to 0 at the anterior lung margin (19). Figure 14 shows average results from four subjects and what appears as a dramatic increase in flow in the posterior lung. A modification of the technique was used to measure blood flow per alveolus rather than per unit lung volume so as to correct for variations in alveolar size. Injections of ^{133}Xe were made while under G and at end-inspiration as before, but after a few seconds to allow

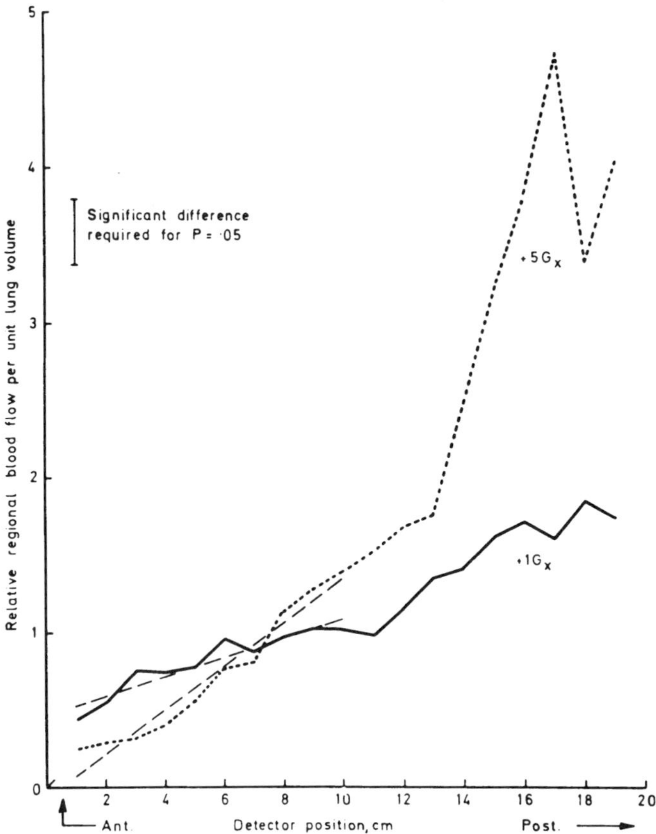

Figure 14 Effect of $+G_x$ acceleration on the regional distribution of blood flow per unit lung volume. Average data from four subjects. The slopes of linear regressions obtained from the upper 10 cm of the scans (dashed lines) are highly significantly different ($p = 0.001$). (From Ref. 19.)

the injectate to be distributed to the lungs, the centrifuge was brought to a stand-still while the subject inspired to TLC so as to achieve uniform alveolar expansion before the scan was made. A single rebreathing scan was then used to correct for the varying field of view of the collimator and a quite different picture emerged as illustrated in Figure 15. Blood flow over the upper 10 cm of lung increases linearly and at a rate precisely proportional to acceleration and, therefore, to the hydrostatic pressure gradient within the pulmonary circulation. The pattern is somewhat different from that predicted from the simple three-zone model, however. Thus, at $+5$ G_z, flow increased linearly over a vertical extent of some 10 cm, equivalent to a pressure increase of 50 cmH$_2$O, too great to be accounted for solely by zone II pressure relationships. It must be assumed that the rate of increase caused by passive capillary distension (zone III) is similar to that of zone II and that there is a further, fourth, zone of decreasing blood flow, which occurs despite continuing increases in intravascular pressures.

West et al. (27) accounted for a similar zone of decreasing flow seen in the isolated dog lung on the basis of increasing extravascular pressure and alveolar edema. Such a mechanism could explain a leveling off in flow, but since the densities of blood, tissue, and tissue fluid are effectively the same, it is difficult to see how it can explain a *decrease* in flow with distance down the lung. Also, the acceleration exposure times were probably too brief for significant edema fluid to form; indeed, pulmonary edema has never been noted in human centrifuge

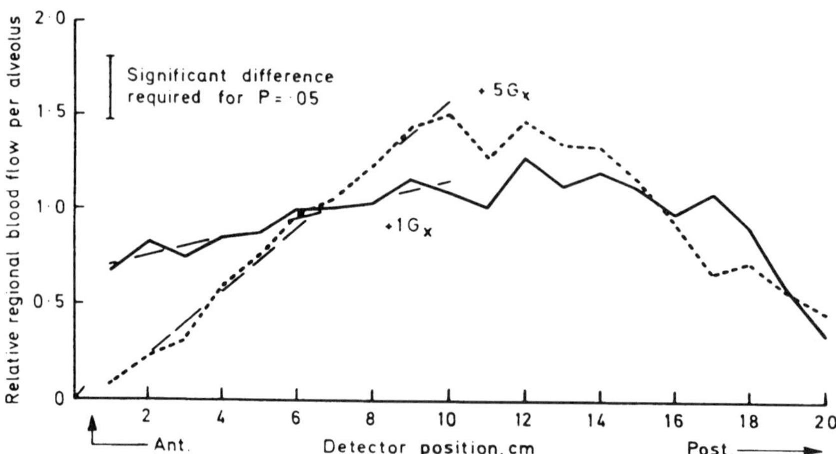

Figure 15   Effect of $+G_x$ acceleration on the regional distribution of blood flow per alveolus. Average data from three subjects. The slopes of linear regressions obtained from the upper 10 cm of the scans (dashed lines) are highly significantly different ($p = 0.001$). (From Ref. 19.)

exposures. The $+G_x$ ventilation studies referred to earlier (Sec. III.E) demonstrated a nonventilated zone at the back of the lung, which was presumed to indicate regional closure of airways. A further measure of regional alveolar size can be obtained by dividing regional flow per unit volume (see Fig. 14) by regional flow per alveolus (see Fig. 15). When allowance was made for the different vertical extent of the lungs at the time of the scans (TLC at $+1$ G_x and end-expiration at $+5$ G_x, respectively) the result was as shown in Figure 16 with the relative alveolar volume falling to 20% of the regional TLC at the back of the lung. Again, this is evidence that these alveoli were at their regional RV and that their airways would have been closed, and offers a possible explanation for the decreased blood flow noted earlier. First, alveolar pressure will no longer be atmospheric, but will rise to that of the surrounding tissues, so offering a zone of potentially constant pressure/flow characteristics; second, through reflex or mechanical constriction of the vascular bed, blood flow to closed off alveoli is reduced. Such a mechanism makes good sense, teleologically, since it will act to reduce ventilation/perfusion inequalities. A fourth zone of decreasing blood flow with distance down the lung was not seen in the $+G_z$ studies (see Fig. 13), presumably because the position of the lowermost lung within the scanning field of view changed significantly with the varying level of the diaphragm, so making data imprecise in the region of concern.

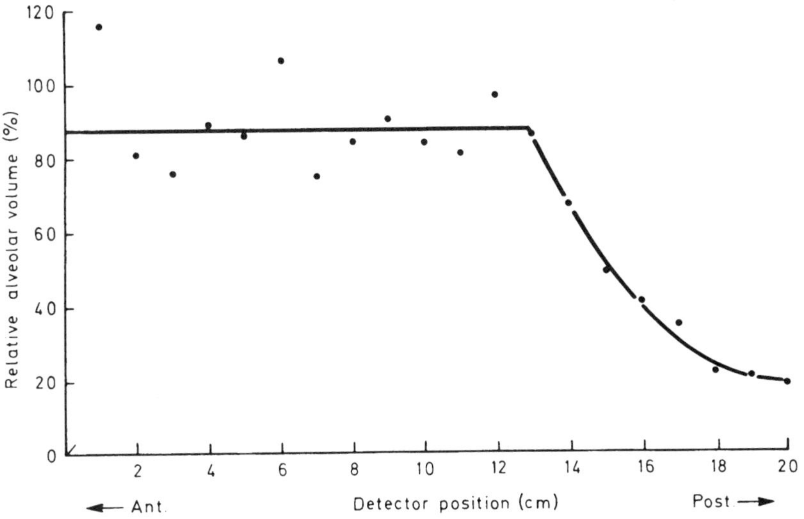

Figure 16 Regional distribution of relative alveolar volume at FRC and $+5$ G_x expressed as a percentage of that at TLC and $+1$ G_x. (From Ref. 15.)

V. Consequences of Acceleration-Induced Ventilation and Perfusion Inequalities

A. Ventilation/Perfusion Ratio

Taking values of unity for the overall ventilation and perfusion of the lung, the relative ventilation/perfusion ratio distributions can be calculated from the relevant scans. $+G_z$ acceleration causes an increasing volume of unperfused, but ventilated alveoli in the upper lung (a simple addition to the anatomical dead space), with the ventilation/perfusion ratio then falling more steeply to reach a value of some 0.5 at the lung base (12). Lower values were not seen since a zone of nonventilated lung was not identified in the scans for the reason given above. The more precise $+G_x$ data did show such a zone, and the resulting ventilation/perfusion ratios are shown in Figure 17. Under normal gravity with the subject supine $(+1\ G_x)$, most of the lung has a ventilation/perfusion ratio close to unity (values between 0.5 and 2.0), but at $+5\ G_x$, only a 5-cm slice in the center of

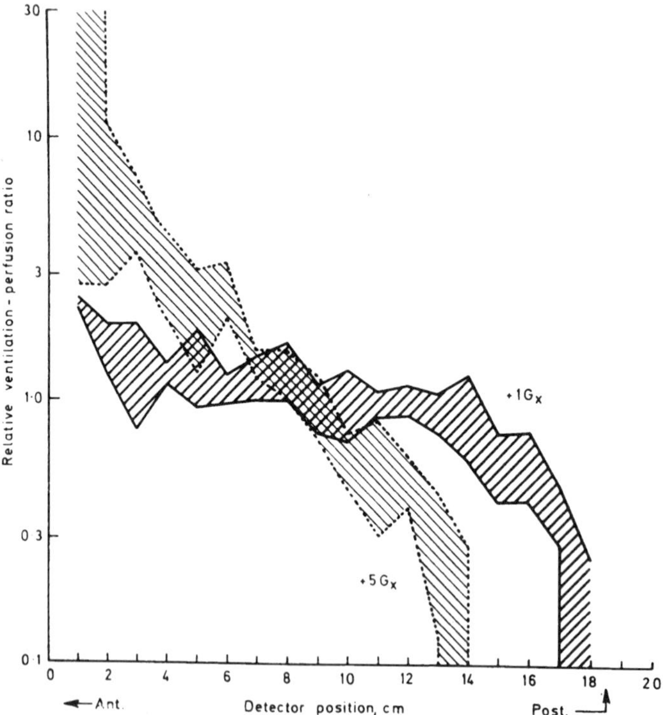

Figure 17 Relative ventilation/perfusion ratios plotted against distance down the anteroposterior axis of the lung at $+1\ G_x$ and $+5\ G_x$. The shaded areas outline the range of values found in three subjects. (From Ref. 19.)

the lung has such values, with ratios of greater than 3 seen in the anterior lung and of 0 throughout the most posterior few centimeters.

B. Arterial Desaturation

The perfusion of unventilated lung units will initially promote further gas exchange until equilibrium is reached with the pulmonary arterial blood. Thereafter, continued perfusion will constitute a right-to-left shunt with ensuing arterial desaturation. Figure 18 shows values for arterial saturation published from divers centrifuge studies, each point representing the average value found from at least three, and up to 31 subjects, with the acceleration axes as indicated. For the obvious reason of human tolerance, the higher values are all from $\pm G_x$ measurements, but saturations start to fall at 4 G independent of posture, with values down to 70 to 80% at the higher levels of acceleration. The single normal value for a prone subject at 6 G (open circle) stands out from the others, the explanation being that the FRC is increased by $-G_x$ acceleration so that the subjects will be breathing with their end-inspiratory volumes greater than their CVs, and dependent lung units will remain ventilated, at least during part of each breath.

Transient changes in arterial O_2 tension were recorded in anesthetized dogs during $+2\,G_z$ to $+5\,G_z$ exposures lasting 1 to 2 min using a microelectrode with

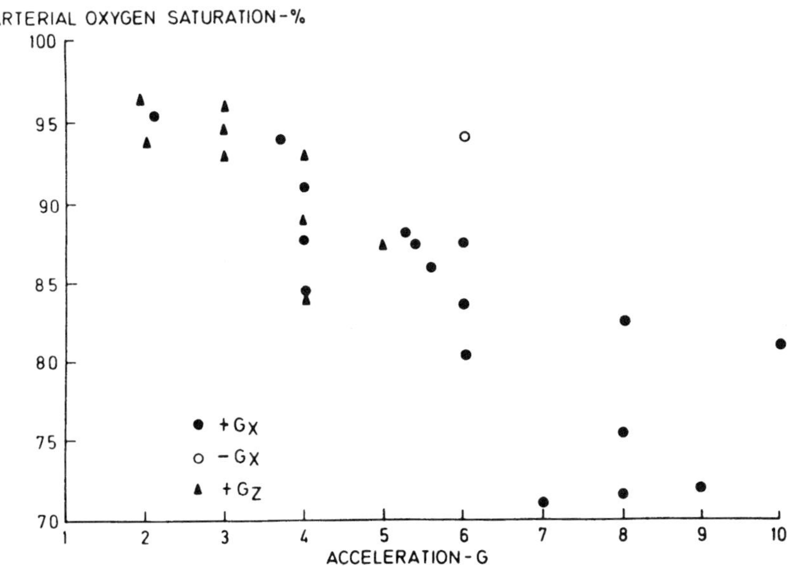

Figure 18 Arterial oxygen saturations reported for human subjects exposed to the varying axes of acceleration as indicated. Points represent average values from 3 to 31 subjects during centrifuge exposures of from 50 s to 6 min, breathing air. (From Ref. 6.)

its tip lying in the femoral artery (28). Changes took place in three phases, with a progressive fall during the exposure followed by a transient recovery upon return to 1 G and a long delayed final recovery to normal. Figure 19 illustrates an experiment in which a dog was exposed to $+3$ G$_z$ for 60 s while spontaneously breathing air and the three phases are clearly seen. It was concluded from these and other measurements that the initial fall was caused by acceleration-induced shunting and reduced cardiac output, the transient recovery was due to a redistribution of pulmonary blood flow and possibly also of ventilation upon return to 1 G, and the delayed final recovery was caused by a persistence of shunting until closed-off and/or atelectatic alveoli were reventilated, final recovery often being delayed until the lung was mechanically hyperinflated. Figure 20 illustrates a further experiment in which an anesthetized dog was artificially ventilated at a constant rate and depth and exposed to $+4$ G$_z$ for 60 s. The low initial value for P_{O_2} is explained by residual atelectasis following an earlier centrifuge run, but further falls follow the onset of acceleration with the greatest fall actually occurring after the transient recoveries seen in relation to changes in systemic blood pressure during the $+4$ G$_z$ exposure and upon return to 1 G. The fall in expired CO_2 despite constant mechanical ventilation is explained by the alveolar dead space that results from the ventilation of unperfused units in the upper lung.

C. Right-to-Left Shunting

It was shown previously (Sec. III.F) that alveoli, when initially containing soluble gases and rendered atelectatic by acceleration, acted as shunts for pulmonary

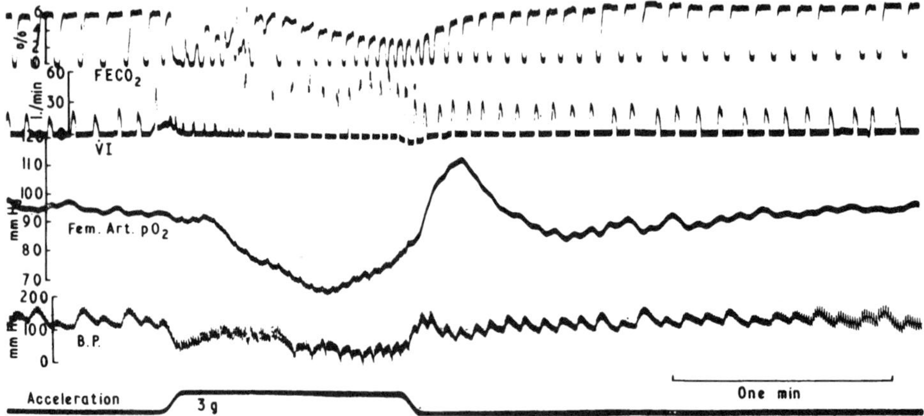

Figure 19   The effect of $+3$ G$_z$ on expired CO_2 concentration, inspired volume, arterial P_{O_2} and systemic blood pressure in an anesthetized dog, supported head-up. (From Ref. 28.)

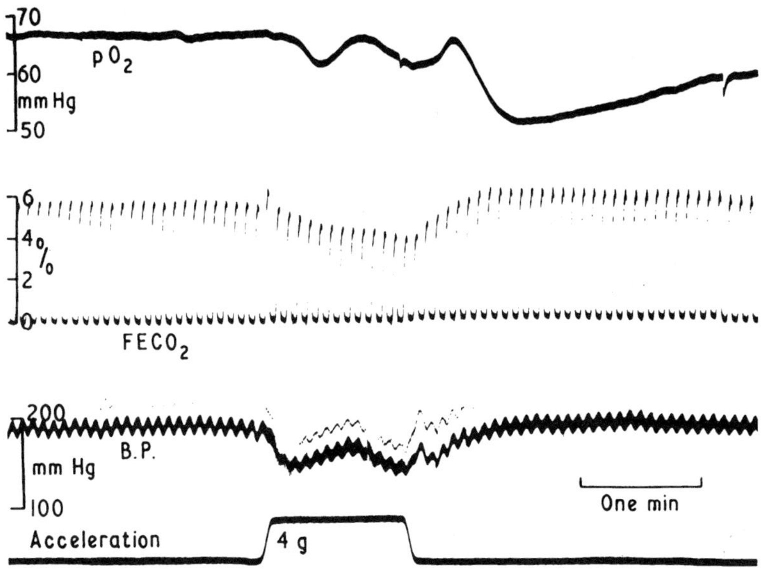

Figure 20 The effect of $+4\,G_z$ on arterial P_{O_2}, expired CO_2 concentration, and systemic blood pressure in an anesthetized dog, supported head-up and mechanically ventilated at a constant rate and depth. (From Ref. 28.)

arterial blood so that an intravenously injected solution of a poorly soluble gas, such as ^{133}Xe, passed through into the systemic circulation. However, the technique used was purely qualitative and gave no indication of the actual fraction of blood shunted in this way. In the normal lung at 1 G, only some 90 to 95% of dissolved Xe is extracted on a single passage through the lungs, but tritium has a blood–gas partition coefficient about one-twelfth that of Xe and should be 99% cleared, even during breathholding at a small lung volume (29). Tritium in saline solution was injected intravenously in four subjects after 2 min of exposure to $+4G_z$, $+4G_x$, and $+7G_x$, and arterial blood sampled during maintained acceleration (6). Since the injectate also contained tritiated water in equilibrium, the fraction of the pulmonary arterial blood passing through atelectatic lung tissue could be calculated by comparing the ratio of $^3H_2$ to ^3H_2O in the injectate with that in the systemic arterial blood, so that absolute concentrations were immaterial. Figure 21 illustrates the fractions obtained in this way before, during, and 1 and 5 min following the three acceleration exposures. Subjects breathed air throughout the procedure and physiological shunts were calculated from the O_2 tension of the arterial blood samples. Figure 22 compares the relevant values, and it is clear that physiological shunts as great as 40% of the cardiac output can

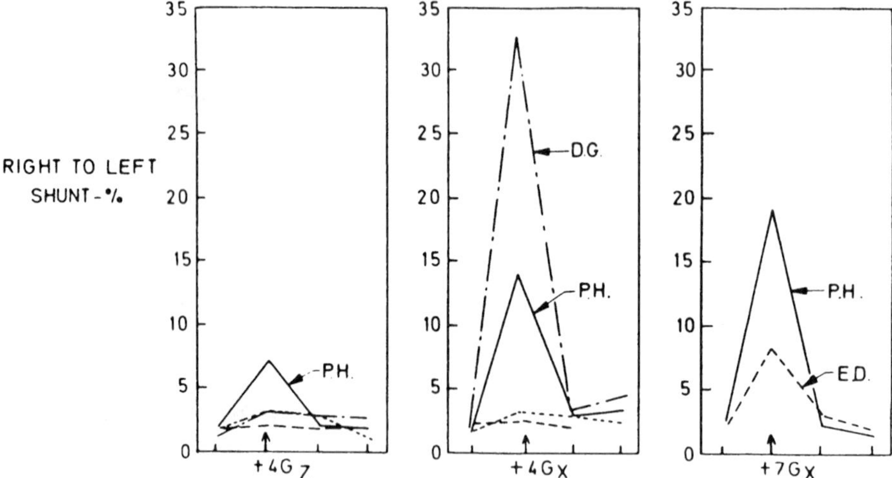

Figure 21   Right-to-left shunts obtained using solutions of tritium injected before, during, and 1 and 5 min following exposure of four subjects to the three conditions of acceleration indicated. (From Ref. 6.)

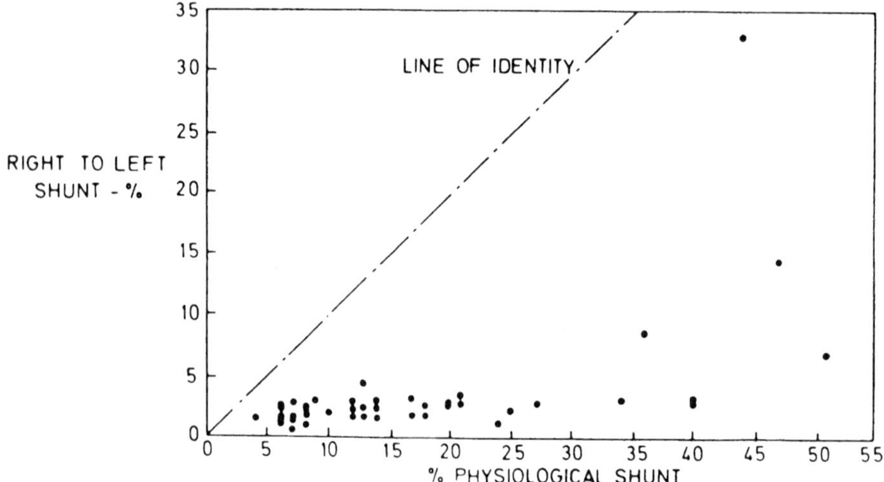

Figure 22   Comparison of the tritium shunts of Figure 22 with the corresponding physiological shunts calculated from arterial O_2 tensions. (From Ref. 6.)

occur in the absence of a significant tritium shunt. Thus, in these cases, the tritium must have diffused into gas-containing, but nonventilated, lung units. However, in subject PH under all three conditions, in subject ED at $+7$ G_x, and most strikingly in subject DG at $+4$ G_x, tritium shunts as large as 33% of the cardiac output occurred during the acceleration exposures. These findings imply that a significant population of lung units had been rendered gas-free by inertial forces, and by a mechanism differing from absorptional atelectasis, since the alveoli had initially contained high concentrations of N_2 and had subsequently reinflated immediately upon return to normal gravity. It is possible that, under some conditions of raised transpulmonary pressure, gas can be squeezed out of alveoli through collateral channels and simply flows back in when pressures return to normal, an observation deserving further study.

Comparable shunt measurements were obtained in anesthetized dogs artificially ventilated on 100% O_2 and exposed to $+G_x$ acceleration (6). Hydrogen in saline solution was used as the indicator and its concentration in the pulmonary arterial and aortic blood measured from indwelling platinum electrodes. In animals breathing air, no more than 5% of the injected dose appeared in the systemic circulation at $+4$ G_x despite physiological shunts calculated in the range of 30 to 50%. As in the human experiments, the injected H_2 must have diffused into gas-containing, but nonventilated, alveoli. However, when the dogs were ventilated on 100% O_2, H_2 shunts of up to 50% were seen. Figure 23 illustrates an example in which a shunt of 7% rapidly increased to 33% and then 58% following exposure to $+4$ G_x, with a concomitant fall in arterial O_2 tension. Thus, nonventi-

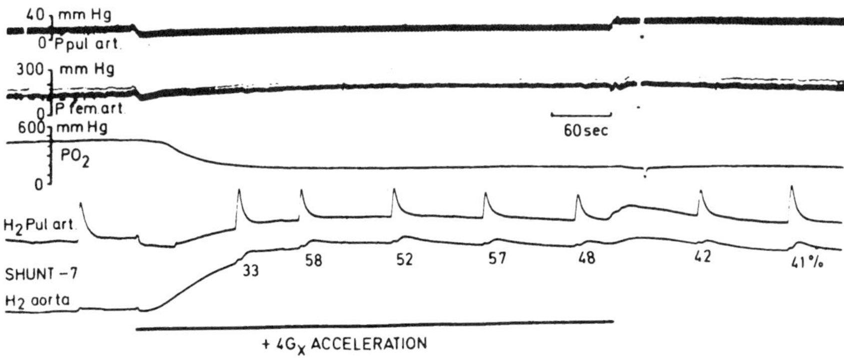

Figure 23 Effect of $+4$ G_x acceleration on pulmonary and femoral arterial pressures, arterial O_2 tension, and potentials recorded from platinum electrodes in the pulmonary artery and aorta of an anesthetized dog artificially ventilated on 100% O_2. Repeat injections of H_2 solution were made into the right atrium and the calculated right-to-left shunts are indicated as percentages of the cardiac output. (From Ref. 6.)

lated alveoli had become atelectatic following absorption of trapped O_2 and the injected H_2 was then able to pass on into the systemic circulation.

D. Effect of Acceleration on Gas Exchange

Figure 24 shows a spirometer tracing and the effect on a relaxed seated subject of a 3-min exposure to $+2$ G_z. Oxygen was added to the circuit at a constant 315 ml/min and its uptake fell during the period of acceleration, only to rise promptly upon return to normal gravity with a small increase overall. This effect was confirmed at $+3$ G_z and Figure 25 illustrates average findings on five subjects for O_2 uptake, CO_2 output, and respiratory gas exchange ratio. During the acceleration exposure (shaded area), the O_2 uptake fell consistently, though only by a small amount, and in the minute following exposure it rose to 160% of the control level ($p = <0.001$). The O_2 debt increased with the duration of the acceleration exposure as well as with the level of acceleration, but was reduced to the level found at a 1G-lower acceleration by the wearing of inflated anti-G trousers. The O_2 uptake could be increased during acceleration by exercise (pulling at a fixed repetition rate on elastic cords with the elbows supported so as to minimize the effect of acceleration on work level), but the uptake was less than when the exercise was carried out at 1 G and the ensuing debt was greater (30). Several factors were considered in trying to account for the decrease in O_2 uptake caused by acceleration, and most were found wanting. However, calculation of the total body O_2 store, taking into consideration the demonstrable fall in the O_2 content of systemic arterial blood, gave a decrease from a resting value of 905 mL at 1 G to 545 mL at $+3$ G_z (6). Thus, during a 3-min centrifuge run, the O_2 stores can fall by up to 120 mL/min, just the value needed to account for the observed O_2 debts and subsequent repayments.

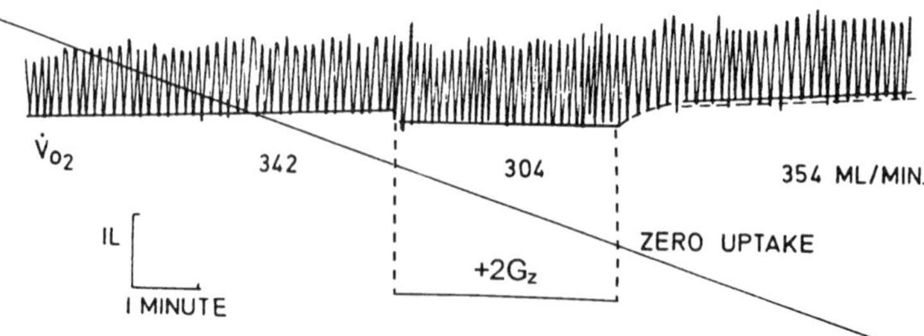

Figure 24   The effect of a 3-min exposure to $+2$ G_z on the uptake of O_2 from a closed-circuit spirometer. A steady inflow of O_2 at 315 mL/min is indicated by the 0 uptake line. (From Ref. 6.)

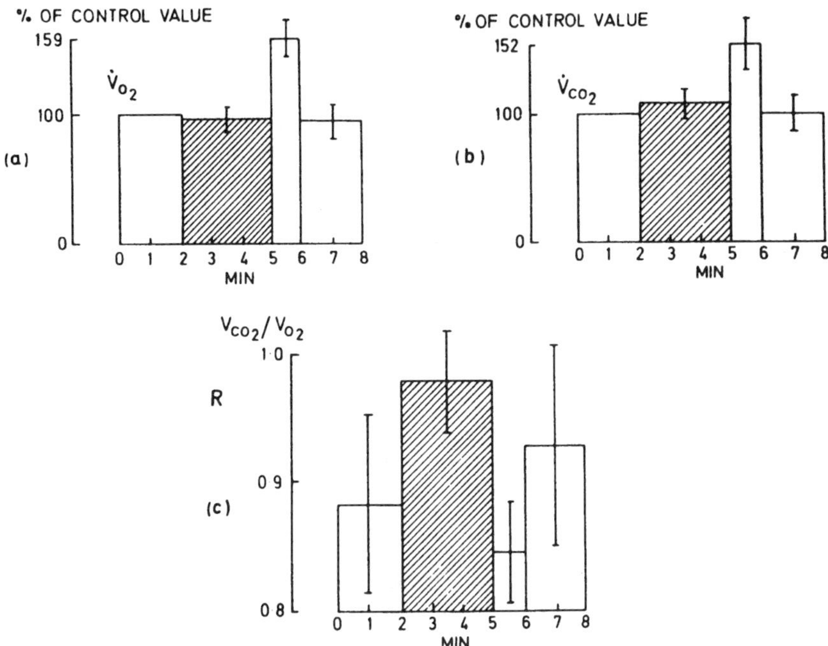

Figure 25 The effect of a 3-min exposure to $+3\ G_z$ on (a) O_2 uptake, (b) CO_2 output, and (c) respiratory gas exchange ratio. Bars indicate ± 1 SD. (From Ref. 6.)

VI. Effect of $\pm G_y$ Acceleration on Ventilation

In the other acceleration axes, $\pm G_z$ and $\pm G_x$, there is continuous lung tissue throughout the vertical extent of the lung, but in lateral decubitus, the left and right lungs are separated by a denser mediastinum. Figure 26 illustrates the specific gravities of the thoracic and abdominal compartments and the transdiaphragmatic pressures that would be present at $-4\ G_y$. This pressure distribution, together with the weight of the mediastinal contents, leads to the left lung being at an overall greater volume than the right so that its density, and resulting vertical gradient in transpulmonary pressure, will be somewhat lower. In lateral decubitus, the interposition of the denser mediastinum displaces the left and right lung pressure–volume curves to produce the result illustrated in the left panel of Figure 27. No change in the lung's mechanical properties is implied, for the simple addition of two identical 1 G pressure–volume curves, with their x axes displaced by up to 30 cmH$_2$O, produces an identical effect (Fig. 28). Also illustrated in the center panel of Figure 27 are expired Ar concentrations following the inspira-

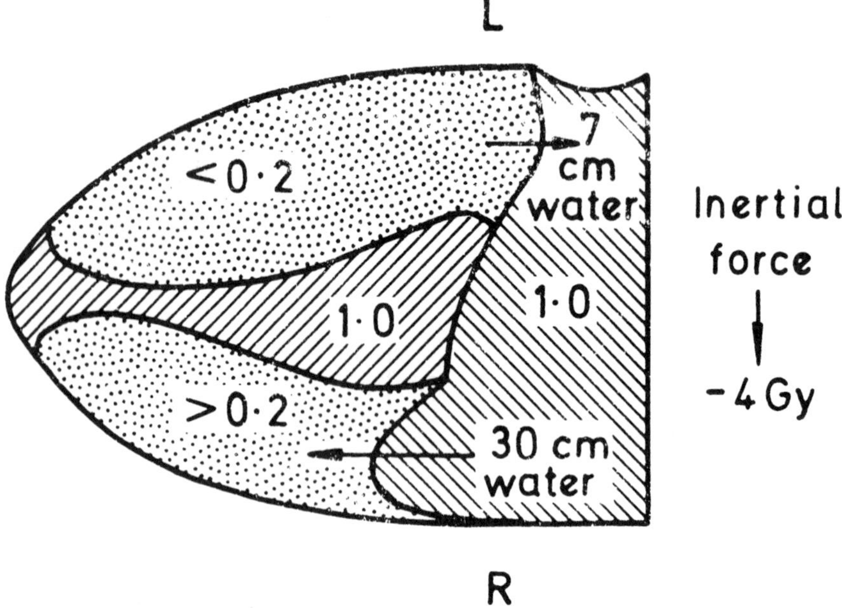

Figure 26 Tissue specific gravity and pressures in the thorax and abdomen at $-4\ G_y$. (From Ref. 11.)

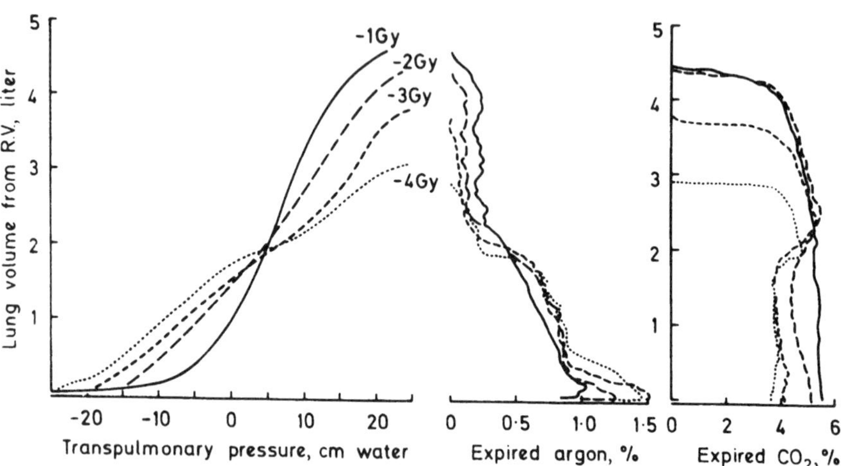

Figure 27 Effect of increasing $-G_y$ acceleration on features of a slow vital capacity expiration after a bolus of Ar had been expired from RV. Plotted against a common volume axis are transpulmonary pressure (left), and expired Ar (center) and CO_2 concentrations (right). (From Ref. 11.)

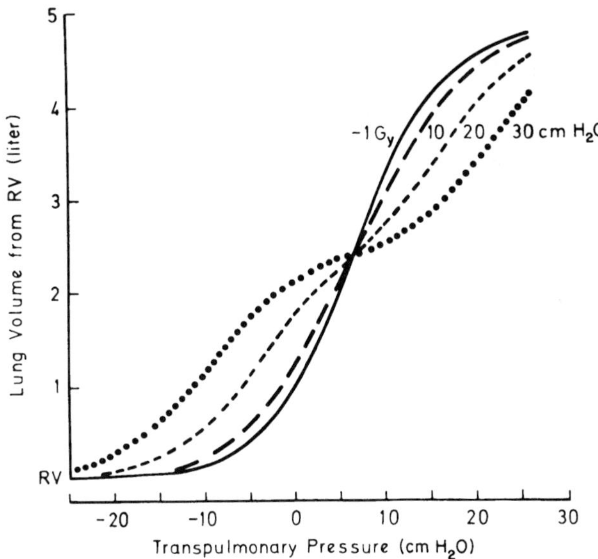

Figure 28 The solid line is the -1 G_y transpulmonary pressure–lung volume curve from Figure 27. The other curves were obtained mathematically by summing two similar curves with their x-axes displaced by up to 30 cmH_2O, and may be compared with the recorded acceleration curves of Figure 27. (From Ref. 15.)

tion of a 50-mL bolus of Ar from RV in the preceding VC inspiration. Inspiratory and expiratory flow rates were 0.5 and 0.3 L/s, respectively, so as to minimize perturbations due to airway resistance effects. At the higher levels of acceleration there are two stepwise increases in Ar concentration, interpreted as the onset of airway closure in dependent lung units, first in the right (lower) lung and then in the left (upper) lung. The steps are coincident with falls in expired CO_2 concentrations (right panel), again indicating two phases of closure of well-perfused lung units and vertical gradients in blood flow distribution across the two lungs in lateral decubitus.

The distribution of ventilation in lateral decubitus was investigated further by inspiring the bolus of Ar at different lung volumes during slow maximal inspirations. Figure 29 shows results from one subject, at $+1$ G_y in the left panel and at $+3$ G_y in the right, with the expired Ar traces placed so as to originate from the lung volume at which each bolus was given. The RV traces are similar to those of Figure 27, but the step increases disappear when the bolus was given after the first 0.75 to 1.0 L of inspirate and are reversed after 1.5 L. These volumes are coincident with the lung volumes at which the steps are seen during expiration

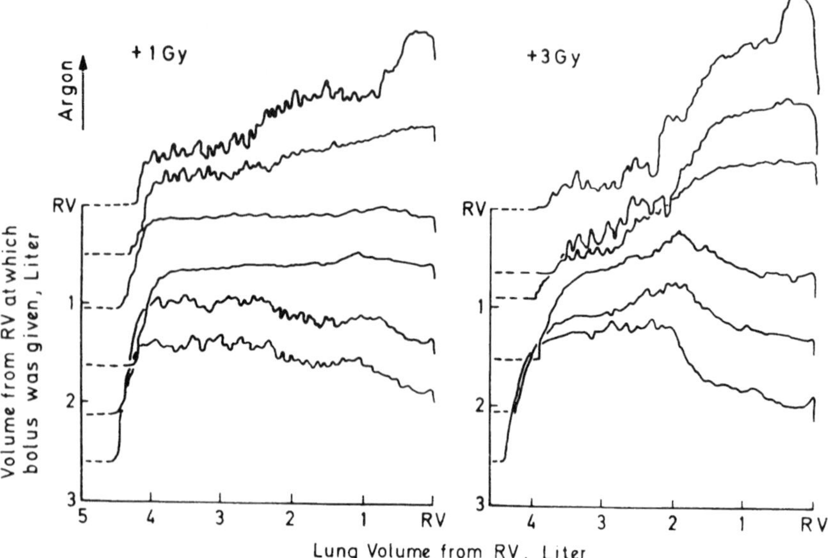

Figure 29 Effect of $+G_y$ acceleration on argon washouts, with boluses of the inert tracer given at the lung volumes indicated by the origin of the traces on the ordinate. (From Ref. 11.)

and suggest that the inspired gas first goes to upper parts of the left lung, then to the entire left lung, followed by its distribution to the upper parts of the right lung, and finally to the entire right lung. This predicted distribution was investigated by using boluses of ^{133}Xe and scanning across the lungs on the centrifuge (31). The subject was supported on a rigid formfitting couch and runs were carried out in right lateral decubitus since in this posture the subject was facing the door of the centrifuge gondola and his face could be seen by the medical observer riding at the center of the centrifuge arm (see Fig. 1). Boluses of ^{133}Xe were given, either at RV or as a sandwich after increasing volumes of air had been breathed in from RV and in every case washed in by a completed slow inflation to TLC. Figure 30 illustrates typical findings and clearly shows the sequential ventilation of the two lungs, with the RV bolus being distributed to the upper parts of the left lung and more than 1.3 L of air needing to be inspired before any gas reaches the right lower lung. At 3.1 L, the bolus becomes preferentially distributed to the right lung. The distributions are not corrected for the volume of lung tissue in the counting field, so tend to show lung shape with the central poorly ventilated area indicating the position of the mediastinum.

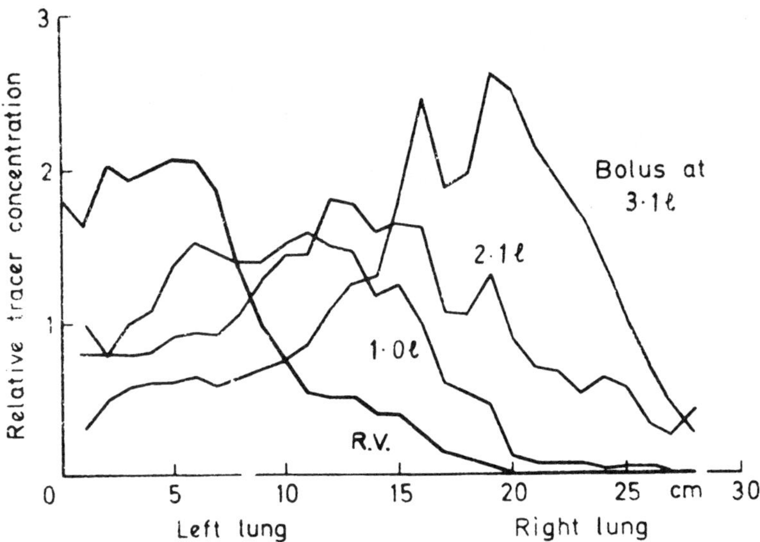

Figure 30   Distributions of radioactivity in the lungs obtained at an acceleration of −4 G_y by vertical scanning of the thorax of a subject lying on his right side after boluses of ^{133}Xe had been inspired at the four different lung volumes indicated. (From Ref. 31.)

VII. Modeling

Salazar and Knowles (32) showed that the pressure–volume characteristics of the human lung could be represented mathematically as a simple exponential function of the form $VC\% = a - be^{-kP}$ and this was confirmed in isolated dog lungs (20), with the inflexion point toward the lower end of the deflation limb indicating the onset of airway closure (see Sec. III.G). While the onset of closure will occur earlier in expiration with increasing acceleration, the RV will be determined only when all units have closed and will, therefore, be unaffected by inertial forces, as observed. A simple two-compartment model has been used to show that acceleration would increase CV and flatten the preceding phase III of bolus washout traces (15), but to model closure occurring asynchronously in the two lungs in lateral decubitus, four compartments are required (31). Each compartment was given the same relative volume, exponential constants, and pressure–volume characteristics as recorded in a test subject at 1 G. Initial tracer gas concentrations were taken from the lateral decubitus ^{133}Xe centrifuge studies and the model then ''computer expired'' to give plots of expired tracer gas concentration against expired volume. When the four compartments were equally separated in terms of their mean transpulmonary pressures, "conventional" washouts with a single phase IV were obtained, but when the first and second, and third and fourth compartments were given transpulmonary pressure differences typical of lung density, but the second and third compartments were separated by a pressure difference to represent the greater density of the mediastinum, the results shown in Figure 31 were obtained. Two phase IVs are computed for the RV bolus and a reversal of the intermediate phase IV occurs when the bolus is given after 3.1 L of inspiration, just as in the experimental washouts illustrated in Figure 29. It is concluded that the complex ventilation distributions and bolus washout patterns

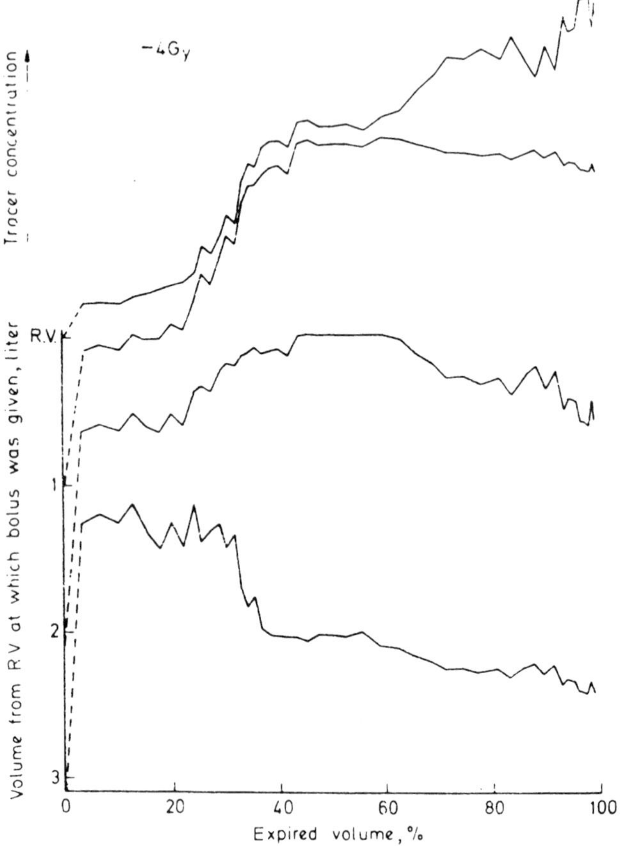

Figure 31 Washouts computed for $-4\,G_y$ using a simple four-compartment lung model with the initial tracer gas concentrations of Figure 30. Transpulmonary pressures for each compartment represent the densities of lung tissue and mediastinum. Compare with the actual washouts of Figure 29. (From Ref. 31.)

that result from exposure to acceleration in lateral decubitus are a simple consequence of the interposition of the denser mediastinum, which lies between the lungs in this body orientation.

VIII. Zero-G Predictions

While the distribution of ventilation was significantly influenced by acceleration, the effect was not truly linear (see Fig. 6) and the gradients in ventilation per unit lung volume could not be extrapolated to 0 G. Furthermore, when the CV was so extrapolated, the 0 G intercepts were at positive volumes of 3 to 9% of the VC (see Fig. 4). These findings suggest that, while an acceleration-dependent gradient in transpulmonary pressure is an important factor in the distribution of pulmonary ventilation, regional differences in mechanical properties play a significant role, particularly at 1 G. It is also important to recognize that all the studies conducted on the centrifuge were done under quasi-static conditions, with flow rates generally set at 0.5 L/s, or less. At higher rates, the influence of airway flow resistance would have been more marked and the effect of acceleration less pronounced.

The effect of acceleration on the distribution of pulmonary blood flow, on the other hand, is much more clear-cut, with the vertical gradients of increasing flow down the lung being, at least in zones II and III, precisely proportional to the level of acceleration, and independent of the actual body posture. These findings indicate that the dominant drivers of regional blood flow in the lung are the hydrostatic pressure gradients in the vasculature and that blood flow, essentially, should be uniformly distributed under conditions of weightlessness.

IX. Summary and Conclusions

With respect to the effects of acceleration, the lung behaves essentially as an isotropic elastic material with ventilatory units sharing common pressure–volume characteristics. Gravitational or inertial forces produce a gradient in transpulmonary pressure that causes units lying at different vertical levels within the lung to operate over different portions of their common pressure–volume curve, leading to a vertical gradient in ventilation. The pressure gradient is a function of lung weight, so will be of the order of 0.2 cmH_2O/cm vertical distance per G. The toe of the deflation limb of the pressure–volume curve is caused by the closure of terminal airways at or close to 0 transpulmonary pressure, so that during a slow maximal expiration, closure will first occur in the lowermost units and then extend upward. This simple model accounts for the effects of $\pm G_z$ and $\pm G_x$ acceleration, but in the case of $\pm G_y$, account must be taken of the greater weight of the mediastinal contents, which causes a discontinuity in the pressure gradient across the two lungs such that their ventilation becomes effectively sequential.

Blood flow distribution is governed by gravity and acceleration-induced hydrostatic pressure gradients that lead to a zone of alveolar dead space in the upper lung tissue (pulmonary arterial pressure $<$ alveolar pressure); two indistinguishable zones of increasing flow in the middle regions due to the waterfall effect (arterial pressure $>$ alveolar pressure $>$ venous pressure) and passive distension/capillary recruitment (arterial pressure $>$ venous pressure $>$ alveolar pressure); and a final zone of decreasing flow in the lowermost nonventilated lung tissue (a combination of increased alveolar pressure and mechanical, or reflex, increases in capillary resistance). The terms "upper" and "lower" refer to the inertial force vectors so that the upper, nonperfused tissue is at the lung apex in $+G_z$ and anterior lung margin in $+G_x$, for example.

The consequence of a dependent zone of perfused, but nonventilated alveoli is the right-to-left shunting of pulmonary arterial blood with consequent arterial desaturation and, if the alveoli close off while containing soluble gases, absorptional atelectasis.

Acknowledgments

On writing this chapter, the author again realizes the debt owed to fellow subjects and centrifuge operating staff at the RAF Institute of Aviation Medicine over many years, and wishes to thank the editors for the opportunity to put this material together. He also wishes to acknowledge the many publications from which the illustrations have been taken, and in particular notes that the original versions of Figures 3, 7, 12, 13, 19, and 22 to 26 were first published by the Advisory Group for Aerospace Research and Development, North Atlantic Treaty Organization (AGARD/NATO) in AGARDograph 133, The Effects of Gravity and Acceleration on the Lung, in November 1970. Figures 27, 28, and 30 are British Crown Copyright/MOD and are reproduced with the permission of Her Majesty's Stationery Office. Other illustrations are reproduced with grateful acknowledgment to the Academic Press, the Aerospace Medical Association, the American Physiological Society, Elsevier Science, and The Royal Society of London. My thanks, finally, to Oxford University Press and its *Oxford Dictionary of Quotations*, for drawing my attention to two prescient poets who had anticipated some aspects of acceleration physiology by several centuries!

References

1. Orth J. Atiologisches und Anatomisches über Lungenschwindsucht. Berlin: Hirschwald, 1887.

2. Gell CF. Table of equivalents for acceleration terminology. Aerosp Med 1961; 32: 1109–1111.

3. Otis AB, Fenn WO, Rahn H. Mechanics of breathing in man. J Appl Physiol 1950; 2:592–607.

4. Keefe JV. Post Flight Respiratory Disorder in Fighter Aircrew. Rep No 20. HQ Fighter Command, Royal Air Force, United Kingdom, 1958.

5. Glaister DH. Lung collapse in aviation medicine. Br J Hosp Med 1969; 2:635–642.

6. Glaister DH, ed. The Effect of Gravity and Acceleration on the Lung. AGARDograph 133. Advisory Group for Aerospace Research and Development, NATO. Slough, England: Technovision Services, 1970.

7. Glaister DH, Farncombe MR. Mass spectrometry with a man-carrying centrifuge. In: Payne JP, Bushman JA, Hill DW, eds. The Medical and Biological Application of Mass Spectrometry. London: Academic, 1979:141–150.

8. Glaister DH. A low weight scintillation counter with focusing collimator. Br J Radiol 1965; 38:625–628.

9. Glaister DH. A flat beam collimator for radioisotope scanning of the lung. Br J Radiol 1967; 40:670–675.

10. Cherniack NS, Hyde AS, Watson JF, Zechman FW. Some aspects of respiratory physiology during forward acceleration. Aerosp Med 1961; 32:113–120.

11. Glaister DH, Ironmonger MR, Lisher BJ. The Effect of Transversely Applied Acceleration on Lung Mechanics in Man. Rep 1340. Flying Personnel Research Committee, MOD (Air Force Dep), London, 1975.

12. Glaister DH. The effect of positive centrifugal acceleration upon the distribution of ventilation and perfusion within the human lung, and its relation to pulmonary arterial and intraoesophageal pressures. Proc R Soc Lond [Biol] 1967; 168:311–334.

13. Glaister DH. The Effect of Positive Acceleration on the Inequality of Ventilation and Perfusion in the Lung. Rep 1231. Flying Personnel Research Committee, MOD (Air Force Dep), London, 1964.

14. Jones JG, Clarke SW, Glaister DH. Effect of acceleration on regional lung emptying. J Appl Physiol 1969; 29:827–832.

15. Glaister DH. Effect of acceleration. In: West JB, ed. Regional Differences in the Lung. New York: Academic, 1977:323–379.

16. Ball WC, Stewart PB, Newsham LGS, Bates DV. Regional pulmonary function studied with xenon 133. J Clin Invest 1962; 41:519–531.

17. Glaister DH. Regional ventilation and perfusion in the lung during positive acceleration measured with ^{133}Xe. J Physiol (Lond) 1964; 177:73–74P.

18. Dollery CT, Gillam PMS. The distribution of blood and gas within the lungs measured by scanning after administration of ^{133}Xe. Thorax 1963; 18:316–325.

19. Glaister DH. Distribution of pulmonary blood flow and ventilation during forward ($+G_x$) acceleration. J Appl Physiol 1970; 29:432–439.

20. Glaister DH, Schroter RC, Sudlow MF, Milic-Emili J. Bulk elastic properties of excised lungs and the effect of a transpulmonary pressure gradient. Respir Physiol 1973; 17:347–364.

21. Glaister DH, Schroter RC, Sudlow MF, Milic-Emili J. Transpulmonary pressure gradient and ventilation distribution in excised lungs. Respir Physiol 1973; 17:365–385.

22. Glaister DH. Effect of acceleration on lung function in man. In: Bhatia B, Chhina GS, Singh B, eds. Selected Topics in Environmental Biology. New Delhi: Interprint, 1975:397–403.

23. Glaister DH. The Relationship Between Closing Volume of the Lung and Transpulmonary Pressure. Rep 1333. Flying Personnel Research Committee, MOD (Air Force Dep), London, 1974.

24. West JB, Dollery CT, Naimark A. Distribution of blood flow in isolated lung: Relation to vascular and alveolar pressures. J Appl Physiol 1964; 19:713–724.

25. Bryan AC, Macnamara WD, Simpson J, Wagner HN. Effect of acceleration on the distribution of pulmonary blood flow. J Appl Physiol 1965; 20:1129–1132.

26. Hoppin FG, Kuhl DE, Hyde RW. Distribution of pulmonary blood flow as affected by transverse ($+G_x$) acceleration. J Appl Physiol 1967; 22:469–474.

27. West JB, Dollery CT, Heard BE. Increased pulmonary vascular resistance in the dependent zone of the isolated dog lung caused by perivascular oedema. Circ Res 1965; 17:191–206.

28. Glaister DH. Transient changes in arterial oxygen tension during positive ($+G_z$) acceleration in the dog. Aerosp Med 1968; 39:54–62.

29. Lassen NA, Mellemgaard K, Georg J. Tritium used for estimation of right-to-left shunts. J Appl Physiol 1961; 16:321–325.

30. Glaister DH. Pulmonary Gas Exchange During Positive Acceleration. Rep 1212. Flying Personnel Research Committee, MOD (Air Force Dep); London, 1963.

31. Glaister DH. Environmental effects on ventilation-perfusion distribution. In: Hutás I, Debreczeni LA, eds. Adv Physiol Sci. Vol. 10. Respiration. Akadémiai Kiadó, Budapest: Pergamon, 1980:215–229.

32. Salazar E, Knowles JH. An analysis of pressure-volume characteristics of the lungs. J Appl Physiol 1964; 19:97–104.

3

Lung Volumes and Chest Wall Mechanics

MARC ESTENNE

Erasme University Hospital
Brussels, Belgium

I. Introduction

The normal respiratory system is exquisitely sensitive to gravity, which causes regional differences in intrapleural pressure, alveolar size, ventilation and perfusion, gas exchange, and parenchymal stresses within the lungs (1–3), and determines the configuration of the relaxed chest wall (4). Despite this susceptibility, the number of experimental observations on the respiratory system under conditions of microgravity has been limited due to the difficulty in creating gravity-free environment, which can only be produced during space flights or aboard high-powered jet airplanes flying through a parabolic trajectory. In this chapter, we consider the effects of changing vertical acceleration (G_z) on lung and chest wall mechanics. Most data in this area have been collected over the last 10 years during parabolic flights and shuttle missions.

II. Lung and Chest Wall Volumes at End-Expiration

Gravity exerts a major influence on lung and chest wall volumes at end-expiration. In an early analysis, Agostoni and Mead (4) suggested that in the upright posture, gravity exerts an inspiratory force on the abdomen that is greater than the expiratory force exerted on the rib cage. In addition, because knowledge of respiratory mechanics at that time indicated that the compliance of the abdomen is greater than the compliance of the ribcage, Agostoni and Mead (4) estimated that the net effect of gravity on the respiratory system should be inspiratory in the upright posture. So, during weightlessness, the diaphragm and the abdomen would shift cranially while the ribcage would only move slightly outward, the net result being a reduction in functional residual capacity (FRC). This prediction has been recently challenged by Liu et al. (5) and Wilson and Liu (6) who reported that the expiratory force on the ribcage and the inspiratory force on the abdomen are similar in magnitude. Because it is now known that, in the upright posture, the compliance of the ribcage is greater than the compliance of the abdomen (7), they concluded that gravity has a net expiratory effect on the system and that FRC should increase in the weightless state. This conclusion was supported by an analysis of the changes in airway opening or pleural pressure in response to accelerations that occur in an elevator or during walking. The following discussion summarizes data obtained in airplanes, shuttle missions, and centrifuges; these data consistently support the prediction of Agostoni and Mead (4) that vertical acceleration has an inspiratory effect in upright humans (8).

A. Microgravity

Early spirometric (9–10) and radiographic (11) studies performed in parabolic flights failed to show a significant change in FRC in microgravity. However, more recent studies by several groups of investigators have consistently demonstrated a 200- to 500-mL decrease in seated FRC during short periods of weightlessness (12–16) (Fig. 1), with a concomitant rise in seated end-expiratory esophageal pressure (14,16,17). Similarly, repeated measurements performed on four to seven astronauts during the SLS-1 mission and on two to four astronauts during the Spacelab D-2 mission using a rebreathing cardiac output test have shown a decrease in inflight FRC compared with preflight and postflight values, with no significant change during the nine-day mission (18,19). On average, inflight FRC was 15% (520 mL) lower than the preflight standing value, and it was intermediate between preflight standing and supine FRC (18). Altogether, these data thus provide unequivocal evidence that both short and prolonged exposure to microgravity produce a substantial reduction in FRC. This reduction may be potentially due to blood volume shifts into the thorax and mechanical factors.

Blood Volume Shifts

Redistribution of blood volume and extracellular fluid due to removal of the hydrostatic pressure gradient is a major response to weightlessness, as will be dis-

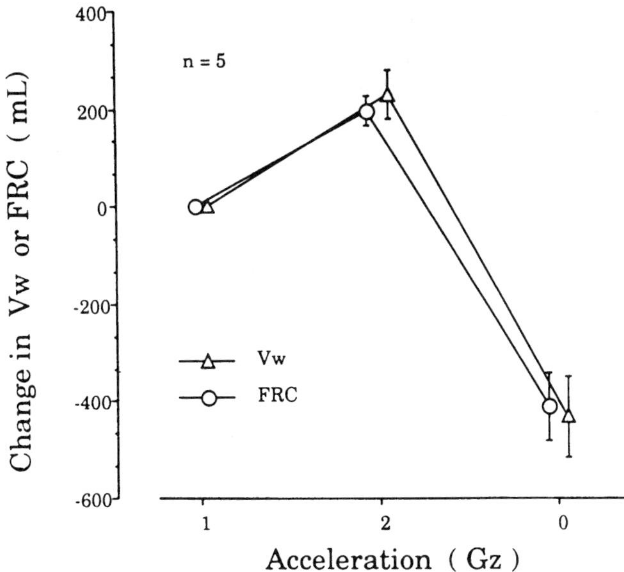

Figure 1 Changes in end-expiratory thoracoabdominal volume (Vw) or functional residual capacity (FRC) as a function of acceleration (G_z) in the sequence observed during parabolic flight paths in five normal seated subjects. Vw and FRC similarly increase at 2 G_z and decrease at 0 G_z relative to 1 G_z. Error bars, $\pm$SE. (From Ref. 15.)

cussed in detail in Chapter 11. Its importance to the cardiorespiratory system is the expected rise in intrathoracic blood volume. The reduction in thoracic impedance that has been demonstrated at 0 G_z during parabolic flights (20) suggests that intrathoracic blood volume actually increases, but the impact of this phenomenon on lung function may be limited by two factors. First, the magnitude of this increase seems to be smaller than suggested by ground-based experiments (21–27). During parabolic flight experiments, Paiva et al. (15) found a similar decrease in thoracoabdominal volume and lung volume at end-expiration after entry into weightlessness (see Fig. 1); this observation suggests that intrathoracic blood volume did not increase substantially because such increase would result in a greater fall in lung than thoracoabdominal volume at 0 G_z. The authors stated that the variability of their data, in particular of ribcage volume at end-expiration (see below), probably limited the resolution of the measurements, but they considered it unlikely that a blood volume shift in excess of ~200 ml could have been missed. This estimate, however, does not necessarily apply to prolonged exposure to weightlessness. Because parabolic flight experiments produce only very brief periods of microgravity that are bracketed by 2 G_z exposures, there may not be enough time to complete blood volume shifts (28).

The second reason why the increase in central blood volume at 0 G_z is expected to have limited effects on the lung is that it is partly accommodated by the chest wall. All other things being equal, changes in blood volume within the thoracoabdominal cavity are expected to cause reciprocal changes in lung and chest wall volumes at end-expiration (4), in much the same way as a pleural effusion (29,30) and a pneumothorax (31) do. The distribution of volume changes between the lung and the chest wall depends primarily on the relative compliance of the two structures, i.e., the more compliant the lung relative to the chest wall, the greater the decrease in lung volume and the smaller the increase in chest wall volume, and vice versa. Studies in normal humans using inflation of fracture splints (24) or inflation of a 1-L balloon in the stomach (32), as well as studies of induced pleural effusions in upright dogs (33), have shown that in these various experimental conditions, the supplementary volume added to the respiratory system is primarily accommodated by the chest wall. The increase in chest wall volume was in general twice as great as the decrease in lung volume, thus indicating a relatively effective protection of intrathoracic gas volume. For these reasons, the increase in intrathoracic blood volume at 0 G_z should only have a limited effect on FRC. Therefore, the reduction in FRC in microgravity would be largely due to mechanical factors.

Mechanical Factors

Using a respiratory inductance plethysmograph, Paiva et al. (15) and Edyvean et al. (13) found that the decrease in FRC in microgravity was primarily due to a decrease in the volume of the abdominal compartment of the chest wall (Fig. 2). They observed that changes in end-expiratory abdominal volume were remarkably reproducible during each parabolic maneuver and that they followed closely the decrease in transabdominal pressure that occurred at 0 G_z (Fig. 3); this indicates that the reduction in abdominal volume reflected unloading of the abdominal contents and was not due to contraction of the abdominal muscles. A change from the upright to the supine posture (1 G_z to 1 G_x) and immersion of seated subjects to the xiphoid process are also associated with an inward displacement of the anterior abdominal wall, a cranial displacement of the diaphragm, and a fall in FRC (4,34,35).

In contrast to changes in end-expiratory abdominal volume, changes in end-expiratory ribcage volume were more variable during parabolic flights. This volume did not change at 0 G_z compared with 1 G_z in one study (see Fig. 2) (15) but it was found to increase by 225 ml at 0 G_z in a subsequent study by the same group of investigators (13). This variability may be related to artifacts due

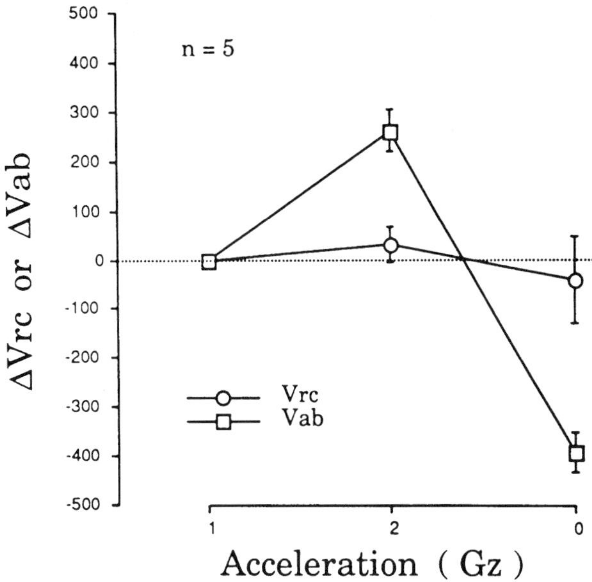

Figure 2   Changes in end-expiratory ribcage (ΔVrc) and abdominal (ΔVab) volumes as a function of acceleration (G_z) in the sequence observed during parabolic flight paths in five normal seated subjects. Vab increases at 2 G_z and decreases at 0 G_z but Vrc is less affected by G_z. Error bars, $\pm$SE. (From Ref. 15.)

to involuntary changes in spinal attitude and differences in experimental conditions, but also to the fact that removal of gravitational stresses does not affect all ribcage dimensions uniformly. Using pairs of magnetometers to measure changes in ribcage dimensions in five normal seated subjects during parabolic flights, Estenne et al. (36) found that microgravity was associated with a consistent motion of the sternum in the cranial direction and an increase in the anteroposterior diameter of the lower ribcage (measured at the level of the fifth intercostal space). In contrast, there was a systematic decrease in the transverse diameter of the lower ribcage, which therefore adopted a more circular shape in weightlessness.

The increase in ribcage anteroposterior diameter at 0 G_z was anticipated. In the upright posture at 1 G_z, gravitational loading of the abdomen produces a hydrostatic gradient of pressure that lengthens and stretches passively the ventral abdominal wall. Because the abdominal muscles insert on the sternum and the lower ribs, this passive tension is applied to the lower ribcage and acts to pull it downward. By removing the weight of the abdominal contents, transition to

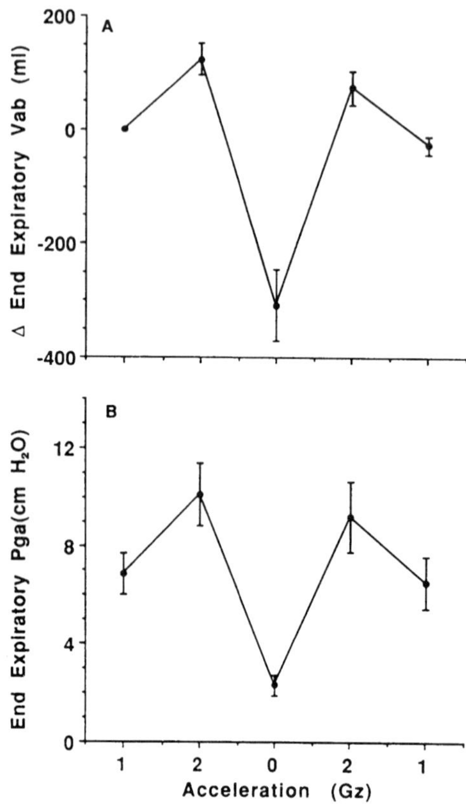

Figure 3 Changes in (A) end-expiratory abdominal volume (ΔVab) and (B) end-expiratory gastric pressure (Pga) as a function of acceleration (G_z) in the sequence observed
during parabolic flight paths in five normal seated subjects. End-expiratory Vab and Pga
increase both at 2 G_z and decrease at 0 G_z. Error bars, $\pm$SE. (From Ref. 13.)

weightlessness removes the tension in the abdominal wall and transabdominal
pressure becomes nearly 0 (see Fig. 3) (13). As a result, the sternum and the ribs
move upward and the ventral ribcage expands. A similar change is observed
during immersion of seated subjects from hips to lower sternum (37). It is more
difficult to understand why the transverse diameter of the lower ribcage did not
similarly increase at 0 G_z. The more likely explanation is that this dimension is
primarily driven by transabdominal pressure. To the extent that the transverse
diameter was measured in the abdomen-apposed portion of the ribcage (36), its
dimension at end-expiration should be primarily influenced by transabdominal

rather than transthoracic pressure. The observation that the former consistently decreased on going from 1 to 0 G_z (13) might thus account for the response of the transverse diameter. The much smaller length of diaphragm apposed to the ventral as compared with the lateral aspects of the cage (38–41) presumably explains why the fall in transabdominal pressure did not act to decrease the anteroposterior diameter.

B. Hypergravity

Despite the greater facility for studying the effects of increased acceleration, experimental studies of the effect of hypergravity on FRC have provided conflicting results. Although Glaister (42) observed a descent of the diaphragm dome at 3 G_z, suggesting an increased abdominal volume, Grassino et al. (43) did not find any consistent change in end-expiratory abdominal cross section or FRC between 1 and 3 G_z. Similarly, Boutellier et al. (44) and Arieli et al. (45) found no change in FRC with a threefold increase in head-to-foot acceleration. On the other hand, four studies during parabolic flights showed a 50- to 200-mL increase in FRC during the 1.8 G_z period preceding the onset of weightlessness (see Fig. 1) (12–15). In the study by Kays et al. (14) in airplanes and a recent study by Estenne et al. (46) in a human centrifuge, end-expiratory esophageal pressure in the seated posture decreased with increasing G_z. This change cannot be accounted for by a direct effect of acceleration (47); it suggests, therefore, a true increase in FRC.

These discrepant results might be due to at least two factors. First, measurements of FRC were obtained using different techniques. For example, Grassino et al. (43) inferred changes in FRC from measurements of vital capacity and inspiratory capacity, Boutellier et al. (44) and Arieli et al. (45) used a rebreathing technique, and Kays et al. (14) used a body plethysmograph. Second, changes in FRC should be affected by the response of the abdominal muscles. In normal subjects breathing at rest, going from supine to upright elicits tonic activation of the abdominal muscles that is more pronounced in the caudal than in the cranial portion of the ventral abdominal wall (48). This response, which is expected to similarly occur in subjects placed in hypergravity conditions (46), should oppose the effects of acceleration on the abdomen and minimize or even prevent the increase in volume that would otherwise occur. Therefore, depending on the level of abdominal muscle activity, increasing gravitational stresses may have different effects on abdominal volume and FRC. When seated subjects are explicitly instructed to relax the abdominal muscles, end-expiratory abdominal volume linearly increases with G_z; this increase faithfully follows the increase in end-expiratory gastric pressure, which reflects the increasing weight of the hydrostatic column within the abdominal cavity (Fig. 4) (46). However, changes in end-

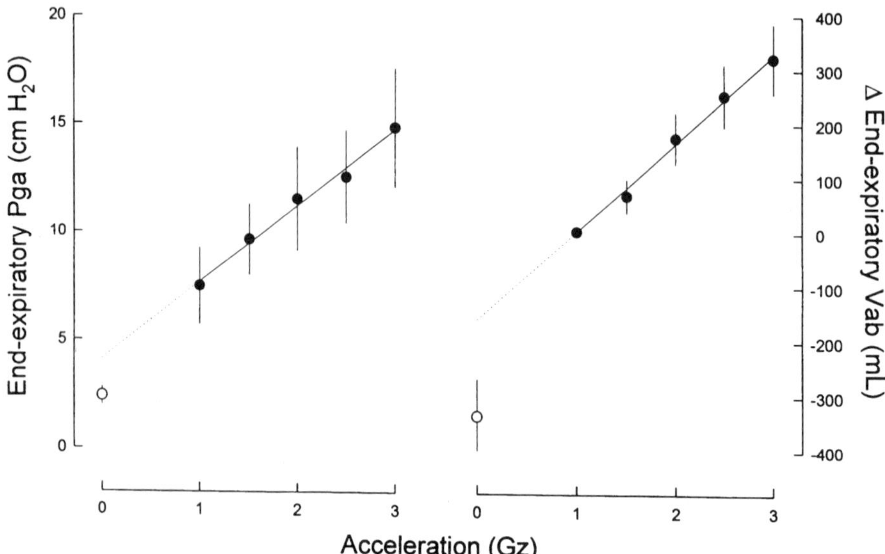

Figure 4   Effects of increasing acceleration from 1 G_z to 3 G_z on end-expiratory gastric pressure (Pga) and changes in end-expiratory abdominal volume (ΔVab) in five normal subjects studied in a centrifuge (solid circles). Dashed lines and open circles show extrapolation to 0 G_z and actual values obtained at 0 G_z during previous parabolic flight experiments in the same five subjects. Extrapolation to 0 G_z is close to actual value measured at 0 G_z for Pga but differs from actual value measured at 0 G_z for ΔVab. Data are means $\pm$SE. (From Ref. 46.)

expiratory abdominal volume per unit change in acceleration are less during increases from 1 G_z to 3 G_z than during decreases from 1 G_z to 0 G_z (see Fig. 4) (15,46).

III.   Static Lung Volumes

A.   Vital Capacity

During early measurements in parabolic flights, several investigators found no consistent changes in vital capacity (VC) between 1 and 0 G_z in seated normal subjects (49,50). Similarly, chest radiographs taken at residual volume (RV) and total lung capacity (TLC) after 10 sec of microgravity showed no significant effect of weightlessness (11). On the other hand, Paiva et al. (15) noted an 8% decrease in slow expiratory VC at 0 G_z in four subjects seated in an aircraft flying parabolic trajectories. The first measurements of VC during prolonged exposure to weightlessness were obtained during the last 2 weeks of the Skylab III mission in 1973 and at various intervals throughout the 84-day Skylab IV mission in 1973/4. Of the six astronauts studied, four showed a decrease in inflight VC approaching 10% of the preflight control value (51). Although this change has been attributed in part to the low barometric pressure inside the spacecraft (equivalent to 27,000 ft. altitude), recent measurements performed during the SLS-1 mission where ambient pressure was 759 mmHg have shown that VC was reduced by about 5% (~200 ml) on flight day 2 compared with the standing 1 G_z preflight value (17,52). This reduction, however, was no longer present on subsequent flight days (Fig. 5). The authors proposed that the initial fall in VC was due to an increase in intrathoracic blood volume, a reduction in the subjects' efforts due to symptoms of space motion sickness, or a combination of the two factors.

Altogether, these data thus indicate that gravitational unloading produces a very slight and transient reduction in VC. Although smaller in magnitude, this change is qualitatively similar to that elicited in normal subjects by water immer-

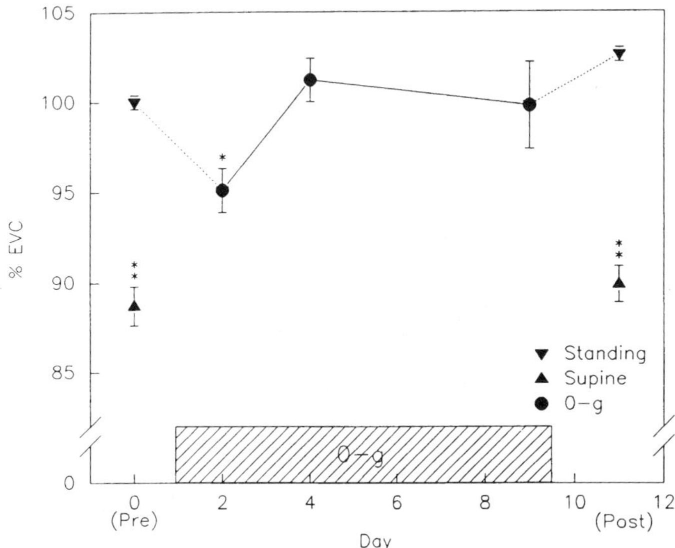

Figure 5   Means (±SE) of individually normalized expired vital capacity (EVC) obtained from four payload crewmembers during the SLS-1 mission. Means from all the preflight sessions are pooled and displayed at day 0. Expired vital capacity on flight day 2 was intermediate between standing and supine values preflight and was reduced compared with flight days 4 and 9. Data that differ significantly from the standing control value are marked: * p <0.05; ** p <0.001. (From Ref. 52.)

sion to the neck (34,53,54), simultaneous inflation of an abdominal bladder and of antishock trousers to lower limbs (55), and postural changes such as head-down tilt and going from upright to supine (4). In contrast to these experimental conditions and to weightlessness, increases in gravitational stresses up to 3 G_z do not produce consistent changes in VC (43,49,56).

B. Residual Volume

Two studies during parabolic flights have shown that RV decreased by 140 to 310 mL at 0 G_z compared with 1 G_z (14,15), but these changes did not reach statistical significance due to a large intersubject variability. On the other hand, measurements during the SLS-1 mission demonstrated a significant reduction in RV compared with preflight standing and supine values (Fig. 6); on average, RV in weightlessness was 430 mL (22%) smaller than 1 G_z standing values and 320 mL (17%) smaller than 1 G_z supine values (18). This observation is intriguing, as RV seems to be a fairly stable volume. For example, it does not change significantly with water immersion (57–59), transition from upright to supine (4), and inflation of an anti-G suit (58). Because the last two conditions increase intratho-

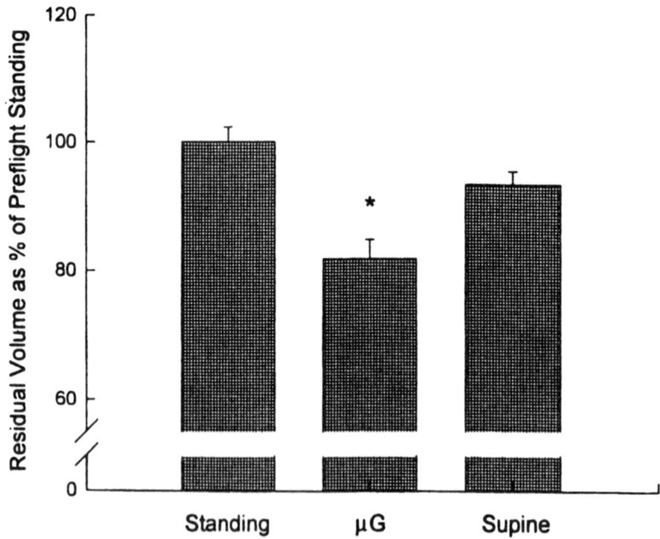

Figure 6 Mean ($\pm$SE) values of residual volume (RV) obtained from four payload crewmembers during the SLS-1 mission. During microgravity, RV was significantly reduced compared with preflight standing and supine values. Significantly different from standing (p < 0.05). (From Ref. 18.)

racic blood volume without changing RV, the increase in intrathoracic blood volume occurring in microgravity is unlikely to be the major reason for the decrease in RV. Another possible mechanism is related to regional differences in alveolar volume. At 1 G_z in both upright and supine postures, the weight of the lung determines large apicobasal gradients of regional lung volume and pleural pressure such that the apical alveoli are relatively overexpanded whereas the basal alveoli are relatively compressed. In microgravity, upper lung regions are no more subjected to expansion forces because of the weight of dependent lung regions, and apicobasal gradients of regional lung volume are presumably abolished. This will reduce the regional RV of upper lung regions. As a result, total RV in microgravity may be reduced compared with either standing or supine 1 G_z values (18).

C. Total Lung Capacity

Total lung capacity decreases slightly during both short and prolonged exposure to microgravity. During parabolic flight experiments, Paiva et al. (15) reported a 470-mL decrease in TLC at 0 G_z compared with 1 G_z; during the SLS-1 mission, the decrease in TLC (which was nonsignificant due to the increased inflight variability and reduced number of measurements) averaged 400 mL compared with preflight standing values (18) and was primarily accounted for by the reduction in RV. A decrease in TLC is also observed on going from upright to supine but it results from a decrease in VC rather than RV.

IV. Forced Expiratory Volumes and Flows

Studies during parabolic flights have shown that short periods of weightlessness generally produce a slight decrease in forced expiratory VC (FVC) (60), forced expiratory volume in 1 sec (FEV$_1$), FEV$_1$/VC ratio, and maximum expiratory flow rates (49,60). In addition, in their study of maximum expiratory flow–volume (MEFV) curves, Guy et al. (60) found that when compared with 1 G_z values, flow rates at 0 G_z were higher at large lung volumes and smaller at low lung volumes, resulting in scooping of the curve. These alterations, however, were small in magnitude and were not present in all subjects. Qualitatively similar changes in the descending limb of the MEFV curve have been reported during head-out immersion (35,53) and in the supine as opposed to the upright posture (60,61). They have been attributed to alterations in lung recoil pressure due to vascular engorgement (35) and subsequent changes in local airway stresses that alter location and motion of airway choke point during forced expiration.

In contrast to data obtained during parabolic flight experiments, data obtained during the SLS-1 mission did not show any consistent changes in midexpiratory flows and in the shape of the descending limb of MEFV curves (62). These curves, however, were not aligned at similar absolute lung volumes, which could introduce a systematic error. On the other hand, peak expiratory flow rate (PEFR) decreased by 12.5% on mission day 2 compared with preflight standing values; PEFR then gradually increased and values on mission day 9 were similar to preflight data. A similar trend was seen for FVC, which was reduced early in the mission but increased above preflight values at the end of the flight. The reduction in FVC, however, was considered to be too small to entirely account for the decreased PEFR. The lack of physical stabilization while performing forced expiratory maneuvers in weightlessness was proposed as a possible mechanism; the importance of this factor is supported by the fact that when MEFV maneuvers were performed during parabolic flights with the subject strapped to the aircraft chair, there was only a slight decrease in PEFR at 0 G_z (60).

V. Pattern of Breathing and Chest Wall Compliance

Although early observations during parabolic flights suggested that breathing frequency may increase in microgravity without concomitant changes in tidal volume (8), more recent studies have shown that 20-s periods of weightlessness produce only small and inconsistent changes in tidal volume, breathing frequency, mean inspiratory flow rate, and inspiratory duty cycle (12–15,36). This is consistent with the effects of posture and head-out immersion, neither of which alter the temporal pattern of breathing at rest (37,63). In contrast, prolonged exposure to microgravity may well have a different effect. Measurements during the SLS-1 mission have shown a 15% reduction in tidal volume inflight compared with preflight standing values; this alteration remained constant for the duration of the orbital flight and was accompanied by a slight increase in breathing frequency (18). Measurements performed by Prisk et al. (64) in the same mission provided evidence that this change was not due to increases in interstitial lung fluid and stimulation of J receptors, which may result in rapid, shallow breathing. The explanation for the different effects of short and prolonged exposures to microgravity remains therefore to be determined.

Separate measurements of ribcage and abdominal volume indicate that the relative contribution of the two chest wall compartments to tidal volume is markedly altered by changes in G_z. In two successive studies, Paiva et al. (15) and Edyvean et al. (13) found that in seated normal subjects, the contribution of the abdomen to tidal volume (ΔVab/V_T) increased from 0.33–0.39 to 0.51–0.57 on going from 1 G_z to 0 G_z (Fig. 7), the value found in microgravity being intermediate between that in the upright and the supine posture (65,66). Data from the Spacelab D-2 mission and from the Euromir-95 mission have also demonstrated a consistent increase in ΔVab/V_T at 0 G_z that remained constant throughout the

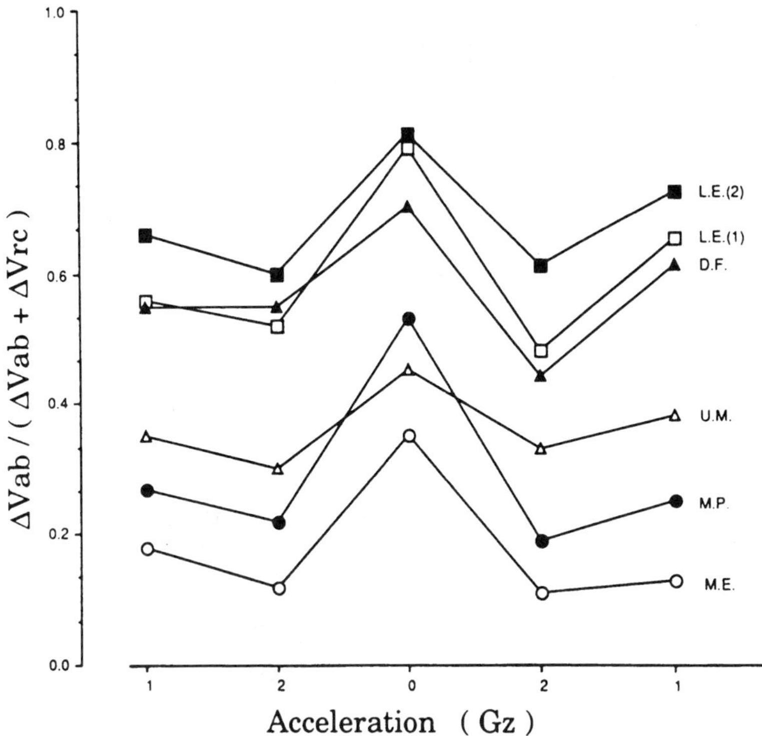

Figure 7   Abdominal contribution to tidal breathing [ΔVab/(ΔVab + ΔVrc)] at different levels of acceleration (G_z) in five normal seated subjects. Data represent average values based on measurements during 4 to 10 parabolas. Results on LE are shown separately for flights 1 and 2, respectively. The contribution of the abdomen to tidal volume increases at 2 G_z and decreases at 0 G_z in all subjects. (From Ref. 15.)

orbital flight (67). The greater contribution of the abdomen to tidal volume at 0 G_z is due, at least in part, to the increased compliance of the abdomen (13,67) that results from removal of passive tension in the ventral abdominal wall (Fig. 8). On transition to weightlessness, abdominal compliance increased on average by ~60% relative to 1 G_z in the five subjects studied by Edyvean et al. (13). Abdominal compliance also increases on going from upright to supine (4,7,68). Conversely, abdominal compliance and contribution to tidal volume decrease gradually when head-to-foot acceleration is increased from 1 G_z to 3 G_z (46).

The increased abdominal contribution to tidal volume at 0 G_z was accompanied by a reciprocal decrease in ribcage contribution. This alteration cannot be

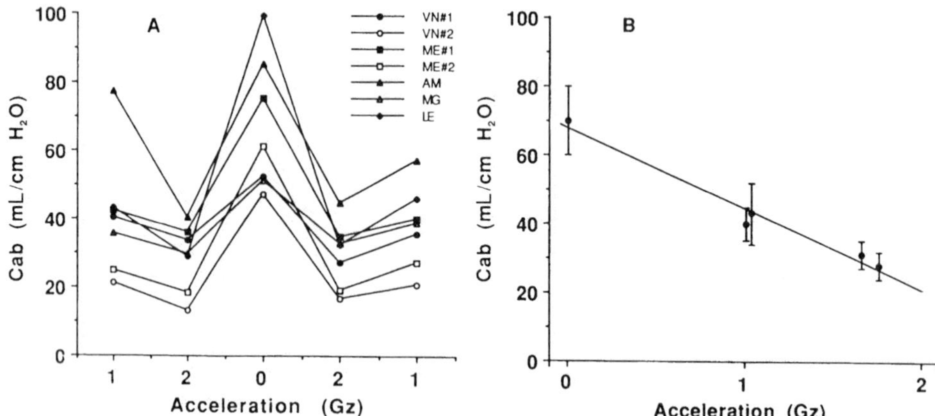

Figure 8   (A) abdominal compliance (Cab) as a function of G_z during parabolic maneuvers in five normal seated subjects, two of whom were studied in duplicate on different flights. Each symbol is the mean of five to eight parabolas. Cab increases at 2 G_z and decreases at 0 G_z in all subjects. (B) Cab vs. G_z. Symbols represent the mean of five subjects. Error bars, $\pm$SE. (From Ref. 13.)

explained by a reduction in the compliance of the ribcage (67), but is related to changes in the mechanical coupling between the lower portion of the cage and the diaphragm, and to derecruitment of ribcage inspiratory muscles. Measurements with magnetometers have shown a reduced tidal expansion of both upper and lower portions of the cage along the anteroposterior diameter at 0 G_z (36). The smaller expansion of the lower ribcage may be explained by the increased abdominal compliance that reduces the insertional and appositional components of the expanding action of the diaphragm (69). This mechanism, however, cannot account for the reduced tidal expansion of the upper ribcage because the expanding forces produced by diaphragm contraction do not apply, or do so only to a limited extent, to the upper portion of the cage (38). This alteration is likely to be related to derecruitment of the scalene muscles at 0 G_z. Direct electromyographic recordings from these muscles, which inflate the upper portion of the ribcage during quiet inspiration at 1 G_z (38,70), have shown a consistent decrease in phasic inspiratory activity in weightlessness (36). This adjustment may be part of the reflex referred to as operational length compensation (71). When the operating length of the diaphragm—and with it, its force-generating capacity— increases, as it presumably does in microgravity, the inspiratory activation of all inspiratory muscles is reduced in order to keep tidal volume constant. This mechanism has been demonstrated during postural changes from upright to supine (65) and during hip-to-xiphoid immersion (37).

VI. Summary

In summary, studies performed in microgravity have provided very consistent results that indicate that decreasing G_z:

1. Reduces FRC and increases end-expiratory esophageal pressure
2. Reduces gastric pressure and abdominal volume at end-expiration
3. Makes the ribcage more circular at end-expiration
4. Has minor effects on VC and TLC but decreases RV
5. Produces a slight reduction in peak and maximum expiratory flow rates
6. Increases abdominal contribution to tidal volume and abdominal compliance without substantially changing tidal volume and temporal pattern of breathing

Reciprocal changes are observed as G_z is increased.

References

1. Engel LA. Effect of microgravity on the respiratory system. J Appl Physiol 1991; 70:1907–1911.
2. West JB, ed. Regional Differences in the Lung. New York: Academic Press, 1977.
3. West JB. Pulmonary function in space. JAMA 1997; 277:1957–1961.
4. Agostoni E, Mead J. Statics of the respiratory system. In: Fenn WO, Rahn H, eds. Handbook of Physiology. Sect. 3. Vol. 1. Respiration. Washington, DC: American Physiological Society 1964:387–409.
5. Liu S, Wilson TA, Schreiner K. Gravitational forces on the chest wall. J Appl Physiol 1991; 70:1506–1510.
6. Wilson TA, Liu S. Effect of acceleration on the chest wall. J Appl Physiol 1994; 76:1242–1246.
7. Estenne M, Yernault JC, De Troyer A. Rib cage and diaphragm-abdomen compliance in humans: Effects of age and posture. J Appl Physiol 1985; 59:1842–1848.
8. Estenne M, Paiva M. Effect of gravity on the respiratory system: Is it inspiratory or expiratory? J Appl Physiol 1995; 78:757.
9. von Baumgarten RJ, Baldrighi G, Vogel H, Thumler R. Physiological response to hyper- and hypogravity during rollercoaster flight. Aviat Space Environ Med 1980; 51:145–154.
10. Wetzig J, von Baumgarten R. Respiratory parameters aboard an aircraft performing parabolic flights. In: Proceedings of the 3d European Symposium on Life Sciences Research in Space (ESA SP-271). 1987:47–50.
11. Michels DB, Friedman PJ, West JB. Radiographic comparison of human lung shape during normal gravity and weightlessness. J Appl Physiol 1979; 47:851–857.
12. Beaumont M, Fodil R, Isabey D, Lofasso F, Touchard D, Harf A, Louis B. Gravity effects on upper airway area and lung volumes during parabolic flight. J Appl Physiol 1998; 84:1639–1645.

13. Edyvean J, Estenne M, Paiva M, Engel LA. Lung and chest wall mechanics in microgravity. J Appl Physiol 1991; 71:1956–1966.

14. Kays C, Choukroun ML, Techoueyres P, Varenne P, Vaida P. Lung mechanics during parabolic flights (abstr). In: Proceedings of the 5th European Symposium on Life Sciences Research in Space, 1993.

15. Paiva M, Estenne M, Engel LA. Lung volumes, chest wall configuration, and pattern of breathing in microgravity. J Appl Physiol 1989; 67:1542–1550.

16. Prisk GK, Elliott AR, Guy HJB, West JB. Lung volumes and esophageal pressures during short periods of microgravity and hypergravity (abstr). Physiologist 1990; 33:A83.

17. Videbaek R, Norsk P. Atrial distension in humans during microgravity induced by parabolic flights. J Appl Physiol 1997; 83:1862–1866.

18. Elliott AR, Prisk GK, Guy HJB, West JB. Lung volumes during sustained microgravity on Spacelab SLS-1. J Appl Physiol 1994; 77:2005–2014.

19. Verbanck S, Larsson H, Linnarson D, Prisk GK, West JB, Paiva M. Pulmonary tissue volume, cardiac output, and diffusing capacity in sustained microgravity. J Appl Physiol 1997; 83:810–816.

20. Lathers CM, Charles JB, Elton KF, Holt TA, Mukai C, Bennett BS, Bungo MW. Acute hemodynamic responses to weightlessness in humans. J Clin Pharmacol 1989; 29:615–627.

21. Arborelius M, Balldin UI, Lilja B, Lundgren CEG. Hemodynamic changes in man during immersion with the head above water. Aerosp Med 1972; 43:592–598.

22. Blomqvist CG, Stone HL. Cardiovascular adjustments to gravitational stress. In: Shepherd JT, Abboud FM, eds. Handbook of Physiology. Sect. 2. Vol. 3. The Cardiovascular System. Washington, DC: American Physiological Society 1983:1025–1063.

23. Chadha TS, Lopez F, Jenouri G, Birch S, Sackner MA. Effects of partial anti-G suit inflation on thoracic volume and breathing pattern. Aviat Space Environ Med 1983; 54:324–327.

24. Kimball WR, Kelly KB, Mead J. Thoracoabdominal blood volume change and its effect on lung and chest wall volumes. J Appl Physiol 1986; 61:953–959.

25. Sjöstrand T. Determination of changes in the intrathoracic blood volume in man. Acta Physiol Scand 1951; 22:114–128.

26. Sjöstrand T. The regulation of blood distribution in man. Acta Physiol Scand 1952; 26:312–327.

27. Weissler AM, Leonard JJ, Warren JV. Effects of posture and atropine on cardiac output. J Clin Invest 1957; 36:1656–1662.

28. Tenney SM. Fluid volume redistribution and thoracic volume changes during recumbency. J Appl Physiol 1959; 14:129–132.

29. Estenne M, Yernault JC, De Troyer A. The mechanism of relief of dyspnea after thoracocentesis in patients with large pleural effusions. Am J Med 1983; 74:813–819.

30. Gilmartin JJ, Wright AJ, Gibson JJ. Effects of pneumothorax or pleural effusion on pulmonary function. Thorax 1985; 40:60–65.

31. Heaf PJD, Prime FJ. Mechanical aspects of artificial pneumothorax. Lancet 1959; ii:468–470.

32. Gilroy RJ Jr, Lavietes MH, Loring SH, Mangura BT, Mead J. Respiratory mechanical aspects of abdominal distension. J Appl Physiol 1985; 58:1997–2003.

33. Krell WS, Rodarte JR. Effects of acute pleural effusion on respiratory system mechanics in dogs. J Appl Physiol 1985; 59:1458–1463.

34. Agostoni E, Gurtner G, Torri G, Rahn H. Respiratory mechanics during submersion and negative pressure breathing. J Appl Physiol 1966; 21:251–258.

35. Prefaut C, Lupi-h E, Anthonisen NR. Human lung mechanics during water immersion. J Appl Physiol 1976; 40:320–333.

36. Estenne M, Gorini M, Van Muylem A, Ninane V, Paiva M. Rib cage shape and motion in microgravity. J Appl Physiol 1992; 73:946–954.

37. Reid MB, Loring SH, Banzett RB, Mead J. Passive mechanics of upright human chest wall during immersion from hips to neck. J Appl Physiol 1986; 60:1561–1570.

38. Estenne M, De Troyer A. Relationship between respiratory muscle EMG and rib cage motion in tetraplegia. Am Rev Respir Dis 1985; 132:53–59.

39. Gauthier A, Verbanck S, Estenne M, Segebarth C, Macklem PT, Paiva M. Three-dimensional reconstruction of the in vivo human diaphragm shape at different lung volumes. J Appl Physiol 1994; 76:495–506.

40. Pettiaux N, Cassart M, Paiva M, Estenne M. Three-dimensional reconstruction of human diaphragm with the use of spiral computed tomography. J Appl Physiol 1997; 82:998–1002.

41. Whitelaw WA. Shape and size of human diaphragm in vivo. J Appl Physiol 1987; 62:180–186.

42. Glaister DH. Ventilation and mechanics of breathing. In: Glaister DH, ed. The Effects of Gravity and Acceleration on the Lung. Slough, England: Technivision Services, 1970:19–36.

43. Grassino AE, Forkert L, Anthonisen NR. Configuration of the chest wall during increased gravitational stress in erect humans. Respir Physiol 1978; 33:271–278.

44. Boutellier U, Arieli R, Farhi LE. Ventilation and CO_2 response during $+G_z$ acceleration. Resp Physiol 1985; 62:141–151.

45. Arieli R, Boutellier U, Farhi LE. Effect of water immersion on cardiopulmonary physiology at high gravity ($+G_z$). J Appl Physiol 1986; 61:1686–1692.

46. Estenne M, Van Muylem A, Kinnear W, Gorini M, Ninane V, Engel LA, Paiva M. Effects of increased $+G_z$ on chest wall mechanics in humans. J Appl Physiol 1995; 78:997–1003.

47. Bryan AC, Milic-Emili J, Pengelly D. Effect of acceleration on the distribution of pulmonary ventilation. J Appl Physiol 1966; 21:778–784.

48. De Troyer A. Mechanical role of the abdominal muscles in relation to posture. Respir Physiol 1983; 53:341–353.

49. Foley MF, Tomashefski JF. Pulmonary function during zero-gravity maneuvers. Aerosp Med 1969; 40:655–657.

50. Michels DB, West JB. Distribution of pulmonary ventilation and perfusion during short periods of weightlessness. J Appl Physiol 1978; 45:987–998.

51. Sawin CF, Nicogossian AE, Rummel JA, Michel EL. Pulmonary function evaluation during the Skylab and Apollo-Soyuz missions. Aviat Space Environ Med 1976; 47:168–172.

52. Guy HJB, Prisk GK, Elliott AR, Deutschman III RA, West JB. Inequality of pulmonary ventilation during sustained microgravity as determined by single-breath washouts. J Appl Physiol 1994; 76:1719–1729.
53. Elliott R, Prisk GK, Guy HJB. Effect of immersion on flow-volume curve configuration (abstr). FASEB J 1989; 3:6240.
54. Hong SK, Cerretelli P, Cruz JC, Rahn H. Mechanics of respiration during submersion in water. J Appl Physiol 1969; 27:535–538.
55. Regnard J, Baudrillard P, Salah B, Xuan TD, Cabanes L, Lockhart A. Inflation of antishock trousers increases bronchial response to metacholine in healthy subjects. J Appl Physiol 1990; 68:1528–1533.
56. Pyszczynski D, Mink SN, Anthonisen NR. Increased gravitational stress does not alter maximum expiratory flow. J Appl Physiol 1985; 59:28–33.
57. Burki NK. Effect of immersion to water and changes in intrathoracic blood volume on lung function in man. Clin Sci Mol Med 1976; 51:303–311.
58. Buono MJ. Effect of central vascular engorgement and immersion on various lung volumes. J Appl Physiol 1983; 54:1094–1096.
59. Robertson CH Jr, Engle CM, Bradley ME. Lung volumes in man immersed to the neck: Dilution and plethysmographic techniques. J Appl Physiol 1978; 44:679–681.
60. Guy HJB, Prisk GK, Elliott AR, West JB. Maximum expiratory flow-volume curves during short periods of microgravity. J Appl Physiol 1991; 70:2587–2596.
61. Castille R, Mead J, Jackson A, Wohl ME, Stokes D. Effects of posture on flow-volume curve configuration in normal humans. J Appl Physiol 1982; 53:1175–1183.
62. Elliott AR, Prisk GK, Guy HJB, Kosonen JM, West JB. Forced expirations and maximum expiratory flow-volume curves during sustained microgravity on SLS-1. J Appl Physiol 1996; 81:33–43.
63. Burki NK. The effects on changes in functional residual capacity with posture on mouth occlusion pressure and ventilatory pattern. Am Rev Respir Dis 1977; 116: 895–900.
64. Prisk GK, Guy HJB, Elliott AR, Deutschmann III RA, West JB. Pulmonary diffusing capacity, capillary blood volume, and cardiac output during sustained microgravity. J Appl Physiol 1993; 75:15–26.
65. Druz WS, Sharp JT. Activity of respiratory muscles in upright and recumbent humans. J Appl Physiol 1981; 51:1552–1561.
66. Sharp JT, Goldberg NB, Druz WS, Danon J. Relative contribution of rib cage and abdomen to breathing in normal subjects. J Appl Physiol 1975; 39:608–618.
67. Wantier M, Estenne M, Verbanck S, Prisk GK, Paiva M. Chest wall mechanics in sustained microgravity. J Appl Physiol 1998; 84:2060–2065.
68. Konno K, Mead J. Static volume-pressure characteristics of the rib cage and abdomen. J Appl Physiol 1968; 24:544–548.
69. De Troyer A. Respiratory muscles. In: Crystal RG, West JB, eds. The Lung. Vol. 1. Scientific Foundations. New York: Raven Press, 1991:869–883.
70. De Troyer A, Estenne M. Coordination between rib cage muscles and diaphragm during quiet breathing in humans. J Appl Physiol 1984; 7:899–906.
71. Green M, Mead J, Sears TA. Muscle activity during chest wall restriction and positive pressure breathing in man. Respir Physiol 1978; 35:283–300.

4

Ventilation Distribution

MANUEL PAIVA

Université Libre de Bruxelles
Brussels, Belgium

G. KIM PRISK

University of California, San Diego
La Jolla, California

I. Introduction

A major breakthrough in the understanding of the distribution of ventilation was achieved in 1966 with the publication of three papers using radioactive gas techniques to measure regional pulmonary ventilation (1–3). Not only were the measurements more reliable than lobar spirometry, but quantitative data could also be obtained on a topographical basis, corresponding to, for example, the vertical distance from the top to the bottom of the lung. Since then, most of the information on regional ventilation distribution has been obtained with radioactive gases. However, these techniques have never been used in space, and most of the experiments used to study microgravity effects on the lung are indirect techniques such as single- and multiple-breath inert gas washouts (4,5). Predictions were made of the results of the experiments performed in microgravity (obtained in space or in aircraft during parabolic trajectories), and from the differences between predictions and observations we gained insight into the behavior of inhaled gas in the human lung. In this chapter we present a number of examples where studies in microgravity have provided insight into the basic physiology.

II. Topographical Ventilation Inhomogeneity

A. Time Constants

One of the first models of gas transport in the lung was an electric analog (6), where pressure, flow, compliance, and flow resistance correspond to voltage, current, capacitance, and electrical resistance, respectively. Using this analog, the volume of gas (ΔV) going to a lung unit at a given flow rate $\dot{v}$ is related to the pressure swing (ΔP), the compliance of the unit (C), and the resistance of its airway (R) by the equation

$$\Delta P = \Delta V/C + \dot{v} R$$

If we consider two parallel units, A and B (Fig. 1, lower left), nonuniformity of ΔP, C, or R may result in convection-dependent inhomogeneities (CDI). Assume first that ΔP is a constant. When $\dot{v} R \ll \Delta V/C$ (or $\dot{v} \ll \Delta V/RC$, where RC is the mechanical time constant), it is apparent from the equation that $\Delta V_A/\Delta V_B \approx C_A/C_B$. This means that for low flows, the distribution of gas is primarily determined by inequalities in compliance. Conversely, when resistance becomes very large (such as might occur at high flows), if $\dot{v} R \gg \Delta V/C$, the equation predicts that $\dot{v}_A R_A \approx \dot{v}_B R_B$. After integration over the inspired volume (assuming time-independent resistances), $\Delta V_A/\Delta V_B \approx R_B/R_A$ and so the distribution of gas is primarily determined by inequalities in resistance.

For normal subjects breathing quietly, $\dot{v} R$ is always much smaller than $\Delta V/C$. This is also the case for slow inspiratory and expiratory vital capacities (VC) and the equation $\Delta V_A/\Delta V_B \approx C_A/C_B$ is valid for any lung volume (see Fig. 1, top). Furthermore, there is evidence that swings in transpulmonary pressure are equal over different lung regions for low flows (7). Therefore, regional inhomogeneity of ventilation, in normal subjects during quiet breathing, can be primarily explained by the static mechanical characteristics of the lung.

In microgravity (μG), the intrathoracic pressure becomes more homogeneous, as all hydrostatic pressures (which are by definition, due to weight) necessarily disappear. Experiments performed during parabolic flights, however, suggest that direct effects of gravity on the mechanical properties of the lung are small (8). In addition, radiographic studies performed in parabolic flight (9) show little change in the conformation of the ribcage and chest wall (see Chapter 3 for more detail). Fluid shifts in microgravity and the modifications of central venous pressure may also play a small role in the mechanical properties of the lung. These subjects are discussed in Chapters 3 and 10, respectively.

B. The "Onion-Skin" Model

Because the first topographical studies of ventilation used external counters, a region corresponded to the counter field of view. This resulted in the terms "re-

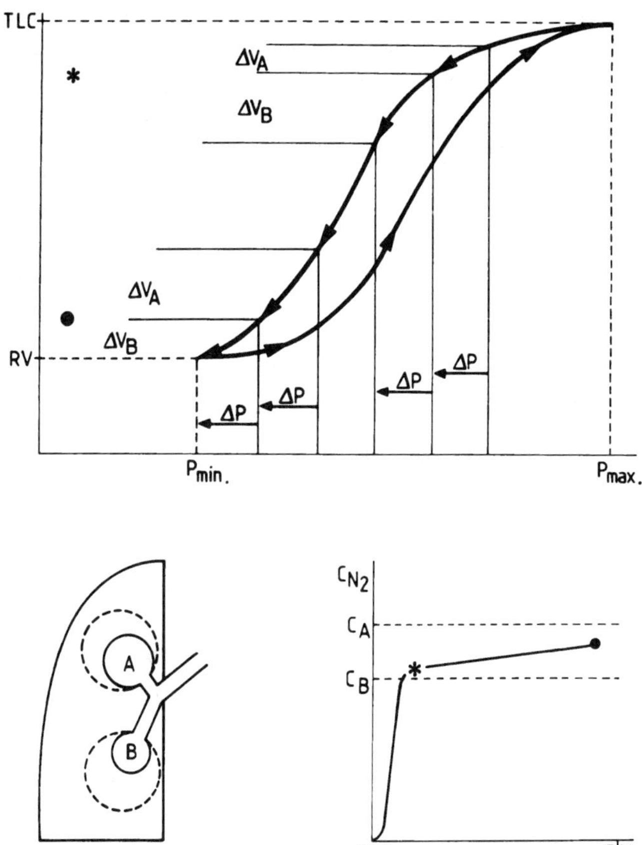

Figure 1   (Lower left) Schematic representation of upper (A) and lower (B) lung units at RV (continuous circles, $V_A > V_B$) and at TLC (dashed circles, $V_A = V_B$). (Lower right) For a VC inspiration of oxygen, the nitrogen concentration in unit A is higher than that in unit B. (Top) The pressure–volume loop for a VC breath (downward arrows for the expiratory limb). The diagram aims at explaining that if the vertical gradient of pleural pressure remains constant during expiration (ΔP between units A and B), unit B contributes more ($\Delta V_B > \Delta V_A$) at the beginning of the expiration (expired concentration represented by a star) and unit A contributes more ($\Delta V_A > \Delta V_B$) at the end (expired concentration represented by a solid circle). This mechanism generates a sloping alveolar plateau. (From Ref. 17.)

gional'' and ''interregional'' distribution of ventilation, the latter being synonymous with topographical distribution of ventilation. The main conclusions obtained in 1966 are still valid today and a summary can be found in the review by Milic-Emili (10).

The essential message is illustrated in the well known ''onion-skin'' diagram (one region of lung corresponds to one skin) shown in Figure 2. On the ordinate, regional lung volume (V_r), is expressed as a percentage of regional TLC (TLC_r). On the abscissa, total lung volume (V), is expressed as percentage of TLC (lower axis) and as a percentage of VC (upper axis). The broken line (line of identity) illustrates the case of homogeneous lung expansion where a region of lung expands to the same degree as the expansion in volume of the entire lung. The vertical distance (D) from the top of lung (in cm) is indicated on each curve. At total lung capacity (TLC), the alveoli are equally and maximally expanded, which gives a convenient volume reference for each region of the lung. Each regional lung volume has its TLC volume as reference and is expressed as regional TLC fraction for any other volume. For volumes above closing volume, regional volumes change almost linearly with lung volume, although at different rates from each other. Airways closure occurs in a particular region when the

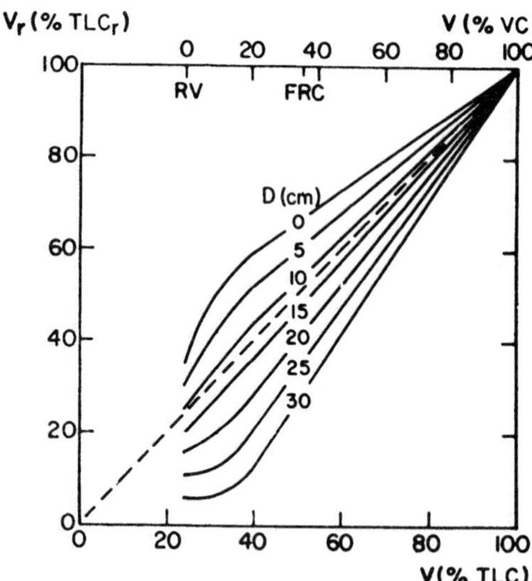

Figure 2 Ordinate: regional lung volume (V_r) expressed as percentage of regional TLC (TLC_r). Abscissa: total lung volume (V) expressed as percentage TLC (lower axis) and as percentage VC (upper axis). The broken line (line of identity) indicates uniform expansion of the regions in equal proportion to that for the entire lung. The vertical distance (D) from the top of lungs (in cm) is indicated on each curve. (From Ref. 3.)

curve for that region becomes flat, and can be seen to occur at greater absolute lung volumes for the lower (gravitational dependent) regions than for the upper. The accepted mechanism for the topographical ventilation inhomogeneity is, primarily, the weight of the lung: the lower zones are, relatively speaking, compressed, and the upper zones, expanded. This model predicts that in μG each region would be equally expanded at the different lung volumes, i.e., all curves coincide with the line of identity. Before the first experiments performed in space, Bryan et al. (1) measured the distribution of ventilation with ^{133}Xe in a human centrifuge at different levels of G. Based on extrapolations to 0 G, the authors suggest that the ventilation should be uniform in μG.

C. Cardiogenic Mixing

The heart shakes the lung parenchyma at each beat, but its quantitative role in gas mixing remains unknown. West and Hugh-Jones (11) observed pulsatile gas flow in lobar and segmental bronchi in human subjects, synchronous with the heart beat. An interesting observation was that the magnitude of flow pulses measured within the lung was much larger than the pulsatile flow measured at the mouth, suggesting that phase differences between flows in different bronchi tend to cancel out the differences in the common airways.

Fukuchi et al. (12) measured intrabronchial nitrogen concentration during inspirations of oxygen at constant flow in vivo and postmortem. They concluded that the mixing action of the heart resulted in an effective diffusion coefficient more than 5 times greater than that for molecular diffusion. However, this calculation is based on a symmetrical model of the lung and may only apply, as a first approximation, for the first generations of the bronchial three. Based on measurements performed before and after the introduction of saline into the pericardial sac, Fukuchi et al. (13) also demonstrated that cardiogenic mixing is due to the mechanical action of the heart on the lung itself, and not to pulsatile capillary blood flow. Engel (14) discussed the mechanisms responsible for cardiogenic mixing, but the unresolved question remains: How much does cardiogenic mixing contribute to the overall mechanisms of gas mixing in the lung? Cardiogenic oscillations persist in μG, and while these may provide some insight into the role of cardiogenic mixing, the action of the heart may be modified in μG, confounding the interpretation somewhat.

D. Single-Breath Nitrogen Washout Test

The first indirect evaluation of topographical ventilation distribution in sustained μG (15) was based on single-breath washouts (SBWs). At residual volume (RV), the subject inspired a 150-ml bolus with 21% oxygen in argon, followed by a controlled inhalation at 0.5 L/sec up to TLC of 100% oxygen. Then, the subject exhaled at the same flow, back to RV. Expired nitrogen and argon plots are displayed in Figure 3 on ground (standing, before the flight) and in space for four of the seven subjects studied during the first dedicated life sciences mission of Spacelab, the nine-day SLS-1 mission, which took place in 1991.

In 1 G (there were no significant differences between preflight and post-flight tests) both nitrogen (upper trace) and argon (lower trace) display the traditional phases of the SBW. From the nitrogen tracing on subject 1, for example, one can see phase I, with 0 nitrogen concentration, corresponding to volume in the upper airways, and the transition (phase II) to purely alveolar gas. Phase III

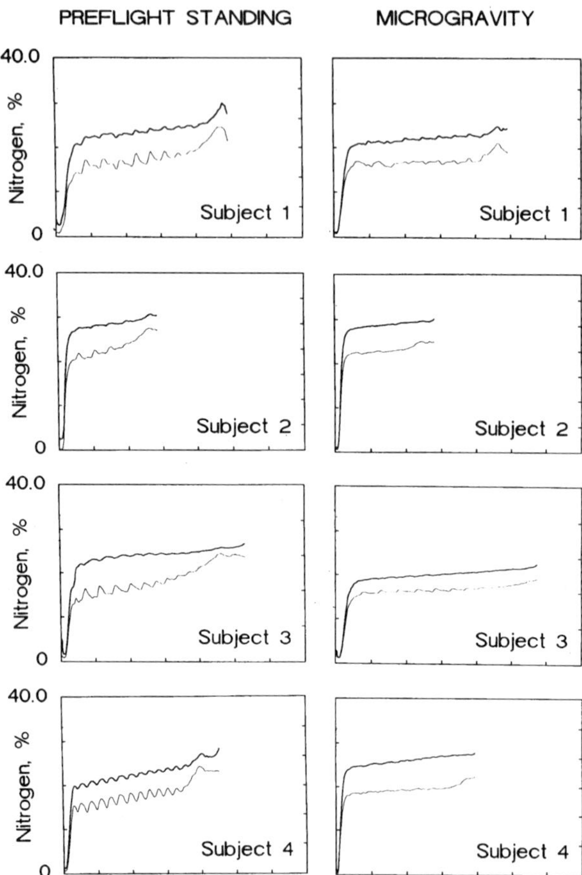

Figure 3   Representative tracings (upper, nitrogen; lower, argon bolus) of SBW from four crew members obtained in standing position in 1 G and in μG. (From Ref. 15.)

or alveolar plateau, i.e., the quasi-linear part with cardiogenic oscillations follows this. A clear transition where the slope changes indicates the onset of phase IV at closing volume, which continues until the maximum of the curve; and a decreasing concentration or phase V (16). Compared with the standing 1 G measurements, there was a marked decrease in ventilatory inhomogeneity during microgravity, as evidenced by the amplitude of the cardiogenic oscillations (44 and 24% of the 1 G values for nitrogen and argon, respectively), height of phase IV (18 and 40%). However, argon phase IV volume was not reduced, showing that considerable ventilatory inhomogeneity, associated with airway closure, remained in the absence of gravity.

Although theoretical studies have suggested a minor role of topographical inhomogeneities on the slope of phase III (17), the experiments performed in space gave the first direct information on the gravity dependence of the slope. The gravitational contribution to phase III slope is illustrated in Figure 1. In the lower panels, the inhomogeneous preinspiratory lung volume due to the weight of the lung is represented at RV in the lower left panel by two units, A and B, with different volumes. After a full inspiration of oxygen (dashed circles), the nitrogen concentration is larger in unit A than in unit B (horizontal dashed lines, lower right). The genesis of the sequential emptying during the subsequent expiration may be explained using a sigmoid-shaped pressure–volume (P–V) curve (top). Consider the expiratory limb (downward arrow) of this curve with a constant pressure gradient ΔP (horizontal arrows) applied to units A and B. The P–V curve represents any part of the lung (i.e., we assume uniform mechanical properties). However, because of the weight of the lung, the pressure applied to unit A is more negative (strictly speaking, more subatmospheric). As a consequence, its compliance is lower at the beginning of expiration. For the same variation of pressure, Figure 1 shows that the total flow is initially weighted preferentially by unit B where the nitrogen concentration (asterisk) is lower. Toward the end of expiration, because of the sigmoid shape of the P–V curve, the ratio of relative flows from each unit is reversed and the end-expired concentration (solid circle) approaches that in unit A (see also lower right). This was shown by Anthonisen et al. (18), who demonstrated that gravity caused the apical lung regions to contribute progressively greater proportions of their initial volume as the expiration progressed. If these were the only mechanisms involved, phase III should have 0 slope in μG. However, in μG, the phase III slope for nitrogen was reduced by only 22% compared with that measured standing in 1 G, demonstrating conclusively that the genesis of phase III is primarily nongravitational (15).

The small decrease of the nitrogen phase III slope was consistent with previous results obtained in parabolic flights (19). However, the interpretation of SBWs performed during parabolic flights suffered from the potential perturbation from the 2 G period proceeding the short period in microgravity, as ventilatory inhomogeneities created during the hypergravity phase of flight may persist into the μG phase. It was not until the measurements in sustained μG that the data collected in parabolic flight could be validated.

Although the slope of phase III of the single-breath washout is an accepted index of ventilation inhomogeneities, it gives not only information of spatial, but also of the temporal inhomogeneities, i.e., inhomogeneities occurring at different moments during the inspired and expired VC. However, topographical (or spatial) inhomogeneities described by the "onion-skin" model can only generate a slope if these units also present a sequential pattern of emptying. Therefore, the homogeneous distribution of inspired gases, or the uniform emptying of the lung are sufficient conditions for the elimination of the topographical component of the slope. The relative contribution of each one of these factors to the 22% decrease of slope in μG remains a matter of speculation.

E. Single-Breath Bolus Washins

While the phase III slopes for nitrogen were reduced by only about 22% in μG compared with 1 G, phase III slopes for the argon bolus inhaled at RV were reduced by 71% to only 29% of their control value (15). However, the interpretation in microgravity is difficult, because of our ignorance of airway closure under these circumstances.

Single-breath bolus washins were performed in μG (20). In an ideal experiment, where successive boluses of a given inert gas are inspired at different points of an inspired VC (free of this gas), adding all the individual expired concentrations should be equivalent to a continuous single inspiration of the same inert gas. For example, assume a VC of 5 L. Theoretically, performing 50 single-breath washins of 100 ml each, inspired at equal volume intervals during a VC inspiration and adding up the 50 expired bolus concentration curves should be equivalent to a complete SBW. Performing so many experiments is impractical, but tests performed both on the ground and during parabolic flights showed that phase III slopes of the SBW could be accurately reconstructed from a small number of bolus tests spaced over the VC range (20).

The bolus tests performed in both 1 G and in μG showed that almost all of the phase III slope seen in conventional SBW tests is a result of events below closing volume or near to TLC, with little slope resulting from gas inspired in the middle lung volume range. Consequently, part of the gravity dependence of the phase III slope is associated with airway closure, making its interpretation even more complex. This conclusion is consistent with the larger reduction of argon bolus slopes in μG, compared with those for nitrogen (15). The observation that the SBW phase III slope is sensitive mainly to ventilatory inhomogeneities near RV and TLC makes it a poor index of inhomogeneities in the range of tidal breathing.

F. Insoluble Gas Rebreathing

A rebreathing test consists of inhaling and exhaling into and out of a bag initially filled with a foreign gas mixture. Since the work of Lewis et al. (21), rebreathing tests have often been used to estimate diffusing capacity and cardiac output. Hughes et al. (22) and Jones et al. (23) have shown that rebreathing inert (insoluble) gases in a closed circuit may be a sensitive technique for detecting ventilatory inhomogeneities.

Consider the rebreathing of argon. After a number of breaths, the argon in the lung and in the bag reach the same concentration, CAr_{eq}. The approach of the initial inspired concentration of argon to CAr_{eq} should be exponential for a perfect homogeneous lung. By considering the lung as a series of parallel units each with a different specific ventilation or $\Delta V/V_0$, the differences between the ideal situation and that observed allows the ventilatory inhomogeneity to be estimated. A major advantage of the rebreathing maneuver over the SBW is that the evaluation of ventilatory inhomogeneity does not require that sequencing between parallel units be present in order for the inhomogeneity to be detected. Furthermore, Jones et al. (23) have shown that rebreathing inert gases in a closed circuit may be a more sensitive technique for detecting convection-dependent ventilatory inhomogeneities than SBW techniques.

Rebreathing tests with argon were carried out on four astronauts before, during, and after the 10-day Spacelab D-2 mission (24). Starting from FRC, the rebreathing maneuver consisted of eight reinspirations from a bag filled with 1.8 to 2.2 L of test gas mixtures containing 5.25% argon. The bag was emptied at each breath. The comparisons between tests performed in space and on the ground showed that the degree of gravity-independent inhomogeneity of specific ventilation was at least as large as the gravity-dependent inhomogeneity of specific ventilation. Based on the previous description of the regional distribution of ventilation, the units responsible for these inhomogeneities are likely to be primarily within the same lung region, since gravity is not present to impose top-to-bottom differences in ventilation.

III. Nontopographical Convection-Dependent Ventilation Inhomogeneity

As it became clear that major nontopographical or intraregional inhomogeneities persist in μG, the understanding of the mechanisms involved required the analysis of gas transport in the periphery of the lung.

A. The Diffusion Front

If an inert gas is inspired at constant flow, it advances rapidly in the conducting airways and a time-independent concentration profile is established at a point deep in the airway, the so-called diffusion front (25). The location of the diffusion front depends on ratio of flow to the diffusion coefficient of the gas. Figure 4 represents diffusion fronts for various ratios of $\dot{v}/D$. For simplicity, we have normalized the concentrations (ordinates) to an inspired concentration of one (F = 1) and have imposed a 0 concentration at the alveolar end (x = 0). During inspiration there is, of course, a continuous increase of concentration at the alveolar end of the lung, as diffusion transports gas to the end of the airway. Thus, it is more appropriate to speak of a quasi-stationary diffusion front. Therefore, the lung can be functionally divided in two zones: the purely convective zone where during inspiration F(x) = 1 and a convective-diffusive zone where F(x) < 1. This partitioning is a direct consequence of Fick's law of diffusion, which states that the flow due to diffusion is proportional to concentration gradient. For F(x) = 1, the concentration gradient is 0 and therefore convective transport is the only mechanism of transport. For F(x) < 1, both convection and diffusion must play a role.

Curves 1 to 3 of Figure 4 are the result of theoretical simulations (26) for an overall (whole lung) inspiratory flow rate of $\dot{v}$ = 0.5 L/s and D = 0.1, 0.225, and 0.6 cm^2/s, respectively. These diffusion coefficients correspond to SF$_6$, N$_2$, and He respectively. Curve 4 is the diffusion front for He at a slow inspiration of $\dot{v}$ = 0.125 L/s. Because the diffusion front depends on both the diffusion coefficient and flow, the convection-dependent units are not anatomically defined.

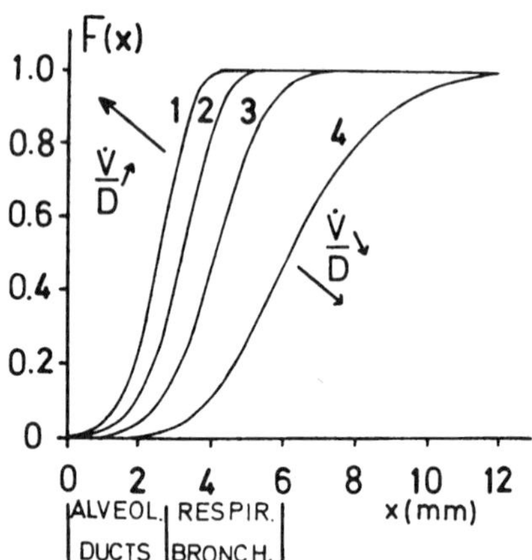

Figure 4 Diffusion fronts for an inspired gas of concentration 1 and for different ratios of flow to diffusion coefficients. The results come from model simulations where the fractional concentration (F) is expressed as a function of distance (x) to alveolar end of the model. Curves 1 to 3 correspond to $\dot{v}$ = 0.5 L/s and D = 0.1, 0.225, and 0.6 cm^2/s respectively. These diffusion coefficients correspond to SF$_6$, N$_2$ and He, respectively. Curve 4 is the diffusion front for He with $\dot{v}$ = 0.125 L/s. (From Ref. 17.)

For the same flow, the convection-dependent units for SF_6 are smaller than those for N_2 or He. For an inspiratory flow of 0.5 L/s, all units larger than the acini are convection dependent for He, whereas within the acini, some of the units are convection dependent for SF_6 but not for He.

B. Inhomogeneous Pressure–Volume Curves

An experimental basis for gravity-independent ventilation inhomogeneities is based on the work of Olson and Rodarte (27). These authors implanted parenchymal markers in isolated dog lobes and measured the deformation of tetrahedrons (the volume defined by four non-coplanar markers) with volumes ranging from 0.2 to 25.1 mL at total lobar capacity. They demonstrated considerable nonuniformity of volume changes. Furthermore, the variability of volume expansions did not follow any topographical orientation. The same behavior was also present in saline-filled lobes, suggesting that the major determinant of the inhomogeneities is related to properties of the parenchyma itself and not to the effects of surfactant. These results are consistent with concentration measurements made by intrapulmonary sampling in open-chest dogs by Engel et al. (28). These authors have shown that large ventilatory inhomogeneities exist within small lung regions and suggested a model based on parallel units with different P–V curves (29), i.e., a model generating convection-dependent inhomogeneities (CDIs).

C. Multiple-Breath Washin and Washout

Figure 5 shows volume and nitrogen concentration tracings, as a function of time, obtained from a typical multiple-breath washout (MBW) test. This test requires a regular breathing pattern with a tidal volume of approximately 1 L, starting from FRC and using pure oxygen for inspiration. The valve switch occurs during an exhalation and the subject was instructed to watch a screen during each inspiration, inhale to an indicator line, and then exhale freely back to FRC. The first and twentieth expirations are represented (inset) as functions of expired volume (30).

If two parallel convection-independent units with different $\Delta V/V_0$ are washed out from their initial nitrogen content, the nitrogen concentration of each one will decrease exponentially. If there is a sequence of emptying between the two units, such that the less-ventilated one preferentially empties late during the expiration, the expired nitrogen will have a positive slope for each expiratory breath, as previously described for topographically distributed units. Dividing this slope by its corresponding mean nitrogen concentration normalizes the slope for the decreasing nitrogen concentration as the washout progresses. It was shown (31) that such a normalized slope (S_n) should increase progressively, ultimately reaching a constant value. Indeed, as the MBW proceeds and the nitrogen concentration decreases in each unit, their relative concentration difference increases and reaches a maximum when the better-ventilated unit is emptied of its nitrogen.

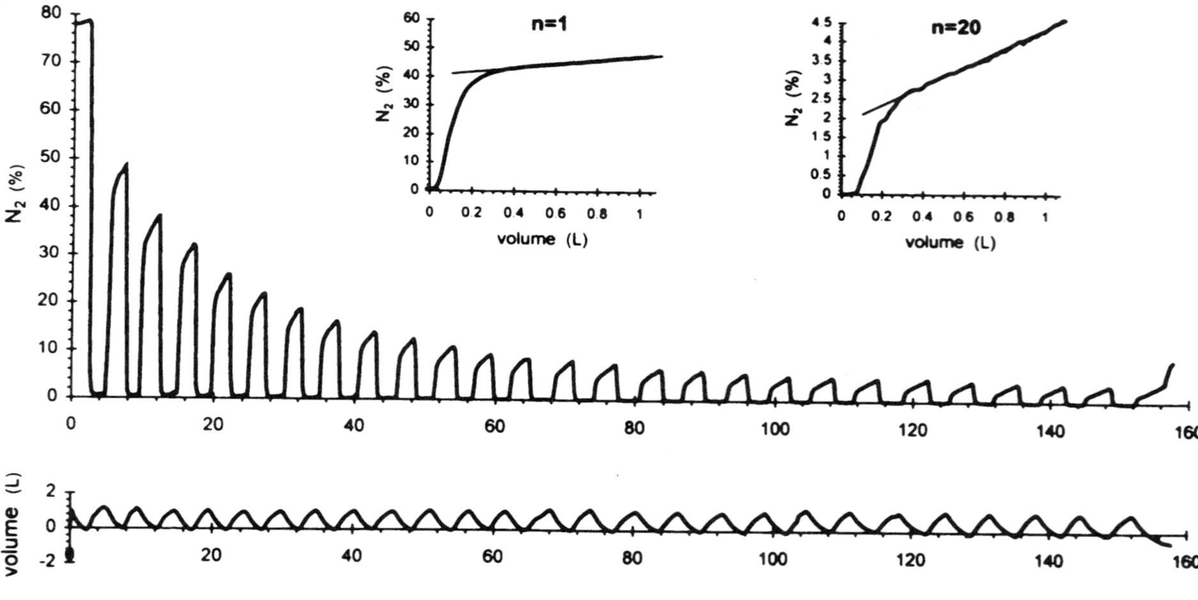

Figure 5 Nitrogen concentration and volume tracings as a function of time during a MBW test from a normal subject during a bronchoprovocation test with histamine, which greatly increases the slopes, particularly at the end of the MBW. (Inset) Illustration of the alveolar slope vs. expired volume for breaths 1 and 20 plotted in an equivalent scaling with respect to mean expired N_2 concentrations of breaths 1 and 20, respectively. (From Ref. 30.)

If the sequence between the units remains constant, this corresponds to a constant S_n. Furthermore, the model predicts S_n values such that the slope on back extrapolation to breath 0 should be $S_0 = 0$ (31).

Figure 6 shows the normalized slopes (mean $\pm$ SD) on one hyperresponsive subject under control conditions (solid symbols) and following bronchoprovocation (open symbols). Data are means $\pm$ SD for three (control) or two (provocation) MBW tests plotted as a function of lung turnover (TO, defined as the cumulative expired volume divided by FRC) (30). These authors defined a convection-dependent index of inhomogeneity, S_{cond}, as the normalized slope difference per

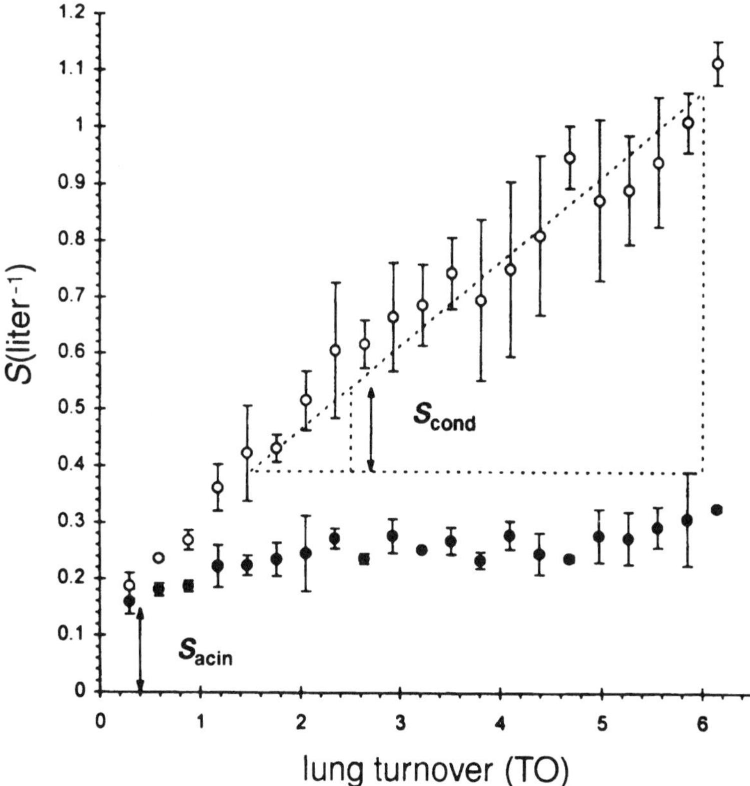

Figure 6 Normalized slopes (slopes divided by mean expired N_2 concentrations) as a function of lung turnover under control conditions (solid symbols) and following bronchoprovocation (open symbols). Data are means $\pm$ SD of three (control) or two (provocation) MBW tests. In the case of the provocation curve, the derivation of the peripheral ventilation index (S_{acin}) and the proximal ventilation index (S_{cond}) is shown. (From Ref. 30.)

TO determined by linear regression between TO = 1.5 and TO = 6. With the exception of S_0, which is clearly different from 0, the experimental points closely follow the predictions of the two-compartment model, and it has been shown that the rate of increase of S_n is a good index of CDI (31).

Figure 7 shows the results obtained in four normal subjects during similar MBW tests, both on the ground (standing and supine) and in μG (32). We will see in the next section that a more refined analysis incorporating diffusion–convection interactions is required for a complete interpretation. However, after the first few breaths, this mechanism (see below) adds a constant to S_n, and there is no change in dS_n/dn among the three different conditions (i.e., the curves—standing, supine, and μG—are essentially linear and parallel to each other). If the parallel units generating the slope were topographically located, one would have expected that in microgravity, S_n would increase much less rapidly with n than in 1 G.

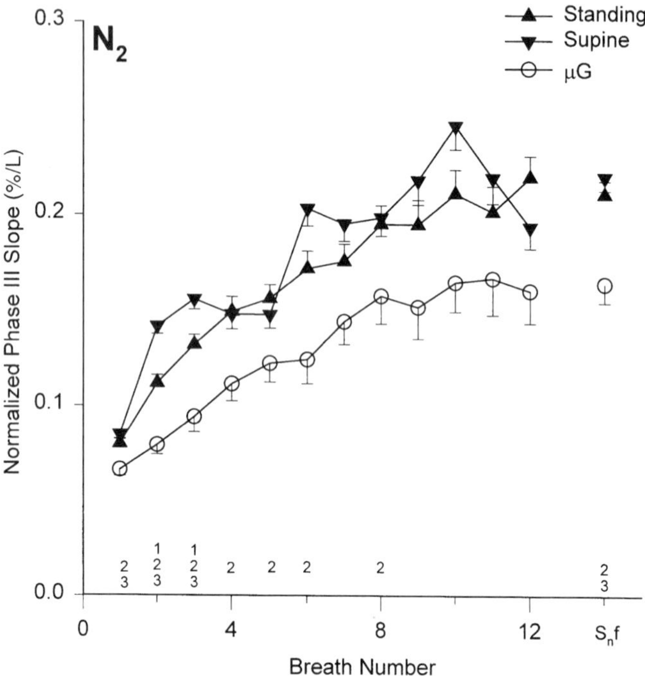

Figure 7 Normalized slopes of nitrogen MBW as a function of breath number from four subjects standing, supine, and in microgravity (μG). Values are means $\pm$ SE. Significance between three positions is indicated by a numerical code below results for each breath: standing vs. supine (1), standing vs. μG (2), and supine vs. μG (3). S_{nf} is the average value for the last 2 breaths. (From Ref. 32.)

That this was not the case means that the units with gravity-dependent ventilation inhomogeneities have little or no expiratory sequencing between them.

Of interest is the comparison of MBWs from mammals with large differences in absolute lung volumes, such as steers and rats (Fig. 8). It was expected that in steers, because of the large lung size, topographical ventilatory inhomogeneities would lead to a significant S_{cond}. This is clearly not the case as evidenced by a constant S_n as the washout progresses (33). Tests performed in horses led to similar results (F. Rollin, personal communication). This demonstrates that in these animals, units with different $\Delta V/V_0$, topographically distributed or not, empty synchronously. The reasons for the different behavior of human lungs from those of other species are not clear.

During the next decade, the experiments performed in μG to study lung ventilation will most probably still be limited to indirect measurements of gas concentrations at the mouth. This will not allow the determination of the anatomi-

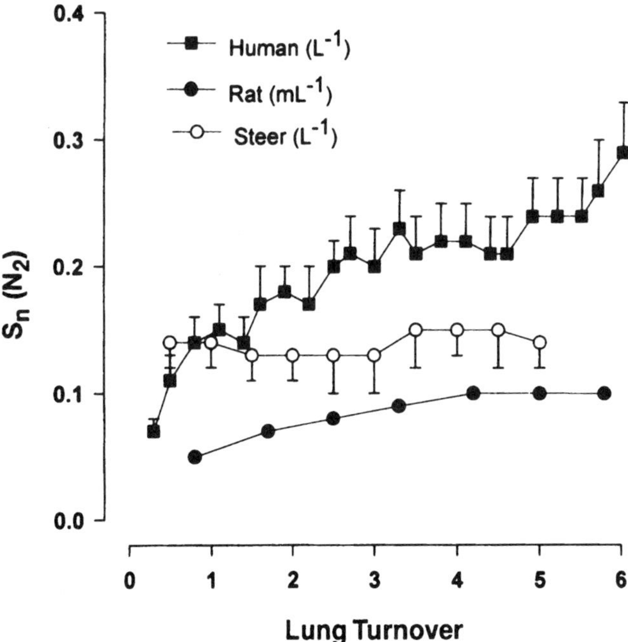

Figure 8 Normalized slopes of nitrogen MBW (means ± SE) as a function of lung turnover in humans, rats, and steers. In rats, SE values are smaller than symbol. Human and rat curves were replotted from Crawford et al. (38) and Verbanck et al. (37) respectively. Note the absence of an increase in S_n as the washout progresses in the curve for steers. (From Ref. 33.)

cal location of the convection-dependent units responsible for the inhomogeneities that persist in μG. Indeed, the expired concentration profiles generated by two large units can be identical to the concentration profile generated by a large number of small units with the equivalent parallel inhomogeneity.

IV. Nontopographical Diffusion-Convection-Dependent Ventilation Inhomogeneity

A. Diffusion-Convection-Dependent Inhomogeneity

Gas mixing in the lung periphery was described in detail in a previous book in this series (17), and we will only address the main aspects here. Diffusion-convection-dependent inhomogeneity (DCDI) is a consequence of interaction of diffusive and convective transport at branch points of asymmetrical structures of ducts (34). Diffusion-convection-dependent inhomogeneity goes against common intuition, in which diffusion is usually thought of as equalizing of gas concentrations, and results in the generation of concentration differences between adjacent lung units.

The branch points leading to DCDI are located at the position of the diffusion front. From Figure 4, for a 0.5 L/s inspiratory flow, these branch points correspond roughly to between 2 and 4 mm from the alveoli for SF_6 and between 2 and 6 mm from the alveoli for He. At the corresponding branch points, diffusion and convection are not independent (they interact through a mass balance constraint) and inhomogeneous gas concentration results if the volumes of the two distal units differ from each other, even in a structure expanding homogeneously and isotropically. Consider an inspiration of oxygen into a nitrogen- (or air-)filled lung. At a branch point leading to units of unequal volume but with airways of comparable airway cross section, the incoming gas is convected in proportion to the volume of the units, but diffused in proportion to the cross sections and diffusion gradient (Fick's law). However, because the smaller and larger units are subtended by airways of comparable cross section and initially the concentration gradient is the same, the amount of gas transported by diffusion is similar for each unit. Because of the difference in the volumes of the units, the result is a higher concentration of inspired gas (oxygen) at the end of inspiration in the smaller unit. On expiration, convection from the larger unit first brings nitrogen-rich gas to the branch point where it diffuses back (in the opposite direction to the convective flow) into the better-ventilated (oxygen-rich) smaller unit. However, oxygen does not diffuse back into the larger unit as effectively because of the larger convective flow out of this unit. Because of the back diffusion of nitrogen into the smaller unit (diffusive pendeluft), nitrogen appears later in the expiration as measured at the mouth than would otherwise have been the case, producing an upward sloping nitrogen plateau (phase III). This mechanism, which generates the sloping alveolar plateau, depends on the dimensions of the units peripheral to those branch points. For He and a 0.5 L/s flow, the units peripheral to the branch points correspond approximately to the acini. In humans, the He slope is usually smaller than the SF_6 slope. In fact, only recently was a situation observed, in lung transplant recipients during episodes of acute rejection, in which the He slope was greater than that of SF_6 (35).

The most convincing experimental demonstration that the theoretical DCDI concept describes the temporal and spatial concentration inhomogeneities in the lung, came from experiments performed in rats, where lung morphometry is known with more details than in any other species (36). The simulations reproduced the experiments remarkably well, including a larger slope for He with respect to SF_6 (37) in these small animals.

For a nitrogen MBW, the slope of the first breath is due mainly to DCDI, which is why back-extrapolation of the S_n curve to breath 0 results in a non-0 intercept (see Fig. 6). It can also be shown that this mechanism, which is mainly intraacinar, has a contribution that is almost constant during a MBW (38). This is the basis for the computation of acinar and conductive components of ventilatory inhomogeneity (S_{acin} and S_{cond}) suggested by Verbanck et al. (30) (see Fig. 6). S_{acin} is determined by subtracting that part attributed to the conducting airways from the slope of the first breath, i.e., S_{cond} multiplied by the TO value of the first breath (≈ 0.3 in the case of Fig. 6).

B.   Helium and Sulfur Hexafluoride Single-Breath Washout

In μG, the phase III slope for nitrogen of the SBW was 78% of its 1 G value, which suggests large ventilatory inhomogeneities in the lung periphery (see Sec. II.D). If these inhomogeneities were intraacinar, it would be expected that gravity would have no effect at this anatomical level. Therefore, $He–SF_6$ SBW performed in μG would be expected to generate slopes decreased by the same amount with respect to their 1 G values. Very surprising results were observed (5). The main findings are summarized below.

Experiments in Sustained Microgravity

In sustained μG, the SF_6 slope decreased more than the He slope, with both slopes becoming similar (Fig. 9). After a 10-s breathhold, the SF_6 slope was actually flatter than that for He (39). It was as if conformational changes in the lung geometry at the level of the acinus or between a few acini had occurred in sustained μG. This was the more surprising as the $He–SF_6$ slope difference was thought to be one of the more robust indices of ventilation inhomogeneity in the lung periphery, with alterations occurring only in pathological situations. However, the fact that 10 hr after landing, the slopes were the same as preflight values suggests that pathological mechanisms such as inflammation were unlikely to be

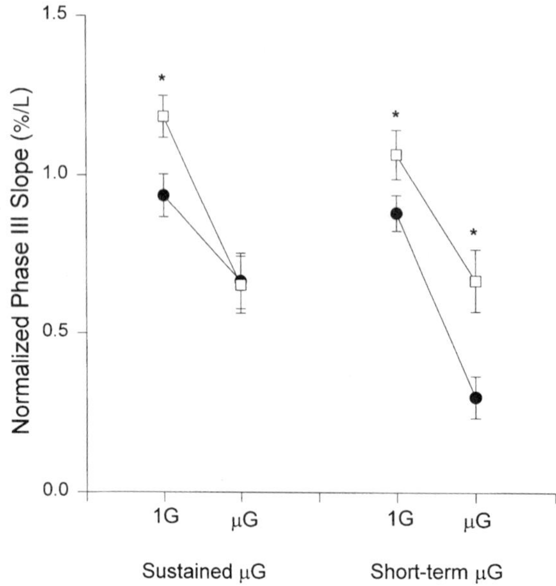

Figure 9 Normalized phase III slopes (by considering pretest concentration as 100% and inspired concentration as 0%) of He (solid circles) and SF_6 (open squares) from SBW tests in a spaceflight study (subjects standing in 1 G and during sustained μG), and from a short-term μG (parabolic flight) study (subjects sitting in 1 G and during short-term μG). * = SF_6 slope significantly different from that of He, $p < 0.05$. (From Ref. 41.)

the cause. While there was a change in RV in μG (~300-ml reduction) (40), tests performed on the ground suggested that this was probably not the cause of the differences between He and SF_6 behavior in μG. However, the more recent observation that much of the phase III slope is a consequence of events in the lung below closing volume makes it impossible to eliminate the change in RV as a potential cause of the changes in the He–SF_6 slope difference.

Experiments in Short-Term Microgravity (Parabolic Flights)

In sharp contrast to the situation in sustained μG, in short-term μG, the He slope decreased more than the SF_6 slope (see Fig. 9), the slope difference being significantly greater in μG (41). This was the more surprising as the results for the phase III slope for nitrogen in sustained μG were consistent to those obtained during parabolic flights (15). These results suggest that the cause of the acinar geometrical changes observed during sustained μG had a long time course (compared with the parabola duration of ~25 s) and thus possibly involved the pulmonary vasculature. Further, these long-term changes occurred in the airways around the acinar entrance, i.e., where the proximal part of the diffusion front stands, because it was the behavior of He that differed between sustained and short-term μG.

Helium and Sulfur Hexafluoride Bolus Washin in Short-Term Microgravity

Diffusion-convection-dependent inhomogeneity resulting from bolus inhalations has a component specifically related to the lung volume when the bolus was inhaled, and simulations of these experiments have shown its potential to understand the slope of the alveolar plateau (42). The bolus tests performed during parabolic flights showed that almost all the phase III slope seen in conventional SBW tests is a result of events below closing capacity or near total lung capacity, with little slope resulting from gas inspired in the middle lung volume range (20). We cannot extrapolate these conclusions to sustained μG because of the differences observed between that environment and parabolic flight (41), but it remains possible that the decreased RV observed in space plays a role. Elliott et al. (40) observed 310- and 220-mL decreases in RV compared with 1 G standing and supine, respectively. Because VC also decreased by 5% (40), the resulting slopes do not correspond to the same absolute lung volumes. The influence of intraregional inhomogeneity of airway closure (43) may also play a role, as closure may take place at different zones of the He and SF_6 diffusion fronts in the gravity-dependent situation. In contrast, airway closure above RV in μG is probably determined by the distribution of mechanical properties of the airways and parenchyma, and may occur at a different anatomical location. Figure 9 also shows that the SF_6 slope behaved similarly when comparing long-term with short-term μG, whereas He behaved differently (a much larger decrease during parabolic flights), suggesting that the corresponding modifications are at the zone of the terminal bronchioles. Finally, to complicate matters, the action of the heart resulting in cardiogenic mixing may have different effects in 1 G and μG.

C. Helium and Sulfur Hexafluoride Multiple-Breath Washout

He and SF_6 MBW seems to be the most informative test regarding ventilation inhomogeneity in the lung periphery. These tests were first performed by Crawford et al. (38) on four subjects over 14 expirations. They showed that after the first five breaths, the He and SF_6 slopes increased similarly with breath number. This result gave experimental support to the concept that S_n increase after the first few breaths is a sensitive index of CDI. Furthermore, because the addition of 1 s of end-inspiratory apnea changed the normalized slopes significantly (44), the S_n increase appeared to correspond to inhomogeneous properties of lung periphery, since the concentration differences must exist between units that are close to each other.

During the Spacelab mission SLS-2, MBWs using He and SF_6 were also performed (32). The results from this MBW study largely confirmed those found previously. Convection-dependent inhomogeneity was largely unaltered between 1 G and μG, in line with the SLS-1 study (45), and the SF_6–He slope difference was reduced (although in these smaller-volume breaths, not abolished) similar to that seen in the SBW study (39). Of particular note, however, was the partitioning of the rise of phase III slope with breath number (the equivalent of the division between S_{cond} and S_{acin}). Although the SLS-2 results were somewhat noisy, partitioning showed that for N_2 and SF_6, both of which have relatively distal quasi-stationary diffusion fronts, there was little difference in the contribution of CDI to overall inhomogeneity between 1 G and μG. However, for He, with a more proximal front, the CDI contribution was abolished in μG. This suggests that the CDI seen to persist in μG must be located between units that are sufficiently close that diffusion of He but not of N_2 or SF_6 was an efficient means of abolishing the concentrations gradients it produces. Thus the nongravitational CDI must exist between acini or between groups of a few acini. This is consistent with the previous observation of Crawford et al. (44), and serves to provide a scale for the convective inhomogeneity that persists in μG.

V. Clinical Applications

Nitrogen SBW was a widespread test of ventilation distribution (46). Its limited success in clinics is likely due not only to its lack of specificity but also to uncertainties on the role of gravity and pathological processes in the derived indices. The experiments performed in μG confirmed that gravity plays a minor role in the slope of phase III. Bolus tests, both on the ground and in μG have also shown that the slope is very sensitive to events occurring both near RV and TLC. As a consequence, this test carries little information on ventilation distribution during normal breathing.

In spite of these limitations, which are valid for normal subjects, He and SF_6 SBW in lung transplant subjects (35) proved to be one of the most interesting applications of this test. For a 1-L inspiration from FRC with the same gas mixture, this test has shown predictive potential for episodes of obliterative bronchiolitis leading to chronic lung rejection (47). During acute rejection episodes, He and SF_6 slopes increase but the He slope becomes larger than that for SF_6, probably due to conformational changes near the entrance of the acinus, possibly as a result of inflammation in that zone.

In terms of the ground applications of MBW performed in μG, a clearer picture has appeared: during normal breathing, inhomogeneities in lung mechanical properties, rather than gravitational distortion, are dominant in determining CDI. The marginal role of gravity simplifies the interpretation of the MBW-derived indices, such as S_{acin} and S_{cond}. Verbanck et al. (48) performed a study in patients with chronic obstructive pulmonary disease (COPD), in whom considerable conductive and acinar impairments of ventilation were expected to exist. The comparison of the relationship of S_{acin} and S_{cond} to standard lung-function indices by means of a principal component factor analysis has shown that S_{acin} and S_{cond} reflect independent alterations to the lung due to pathology. S_{cond} is linked to conductive zone factors such as airway conductance and forced expiratory flows, while S_{acin} is linked to acinar lung-zone factors such as diffusing capacity. Another study in 20 patients with stable asthma, using the same technique, gave evidence of acinar airway involvement (49). In these patients, in contrast to the COPD patients, acinar airway impairment was partially reversible after salbutamol inhalation.

Major lessons learned from μG were a clarification of the functional meaning of indices derived from the MBW and a better understanding of the factors affecting the results of SBW. The results from these studies in μG have assisted in pushing forward the early development of more-sensitive tests of pulmonary function on the ground (30,47).

References

1. Bryan AC, Milic-Emili J, Pengelly D. Effect of gravity on the distribution of pulmonary ventilation, J Appl Physiol 1966; 21:778–784.
2. Kaneko K, Milic-Emili J, Dolovich MB, Dawson A, Bates DV. Regional distribution of ventilation and perfusion as a function of body position. J Appl Physiol 1966; 21:767–777.
3. Milic-Emili J, Henderson JAM, Dolovich MB, Trop D, Kaneko K. Regional distribution of inspired gas in the lung. J Appl Physiol 1966; 21:749–759.
4. West JB, Guy HJB, Elliott AR, Prisk GK. Respiratory system in microgravity. In: Fregly MJ, Blatteis CM, eds. The Handbook of Physiology. Sect. 4. Environmental Physiology. New York: Oxford University Press, 1996:675.
5. Prisk GK. Invited review: Microgravity and the lung. J Appl Physiol 2000; 89:385–396.
6. Otis AB, McKerrow CB, Bartlett RA, Mead J, McIlroy MB, Selverstone NJ, Radford EP Jr. Mechanical factors in distribution of pulmonary ventilation. J Appl Physiol 1956; 8:427–443.
7. Baydur A, Behrakis PK, Zin WA, Jaeger M, Milic-Emili J. A simple method for assessing the validity of the esophageal balloon technique. Am Rev Respir Dis 1982; 126:788–791.
8. Edyvean J, Estenne M, Paiva M, Engel LA. Lung and chest wall mechanics in microgravity. J Appl Physiol 1991; 71:1956–1966.

9. Michels DB, Friedman PJ, West JB. Radiographic comparison of human lung shape during normal gravity and weightlessness. J Appl Physiol 1979; 47:851–857.

10. Milic-Emili J. Topographical inequality of ventilation. In: Crystal RG, et al., eds. The Lung: Scientific Foundations. Vol. 2, 2d ed. New York: Raven, 1997:1415.

11. West JB, Hugh-Jones P. Pulsatile gas flow in bronchi caused by the heart beat. J Appl Physiol 1961; 16:697–702.

12. Fukuchi Y, Roussos CS, Macklem PT, Engel LA. Convection, diffusion and cardiogenic mixing: An experimental approach. Respir Physiol 1976; 26:77–90.

13. Fukuchi Y, Cosio M, Kelly S, Engel LA. Influence of pericardial fluid on cardiogenic gas mixing in the lung. J Appl Physiol 1977; 42:5–12.

14. Engel LA. Dynamic distribution of gas flow. In: Macklem PT, Mead J, eds. Handbook of Physiology. Vol. 3. The Respiratory System. Bethesda: American Physiological Society, 1986:575.

15. Guy HJB, Prisk GK, Elliott AR, Deutschman RAI, West JB. Inhomogeneity of pulmonary ventilation during sustained microgravity as determined by single-breath washouts. J Appl Physiol 1994; 76:1719–1729.

16. Nichol GM, Michels DB, Guy HJB. Phase V of the single-breath washout test. J Appl Physiol 1982; 52:34–43.

17. Paiva M, Engel LA. Gas mixing in the lung periphery. In: Chang HK, Paiva M, eds. Respiratory Physiology. Lung Biology in Health and Disease Series. Vol. 40. New York: Marcel Dekker 1989:245.

18. Anthonisen NR, Robertson PC, Ross WRD. Gravity-dependent sequential emptying of lung regions. J Appl Physiol 1970; 28:589–595.

19. Michels DB, West JB. Distribution of pulmonary ventilation and perfusion during short periods of weightlessness. J Appl Physiol 1978; 45:987–998.

20. Dutrieue B, Lauzon A-M, Verbanck S, Elliott AR, West JB, Paiva M, Prisk GK. Helium and sulfur hexafluoride bolus washin in short-term microgravity. J Appl Physiol 1999; 86:1594–1602.

21. Lewis BM, Lin T, Noe FE, Komisaruk R. The measurement of pulmonary capillary blood volume and pulmonary membrane diffusing capacity in normal subjects: The effects of exercise and position. J Clin Invest 1958; 37:1061–1070.

22. Hughes JMB, Jones HA, Davies EE. Applications of multibreath washin measurements in closed circuit using a mass spectrometer. Bull Eur Physiopathol Respir 1982; 18:309–317.

23. Jones HA, Davies EE, Hughes JMB. Modification of pulmonary gas mixing by postural changes. J Appl Physiol 1986; 61:75–80.

24. Verbanck S, Linnarsson D, Prisk GK, Paiva M. Specific ventilation distribution in microgravity. J Appl Physiol 1996; 80:1458–1465.

25. Paiva M. Gas transport in the human lung. J Appl Physiol 1973; 35:401–410.

26. Paiva M. Theoretical studies of gas mixing in the lung. In: Engel LA, Paiva M, eds. Gas Mixing and Distribution in the Lung. New York: Marcel Dekker, 1985: 221.

27. Olson LE, Rodarte JR. Regional differences in expansion in excised dog lung lobes. J Appl Physiol 1984; 57:1710–1714.

28. Engel LA, Utz G, Wood LDH, Macklem PT. Ventilation distribution in anatomical lung units. J Appl Physiol 1974; 37:194–200.

29. Fukuchi Y, Cosio M, Murphy B, Engel LA. Intraregional basis for sequential filling and emptying of the lung. Respir Physiol 1980; 41:253–266.

30. Verbanck S, Schuermans D, Van Muylem A, Paiva M, Noppen M, Vincken W. Ventilation distribution during histamine provocation. J Appl Physiol 1997; 83:1907–1916.

31. Paiva M. Two new pulmonary functional indexes suggested by a simple mathematical model. Respiration 1975; 32:389–403.

32. Prisk GK, Elliott AR, Guy HJB, Verbanck S, Paiva M, West JB. Multiple-breath washin of helium and sulfur hexafluoride in sustained microgravity. J Appl Physiol 1998; 84:244–252.

33. Rollin F, Desmecht D, Verbanck S, Van Muylem A, Lekeux P, Paiva M. Multiple-breath washout and washin experiments in steers. J Appl Physiol 2000; 81:957–963.

34. Paiva M, Engel LA. Pulmonary interdependence of gas transport. J Appl Physiol 1979; 47:296–305.

35. Van Muylem A, Antoine M, Yernault J-C, Paiva M, Estenne M. Inert gas single-breath washout after heart-lung transplantation. Am J Crit Care Med 1995; 152:947–952.

36. Rodriguez M, Bur S, Favre A, Weibel ER. The pulmonary acinus: Geometry and morphometry of the peripheral airway system in rat and rabbit. Am J Anat 1987; 180:143–155.

37. Verbanck S, Weibel ER, Paiva M. Simulations of washout experiments in postmortem rat lung. J Appl Physiol 1993; 75:441–451.

38. Crawford ABH, Makowska M, Paiva M, Engel LA. Convection- and diffusion-dependent ventilation maldistribution in normal subjects. J Appl Physiol 1985; 59:838–846.

39. Prisk GK, Lauzon A-M, Verbanck S, Elliott AR, Guy HJB, Paiva M, West JB. Anomalous behavior of helium and sulfur hexafluoride during single-breath tests in sustained microgravity. J Appl Physiol 1996; 80:1126–1132.

40. Elliott AR, Prisk GK, Guy HJB, West JB. Lung volumes during sustained microgravity on spacelab SLS-1. J Appl Physiol 1994; 77:2005–2014.

41. Lauzon A-M, Prisk GK, Elliott AR, Verbanck S, Paiva M, West JB. Paradoxical helium and sulfur hexafluoride single-breath washouts in short-term vs. sustained microgravity. J Appl Physiol 1997; 82:859–865.

42. Engel LA, Paiva M. Analysis of sequential filling and emptying of the lung. Respir Physiol 1981; 45:309–321.

43. Engel LA, Grassino A, Anthonisen NR. Demonstration of airway closure in man. J Appl Physiol 1975; 38:1117–1125.

44. Crawford ABH, Makowska M, Kelly S, Engel LA. Effect of breath holding on ventilation maldistribution during tidal breathing in normal subjects. J Appl Physiol 1986; 61:2108–2115.

45. Prisk GK, Guy HJB, Elliott AR, Paiva M, West JB. Ventilatory inhomogeneity determined from multiple-breath washouts during sustained microgravity on Spacelab SLS-1. J Appl Physiol 1995; 78:597–607.

46. Cotes JE. Assessment and Application in Medicine. In: Leathart GL, ed. Lung Function. Vol. 5. Oxford: Blackwell, 1993:251.

47. Estenne M, Van Muylem A, Knoop C, Antoine M. The Brussels Lung Transplant Group. Detection of obliterative bronchiolitis after lung transplantation by indexes of ventilation distribution. Am J Respir Crit Care Med 2000; 162:1047–1051.
48. Verbanck S, Schuermans D, Van Muylem A, Melot C, Noppen M, Vincken W, Paiva M. Conductive and acinar lung-zone contributions to ventilation inhomogeneity in COPD. Am J Respir Crit Care Med 1998; 157:1573–1577.
49. Verbanck S, Schuermans D, Noppen M, Van Muylem A, Paiva M, Vincken W. Evidence of acinar airway involvement in asthma. Am J Respir Crit Care Med 1999; 159:1545–1550.

5

Aerosol Transport in the Lung

CHANTAL DARQUENNE

University of California, San Diego
La Jolla, California

I. Introduction

The deposition of aerosols in the human lung is primarily due to the mechanisms of inertial impaction, gravitational sedimentation, and Brownian diffusion. Inertial impaction causes most of the particles larger than 5 μm to deposit in the upper airways. Brownian diffusion affects smaller particles (<1 μm), which deposit mainly in the alveolar region. Sedimentation is the gravitational settling of particles and mainly affects particles in the size range 1 to 8 μm. Of these three mechanisms, sedimentation is the only gravity-dependent process and is changed in altered gravity (G) environments.

It is also well known that the lung is highly sensitive to gravity, which is responsible for regional differences in ventilation, blood flow, intrapleural pressure, alveolar size, and gas exchange (1). Because the lung distorts under its own weight, the alveoli at the base of the lung are relatively compressed compared with the apical alveoli, and, because poorly expanded alveoli are more compliant, ventilation is greatest near the bottom of the lung and becomes progressively lower near the top. Changes in the G level (i.e., in lung weight) will affect the distribution of ventilation and are therefore expected to affect the distribution and deposition of aerosols.

The first study on the effect of gravity on aerosol deposition in the human lung was made by Muir in the 1960s (2). His interest was to determine the change in aerosol deposition under the reduced gravitational field on the surface of the Moon compared to that on Earth. In his theoretical analysis, Muir focused on the deposition of particles in the size range 1 to 8 µm for which the main factor controlling deposition is sedimentation. Based on the knowledge of deposition on Earth, Muir predicted a reduction in the overall deposition but an increase in deposition in the alveolar region. He suggested, therefore, that astronauts should be more susceptible to infection by bacteria penetrating deeper into the lung. Around the same time, Beeckmans (3) developed a computer program based on experimental data available at 1 G to predict deposition within the respiratory tract. He computed deposition for particle sizes ranging from 0.2 to 15 µm at decreased gravity levels. While his computations also show a decrease in overall deposition and an increase in alveolar deposition at ⅙ G compared with deposition on Earth, Beeckmans predicted a lower alveolar deposition than that suggested by Muir at ⅙ G.

The first measurements of aerosol deposition in altered gravity were made by Hoffman and Billingham (4) in the mid-1970s during parabolic flights using 2-µm-diameter particles. It was not until the mid-1990s that extensive measurements of aerosol deposition in microgravity (µG) were made during a series of parabolic flights aboard the NASA Microgravity Research Aircraft (5–8). This chapter discusses the data collected during these experiments.

Studying the effect of gravity on the transport and deposition of aerosols in the human lung has potential applications both in space and on Earth. Long-term space flight presents a situation in which aerosol deposition may be an important health consideration. In the spacecraft environment, the potential for significant airborne particle concentrations is high, since the environment is closed, and no sedimentation occurs. Measurements in the shuttle air environment have shown a substantial increase in microbial counts during missions (9,10), and a large variety of airborne particles, including hair, food, paint chips, and synthetic fibers, have been found. While routine spacecraft particle loads may be low for much of the time a crew spends in the vehicle, there is no doubt that unexpected events cause very large particle loads indeed. For example, the fire aboard the Mir space station in February 1997 (11) produced large amounts of airborne particles that took in excess of 6 h to clear. Since no measurements were done on air quality, it is unknown how many particles may have remained in the Mir atmosphere once the air was considered ''clear.'' Certainly the crew continued to wear masks for some nights to sleep following the fire. The absence of gravity affects both the amount and the site of aerosol deposition in the respiratory tract. These changes may have long-term health implications for astronauts exposed to high concentrations of particulates.

Even more importantly, the knowledge acquired from the aerosol studies in μG may have implications for applications on Earth. The deposition of aerosols from the environment in the lung is a health risk, at least in some subjects. The initial evidence that air pollution increases mortality came from dramatic pollution episodes, like that in the Meuse Valley of Belgium in 1930 (12) or that in London in 1952 (13). Since then, more than 200 epidemiological studies have linked exposure from atmospheric particulate matter (PM) to pulmonary diseases. A detailed review of these studies was done by Dockery and Pope (14,15), where an extensive reference list may be found. As the data have developed over the years, the correlation between PM exposure and pulmonary effects appeared to be better when particles less than 10 μm (PM_{10}) were considered rather than the total suspended particulates. An even better correlation was found with fine (<2.5 μm) particulates ($PM_{2.5}$) rather than PM_{10}. As an example, Thurston et al. (16) showed a direct relationship between $PM_{2.5}$ and hospital admissions for asthma and respiratory failure. The important role of fine particulates on respiratory health effects has been recognized by the U.S. Environmental Protection Agency (17), which recently established more stringent regulations in the generation of $PM_{2.5}$. The mechanism of the increase in risk factor resulting from PM exposure is, however, still the subject of speculation and is generally unknown. At least part of this uncertainty results from the poor understanding of the nature, site, and magnitude of deposition, especially in the alveolar region of the lung.

Many drugs are now also administered in aerosolized form. The list is long, including β_2 agonists, corticosteroids and other medications for the treatment of asthma, pentamadine for *Pneumocystis carinii* pneumonia, and numerous other drugs. Inhalation is also used in many forms of challenge testing. The list of such uses is growing rapidly. It has long been known that the effect of β_2 agonists as bronchodilators is enhanced if they can be delivered directly to their intended site of action (18). This concept of spatial targeting requires knowledge of the nature of the aerosol being delivered and the behavior of such an aerosol in the lung. Poor spatial targeting is associated with lowered efficacy and potential side effects. For example, deposition of inhaled corticosteroids in the upper airways is associated with fungal infections, hoarseness, and cough (19). Other drugs such as pentamidine or ergotamine (used in the treatment of migraine) have systemic effects that are best achieved if they can be delivered into the alveolar region of the lung, with minimum deposition in other regions (20). Thus, it may be possible to obtain optimal results with small quantities of drugs if spatial targeting puts the drug at exactly the right place in the lung, minimizing the use of a potentially expensive drug, and minimizing harm caused by side effects.

Deposition of inhaled particles depends upon several factors, including the airway anatomy, the breathing pattern, and the intrinsic properties of the particles (diffusion, sedimentation, inertia). By performing aerosol studies in altered gravity environments, one can manipulate particle sedimentation and study its effect on overall deposition. In this manner, one can distinguish between gravitational and nongravitational deposition mechanisms in an attempt to better understand the entire process of deposition, which is involved in pulmonary diseases resulting from the inhalation of airborne pollutants as well as in the delivery of aerosolized drugs. A better understanding of the behavior of aerosols in the lung may be fed back to better aerosol generation, targeting more specific sites of the lung.

There have been no aerosol studies in sustained microgravity. This chapter is dedicated to the description of experimental aerosol studies that have been performed in altered gravity environments obtained during parabolic flights aboard high-powered jet aircraft. During such flights (Fig. 1), the aircraft typically climbs to an altitude of ~10,000 m with the cabin pressurized to ~600 Torr. A ''roller-coaster'' flight profile is then performed. The aircraft is pitched up at ~1.6 G_z to a 45° nose-high attitude. Then the nose is lowered to abolish wing lift, and thrust is reduced to balance drag (thus maintaining μG). A ballistic flight profile results and is maintained until the aircraft nose is 45° below the horizon.

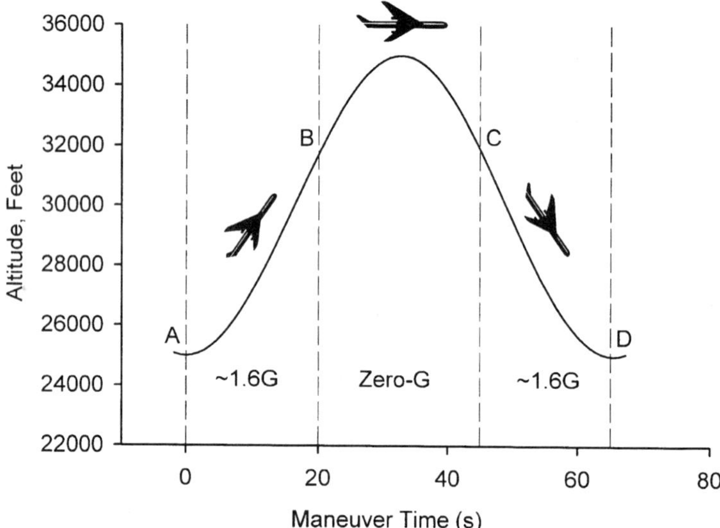

Figure 1 Profile of the maneuver during parabolic flights. (A) The aircraft is pitched up at ~1.6 G to a 45° nose-high attitude. (B) The nose is lowered to abolish wing lift, and thrust is reduced to balance drag maintaining μG. (C) A pullout averaging ~1.6 G is maintained until the aircraft pitches up to a 45° nose-high attitude (D) allowing the cycle to be repeated.

In this manner, μG is maintained for ~25 s. A pullout averaging ~1.6 G_z is maintained for ~40 sec causing the nose to pitch up to a 45° nose-high attitude and allowing the cycle to be repeated.

The chapter is divided in two major sections. The first section deals with total deposition studies and describes measurements made by Hoffman and Billingham (4) with 2-μm-diameter particles and by Darquenne et al. (5) with 0.5- to 3-μm-diameter particles at various G levels. The second section focuses on aerosol bolus studies. Measurements made with 0.5-, 1-, and 2-μm-diameter particles are discussed in an attempt to elucidate the effect of gravity on aerosol deposition and dispersion processes.

II. Total Deposition Studies

In 1975, Hoffman and Billingham (4) published the first experimental results of aerosol deposition in the human lungs at various G levels. They exposed three subjects to 2-μm-diameter polyvinyltoluene particles in μG, 0.5 G, 1 G, and 2 G and measured total deposition at each G level. They showed an almost linear increase in deposition with increasing G (Fig. 2). More recently, Darquenne et al. (5) obtained similar results with 2- and 3-μm-diameter polystyrene latex particles in four subjects in the range 0 to 1.6 G (Fig. 3).

The linear increase in deposition with increasing G level can be explained by the independence between the different deposition mechanisms for these particle sizes. The change in G level affects only deposition by sedimentation, which is a gravitational process. According to Stokes' law, spherical particles sediment with a velocity

$$v_s = \frac{\rho_p d_p^2}{18\,\mu} G \tag{1}$$

where ρ_p is particle density, d_p is particle diameter and μ is gas viscosity. The settling velocity, v_s, is therefore directly proportional to G and one would expect an approximately linear variation of deposition by sedimentation with altered G level. For 2- and 3-μm-diameter particles, diffusion may be neglected and deposition is mainly due to sedimentation in the distal airways and to impaction in the upper airways. Impaction is gravity independent and remains therefore constant with increasing G level for a given particle size. Therefore, to a first approximation, total deposition increases linearly with G. It should be noted that this reasoning ignores secondary effects like change in airspace geometry between G levels.

Darquenne et al. (5) also performed the same type of measurements with 0.5- and 1-μm-diameter particles. Unlike 2- and 3-μm-diameter particles, deposition of the smaller particles did not increase linearly with G (see Fig. 3). The results showed that deposition increased less between μG and 1 G than between

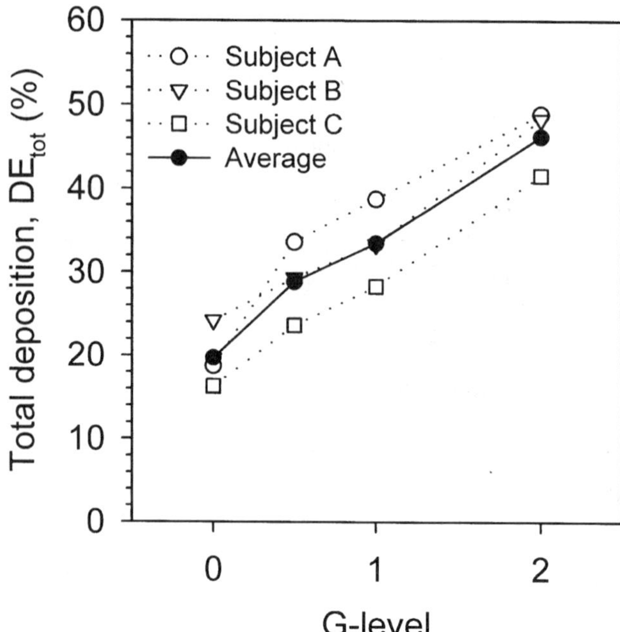

Figure 2 Mean total deposition of 2-μm-diameter particles at various G levels in three subjects. Deposition averaged over the three subjects is also shown. Note the linear increase in deposition with increasing G level. (From Ref. 4.)

1 G and 1.6 G. Assuming that the relationship between 1 and 1.6 G defines the gravitational influence, the extrapolation to μG predicts a deposition less than that measured, especially for 1.0-μm-diameter particles where deposition in μG would be expected to approach 0. As the deposition by impaction is negligible for these small particles, a plausible explanation for the μG results might be a larger deposition by diffusion since sedimentation is absent in μG. The absence of deposition by sedimentation increases the aerosol concentration in the small airways leaving a larger number of particles available for deposition by diffusion and for being transported deeper in the lung, where deposition by diffusion may also occur.

These observations are supported in the results of a simulation obtained with a one-dimensional (1D) model developed by Darquenne and Paiva (21). In that model, a 1D equation describing the transport and deposition of aerosols is solved in a lung structure based on data of Haefeli-Bleuer and Weibel (22). The results of the simulation are shown in Figure 4 as a function of particle size and detail the contribution of inertial impaction, gravitational sedimentation, and

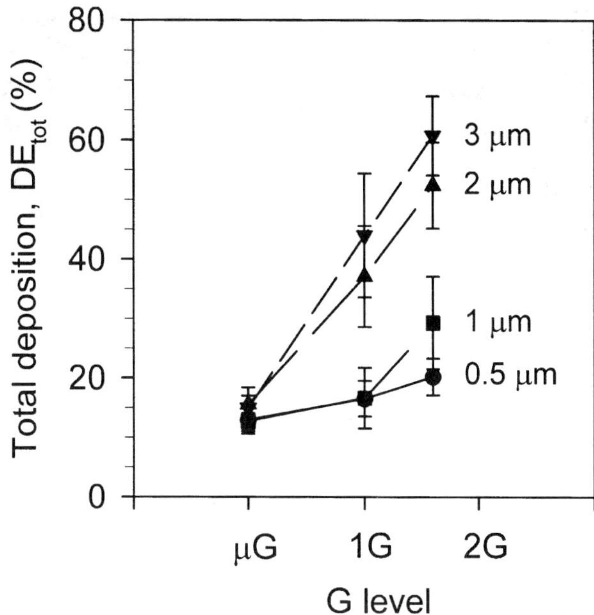

Figure 3 Total deposition of aerosol particles (mean $\pm$ SD) as a function of G level averaged over four subjects. There is a linear increase in deposition with G levels for both 2- and 3-μm-diameter particles, while for 0.5- and 1-μm-diameter particles, deposition increased less between μG and 1 G than between 1 G and 1.6 G. (From Ref. 5.)

Brownian diffusion to the overall deposition. For each particle size, deposition is shown at the different G levels: the left, central, and right bars represent the predicted values in μG, 1 G and 1.6 G, respectively. In μG, sedimentation is absent and deposition results from impaction and Brownian diffusion. While impaction is negligible for 0.5- and 1-μm particles, it contributes to more than 75% of overall deposition for 3-μm particles. The model also predicts that deposition by Brownian diffusion is the main contributor to total deposition for 0.5- and 1-μm particles in μG. In 1 G and 1.6 G, gravitational sedimentation contributes to total deposition and its effect becomes larger with increasing particle size.

The comparison between experimental and numerical data is shown in Figure 5. The experimental data are plotted as a percentage of the predicted values from the model in μG, 1 G and 1.6 G, respectively. A value of 100% means that the model exactly predicted the experimental data. In 1 G and 1.6 G, the experiments are not significantly different from the numerical predictions, while in μG, the measured deposition is significantly larger than the numerical data for each particle size except 3 μm. For 1-μm particles, the measured deposition is more

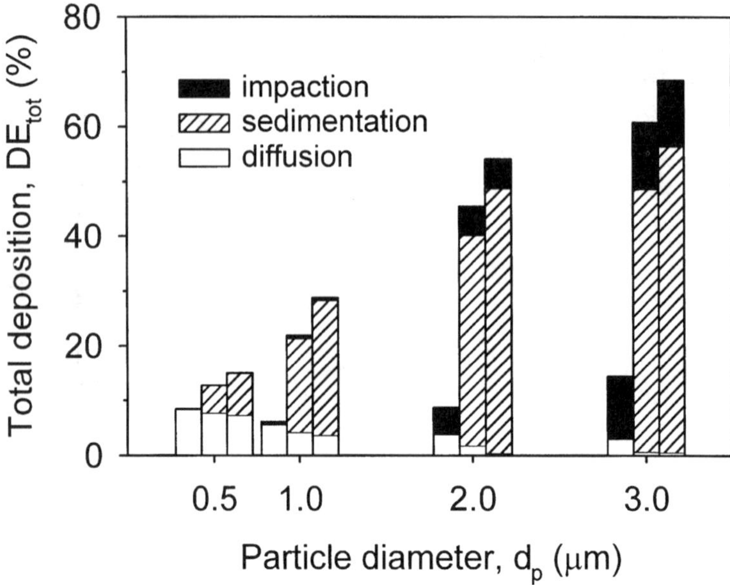

Figure 4 Numerical predictions obtained with a 1D model (21) at different G levels. The model details the contribution of inertial impaction, gravitational sedimentation, and Brownian diffusion to overall deposition. For each particle size (d_p), the left, central and right bars represent the predicted values in μG, 1 G and 1.6 G, respectively.

than twice the predicted value. These results show that the model fails to accurately predict aerosol deposition in μG.

Although the model underestimates deposition in μG (see Fig. 5), the numerical results show a much increased deposition component due to diffusion in μG compared with 1 G and 1.6 G (see Fig. 4). Deposition by diffusion increases from 7.3% at 1.6 G to 8.4% in μG for 0.5-μm particles, from 3.6 to 5.6% for 1-μm particles, from 1.4 to 3.8% for 2-μm particles and from 0.6 to 3.8% for 3-μm particles. This increase is a direct consequence of the absence of sedimentation in μG, leaving more particles available to deposit by diffusion.

However, deposition by Brownian diffusion on its own cannot explain the high deposition measured in μG. An effect that might contribute to the aerosol deposition is the nonreversibility of flows in the airways of the human lung (23). The branching structure of the bronchial tree imparts a directional asymmetry to the flow field, manifested by a difference in the shape of the velocity profile during inspiration and expiration. During inspiration, the airflow is more rapid at the center of the airways than near the walls. Particles in the center of the airways therefore reach more distal generations of the lung than do the particles

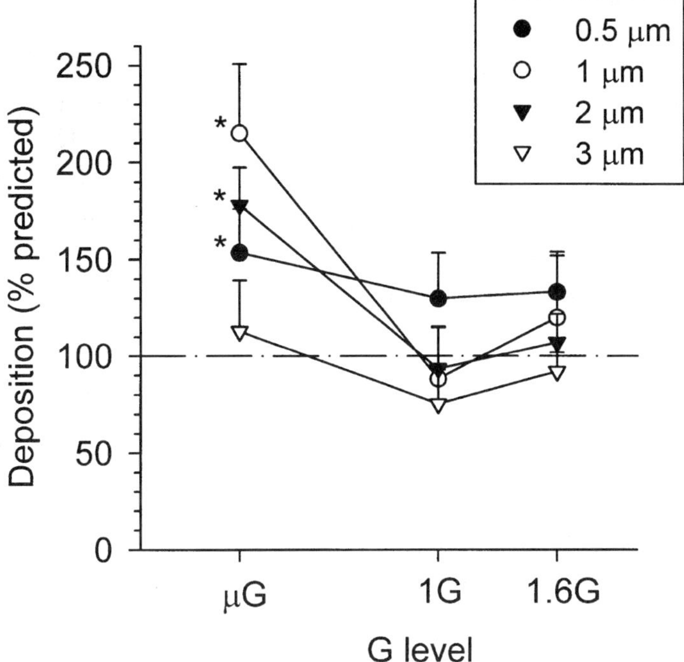

Figure 5 Experimental deposition values displayed as a percentage of the predicted values from the numerical model at different G levels. * Indicates significantly different from 100%. There were no significant differences between experimental values and predicted values in 1 G and 1.6 G. Significant differences were found in μG for 0.5-, 1-, and 2-μm-diameter particles.

located near the airway walls. During expiration, the velocity profile is more blunted, and particles at the center of the airways travel at a slower pace than during inspiration, whereas particles near the airway walls travel faster (23). Other parameters, such as unequal time constants in different parts of the lung during reciprocal tidal breathing and the mechanical pulsations of the heart, will also serve to make flow reversals within the lung asymmetric (24). The nonreversibility of the flows results in additional mixing that serves to move the particles in the direction of the alveoli. This could have the effect of increasing the apparent contribution of diffusional loss of particles in the lung. In the absence of gravity, with the reduction in losses due to sedimentation, this effect will become more obvious than in 1 G.

The increase in the apparent contribution of diffusional loss of particles in the human lung might also be explained by the concept of ''stretch and fold'' described by Butler and Tsuda (25). These authors have shown that the pattern of mixing of convective streamlines in the airways is much more complex than previously thought. In experimental preparations using rat lungs that were filled with low-viscosity silicone fluids, they have shown that the reciprocal motion of the fluid in the airways wraps the streamlines around each other as tidal breathing continues. The effect is like that of the stretching and folding of pastry, with the net effect being to bring previously widely separated streamlines into close apposition to each other and to increase the area over which diffusion can take place. As such, previously long diffusion distances are greatly reduced and this results in an increase in the apparent diffusion coefficient. The mixing associated with convective stretching and folding may also contribute to a greater aerosol deposition by diffusion than that predicted by current models. However, this remains to be tested. Such experiments should be performed in μG to avoid the confounding effects of sedimentation and to therefore directly address the mechanism of "stretch and fold."

Another factor that may explain that deposition in μG is larger than expected is the reduction in the functional residual capacity (FRC) that occurs in the weightless environment (26–29). Elliot et al. (27) found that during sustained periods of microgravity, FRC decreased significantly by 15% ($\sim$500 mL) compared with 1 G standing FRC. Paiva et al. (28) and Edyvean et al. (26) also demonstrated a 200- to 500-mL decrease in FRC during short periods of microgravity. Davies et al. (30) and Heyder et al. (31) showed that deposition increased as the resting expiratory reserve volume was decreased. For 0.5-μm-diameter particles, Davies et al. (30) found a $\sim$2.5% increase in deposition when the tests were performed from FRC $-$ 500 ml instead of FRC. Therefore, the reduction in FRC observed in μG likely contributes to the higher deposition than expected if the tests were performed from the same FRC as in 1 G. However, this increase is much smaller than what was observed in the measurements. So even if the reduction in FRC in μG is corrected by a controlled protocol, deposition would likely still be higher than that predicted by the model.

III. Aerosol Bolus Studies

A. Aerosol Bolus Technique

The aerosol bolus technique has been developed to probe convective gas transport at different depths within the lung (32–36). This technique consists of inserting a small amount of aerosol (a bolus) at a predetermined point in the subject's inspiratory volume and analyzing the distribution of the aerosol bolus in the subsequent exhalation. During its transport through the lung, as the aerosol bolus passes through the airways, it divides into multiple segments that recombine during expiration. If the flow profiles in the bronchial tree were perfectly reversible, the expansion and contraction of the lung perfectly homogeneous, and intrinsic particle motion absent, the expired bolus would be identical to the inspired bolus. However, the expired bolus is always spread over a larger volume than the volume of the inspired bolus. This indicates that during the transport of the bolus through the lung, there is an irreversible transfer of particles from the bolus to the adjacent air. All the mechanisms, except Brownian motion, that are involved in this transfer are collectively referred to as convective mixing (35). Factors contributing to convective mixing include velocity profiles within the airspaces, airway and alveolar geometries, asymmetries between inspiratory and expiratory flows, nonhomogeneous ventilation of the lung, cardiogenic mixing, and possibly the phenomenon of "stretch and fold."

The aerosol bolus technique is a powerful tool to determine the axial dispersion undergone by the particles during their transport into the lung and also to study regional deposition at different lung depths. More particularly, when used at different G levels, the technique allows study of the effect of gravity on both regional deposition and mixing processes in the lung.

B.   Bolus Parameters

Several parameters are typically used to characterize a bolus test. On a graph of aerosol concentration as a function of respired volume (Fig. 6), inhaled and exhaled boluses are characterized by their mode and half-width. The mode of the inhaled bolus measured from the end of the inhalation determines the depth at which the bolus penetrates into the respiratory tract and is referred to as penetration volume (V_p). The mode of the exhaled bolus (M_{ex}) characterizes where the bolus appears in the exhalation. The mode shift (MS) is defined as the difference between the position of the mode of the expired bolus and the penetration volume:

$$MS = M_{ex} - V_p \qquad (2)$$

A negative value of MS indicates that the position of the mode of the expired bolus has shifted to a smaller volume than the location of the inspired bolus, i.e., that the bolus has moved toward the mouth.

The half-width of the inhaled (H_{in}) or exhaled bolus (H_{ex}) is defined as the bolus width (in mL) between the two points of one-half the maximum concentration of the inhaled or exhaled bolus (see Fig. 6). The change in half-width between the inspired and expired bolus reflects the aerosol bolus dispersion and is defined by the following equation:

$$H = (H_{ex}^2 - H_{in}^2)^{0.5} \qquad (3)$$

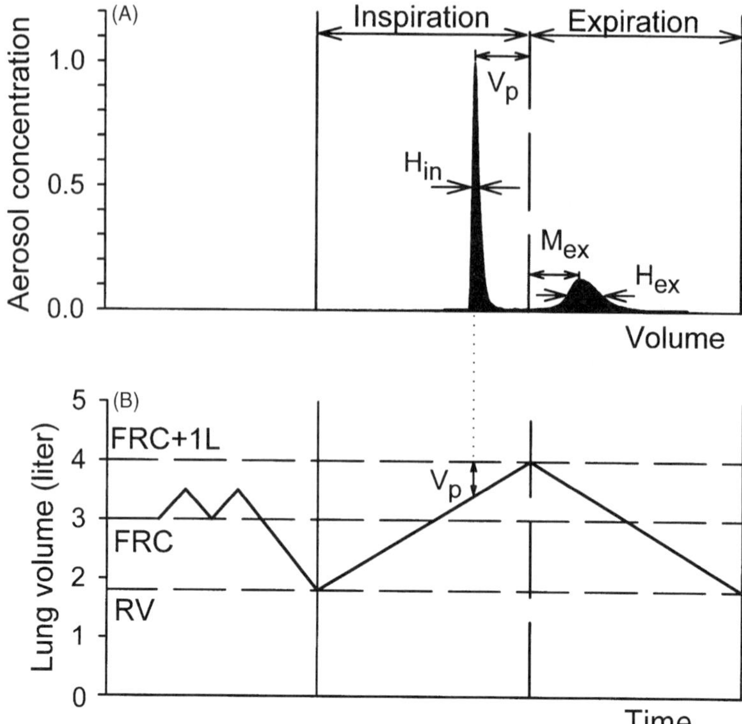

Figure 6 Aerosol bolus test. (A) Aerosol tracing for a penetration volume (V_p) of 500 ml. H_{in} and H_{ex} are the half-widths of the inhaled and exhaled bolus, respectively. M_{ex} is the mode of the exhaled bolus. (B) Volume history. RV, residual volume; FRC, functional residual capacity.

The particle loss in the lungs, i.e., the aerosol deposition (DE) is defined as

$$DE = 1 - \frac{N_{ex}}{N_{in}} \tag{4}$$

where N_{in} and N_{ex} are the number of inhaled and exhaled particles, respectively.

C.   Bolus Inhalations at Various G Levels

Since Altshuler (37) demonstrated in the late 1960s that aerosol bolus inhalations were well suited to study aerosol dispersion in the lung, the aerosol bolus technique has been used by numerous investigators. The technique has been proved useful not only to study convective transport and gas mixing in the lungs (33–

35,38,39) but also to probe deposition at different depths within the respiratory tract (40–42). In normal subjects, both deposition and dispersion have been shown to increase with depth of inhalation of the bolus into the lungs. The technique has also been used as a possible marker for disease-induced change in lung function. In a study on the effect of cystic fibrosis on inhaled aerosol, Anderson et al. (43) showed that aerosol bolus deposition and dispersion were both higher in patients than in healthy subjects and that this difference could be related to lung health. Studies have also been conducted on smokers (44,45) and on patients with other pulmonary diseases like asthma (46) and COPD (47). A detailed review on the use of aerosol bolus inhalations to assess lung pathology can be found elsewhere (48,49).

Several studies have been performed by Darquenne et al. (6–8) during parabolic flights to study the effects of both the G level and particle size on aerosol bolus dispersion and deposition. In these studies, subjects performed a standardized respiratory maneuver consisting of an expiration to residual volume (RV), a controlled inspiration at ~0.45 L/s to approximately 1 L above FRC measured in the seated position in 1 G. Subjects then immediately expired, again at ~0.45 L/s to RV, completing the maneuver. Bolus inhalations of 0.5-, 1-, and 2-μm-diameter particles were performed in μG, 1 G, and 1.6 G, using penetration volumes ranging between 150 and 1500 mL. The results of these studies are discussed in this section.

Aerosol Deposition

The aerosol bolus studies provide insight into the total deposition measurements described in the previous section by probing aerosol deposition at different depths within the lung. As outlined in the total deposition studies, settling due to sedimentation is affected by a change in the G level. For a given particle size, two factors play major roles in the deposition efficiency by sedimentation: time available to particles to sediment and airway size that determines the distance to be covered by the particles before depositing on the walls. Because of this, deposition due to gravitational sedimentation is expected to take place mainly in the distal part of the lung where residence time in airways is high and airway dimensions are small.

At each G level and for each particle size, deposition increased with increasing penetration volume. As an example, Figure 7 displays the deposition of 1-μm-diameter particles at each gravity level. At shallow penetration volumes (V_p = 200 mL), deposition was gravity independent showing that deposition by sedimentation is negligible at these lung depths. In contrast, deeper in the lung, deposition was significantly different from one G level to the other, with the lower deposition being present at lower G levels. Significant differences between μG and 1 G appeared at V_p = 800 mL and between 1 G and 1.6 G at V_p = 400

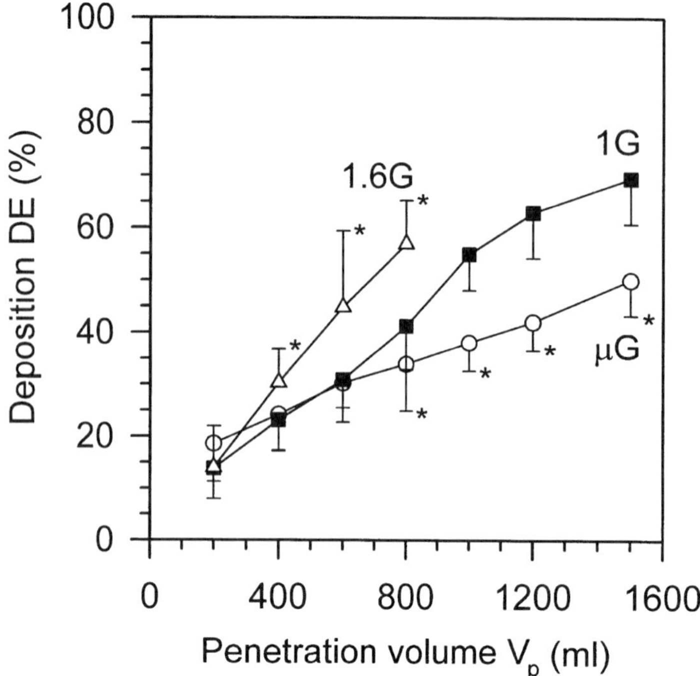

Figure 7 Deposition of 1-μm-diameter aerosol boluses inhaled to different lung penetration volumes collected in μG, 1 G and 1.6 G. * Indicates significantly different from 1 G ($p < 0.05$). Deposition increased with increasing V$_p$ at each G level. There was significant difference in deposition between μG and 1 G for V$_p$ ≥ 800 mL and between 1 G and 1.6 G for V$_p$ ≥ 400 mL. (From Ref. 6.)

mL (see Fig. 7). For 0.5-μm-diameter particles, the sedimentation velocity is smaller than for 1-μm-diameter particles and significant differences between G levels appeared at larger V$_p$ where airway dimensions are smaller and residence time higher. On the other hand, for 2-μm particles, the dependence on the G level was already visible at shallower V$_p$ because of a higher sedimentation rate of these particles compared with the smaller ones.

The effect of particle size on deposition is better shown in Figure 8. Deposition is plotted as a function of the penetration volume for the three different particle sizes in μG and 1 G, respectively. In μG, deposition was not significantly different from one particle size to another (Fig. 8A). In 1 G, deposition increased with particle size reflecting the effect of sedimentation (Fig. 8B). This is consis-

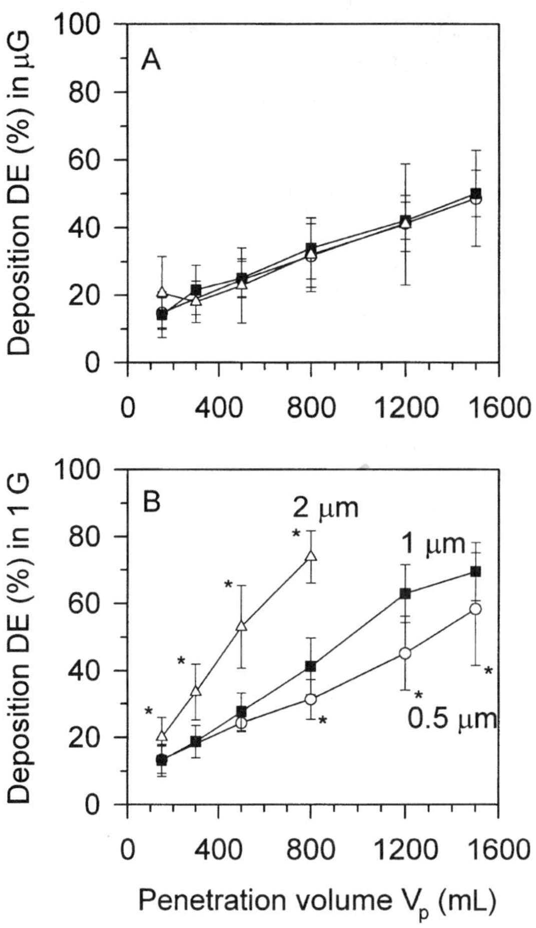

Figure 8 Deposition of aerosol bolus (mean ± SD) averaged over four subjects and plotted as a function of V_p. ○, d_p = 0.5 µm; ■, d_p = 1 µm; △, d_p = 2 µm. * Indicates significantly different from 1 µm ($p < 0.05$). (A) µG. There was no significant difference between particle size. (B) 1 G. There was significant difference in deposition at all V_p between 1- and 2-µm-diameter particles and for $V_p \geq 800$ ml between 0.5- and 1-µm-diameter particles. (From Ref. 7.)

tent with the total deposition studies for which total deposition did not significantly differ in µG between particles in the size range 0.5 to 3 µm (see Fig. 3).

Because the bolus inhalations were performed with the same breathing conditions for each particle size, differences in deposition at a given penetration volume should be explicable by differences in intrinsic particle motions and G level. Table 1 summarizes the intrinsic motions of the particles (diffusion, sedimentation, and inertia) as a function of their size. Particle sedimentation is proportional to the G level (see Eq. 1) and is therefore absent in µG. The inertia of the particles is measured by their stopping distance (50), which is a function of the air velocity within the airways. Details of the calculation can be found elsewhere (7). Comparing 0.5- and 2-µm particles, displacements by diffusion varies by a factor of ~2.2, sedimentation rate by a factor of ~13, and particle inertia as inferred from the stopping distance by a factor of ~15.

Based on these values, if inertial impaction and Brownian diffusion were the main causes of deposition in µG, we would expect different deposition profiles as a function of penetration volume for the three particle sizes. For the smallest particle size, we would expect low deposition by inertial impaction and therefore lower deposition in the upper airways than for larger particle sizes. Conversely there should be higher deposition deeper in the lung because of Brownian diffusion for the smallest particles. The similarity in the deposition profiles between particle size in µG (see Fig. 8A) suggests, therefore, the contri-

Table 1   Intrinsic Particle Motion Values[a]

	Particle diameter (µm)		
	0.5	1	2
Diffusion:			
$\sqrt{\overline{\Delta X^2}}$ (µm)	20	13	8.8
Sedimentation:			
$v_{s,\,\mu G}$ (µm/s)	0	0	0
$v_{s,\,1G}$ (µm/s)	9.5	33	125
Inertia:			
$\theta_{trachea}$ (µm)	1.5	6	23
$\theta_{vp\,=\,1200\,mL}$ (µm)	0.002	0.008	0.032

[a] $\sqrt{\overline{\Delta X^2}}$, Brownian root-mean-square displacement in 1 s; v_s, sedimentation velocity in micro-(µG) and normal gravity (1 G); θ, stopping distance as a measure of particle inertia in the trachea and deep in the lung (V_p = 1200 mL). Data for v_s and $\sqrt{\overline{\Delta X^2}}$ are from Ref. 62. θ is the stopping distance and is calculated from Ref. 50.
Source: Ref. 7.

bution of another mechanism to deposition like additional mixing resulting from the nonreversibility of the flows as discussed in the previous section.

The effect of residence time on aerosol deposition is illustrated in Figure 9. In these experiments, the inspiration of the bolus was followed by different breathholding times before the subjects exhaled to RV (8). Figures 9A and B display the deposition of 1-μm-diameter particles (mean $\pm$ SD) as a function of breathholding time in μG and 1 G for a penetration volume of 150 and 500 mL, respectively. While there were no significant differences in deposition between μG and 1 G at either penetration volume when the tests were performed with no breathholding, deposition in 1 G became significantly and progressively larger than in μG with increasing breathholding time. The increase in deposition between μG and 1 G is reflective of an increase in deposition by gravitational sedimentation. As there is no airflow in the airways during the breathholding, inertial impaction cannot cause any deposition and therefore does not contribute to the increase in deposition. Deposition by Brownian diffusion happens both during μG and 1 G, while gravitational sedimentation only occurs in 1 G. For a given breathholding time, the difference in deposition between μG and 1 G is larger at the larger penetration volume ($V_p = 500$ mL) because of smaller airway dimensions deeper in the lung (8). This observation is consistent with the supposition that gravitational sedimentation is the mechanism responsible for the increase in deposition between μG and 1 G.

Aerosol Dispersion

The measurement of bolus dispersion is a means of probing convective mixing at different depths within the lung. The use of boluses of inert gases (51,52) only permitted study of convective mixing in the proximal airways but not in the periphery of the lung where mixing by diffusion becomes increasingly dominant. Particles have negligible diffusive properties compared with gases and act as a "nondiffusing" gas (32). They are therefore well suited to trace convective gas transport within the lung (35).

Figure 10 illustrates the lack of diffusion of particles compared with gases. In these experiments, Schulz et al. (38) performed bolus inhalations on dogs using both 0.86-μm-diameter particles and sulfur hexafluoride (SF_6) where the diffusion coefficient of the particles is about 5 to 6 orders of magnitude smaller than that of SF_6. While aerosol and SF_6 behave similarly in the conducting airways (volumetric lung depth, VLD $\equiv V_p < 110$ mL), SF_6 boluses expired deeper in the lung showed a more and more developed tail compared to expired particle boluses indicating that gas diffusion is significant in the distal part of the lung contrary to particle diffusion.

As for deposition, aerosol bolus dispersion increases with increasing penetration volume and is gravity dependent, with the greatest dispersion occurring

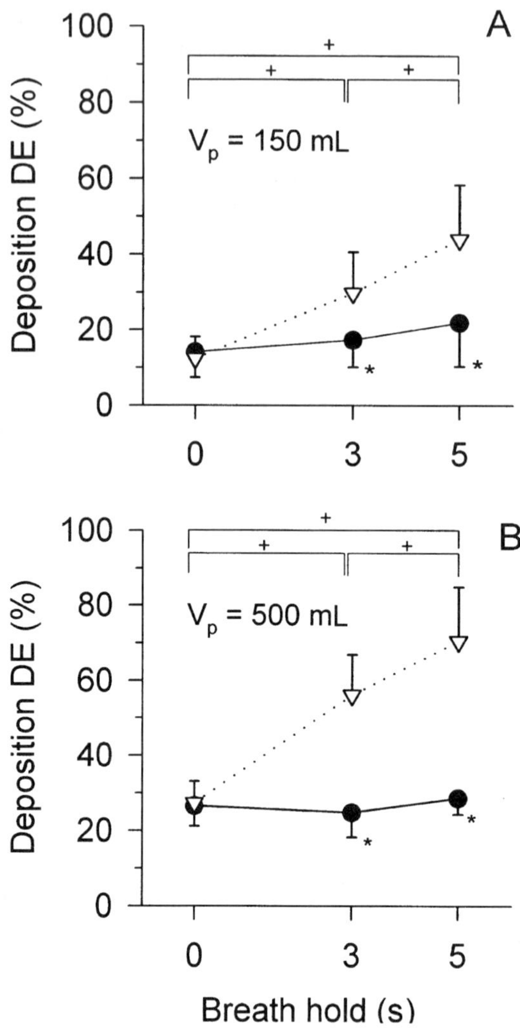

Figure 9 Deposition of 1-μm-diameter aerosol bolus as a function of breathholding time following the end of the bolus inspiration for V_p = 150 mL (A) and V_p = 500 mL (B). ●, μG; ▽, 1 G. * Indicates significantly different (p < 0.05) from 1 G data. + Indicates significant differences between breathholding time in 1 G. There are no significant differences between breathholding times in μG. (From Ref. 8.)

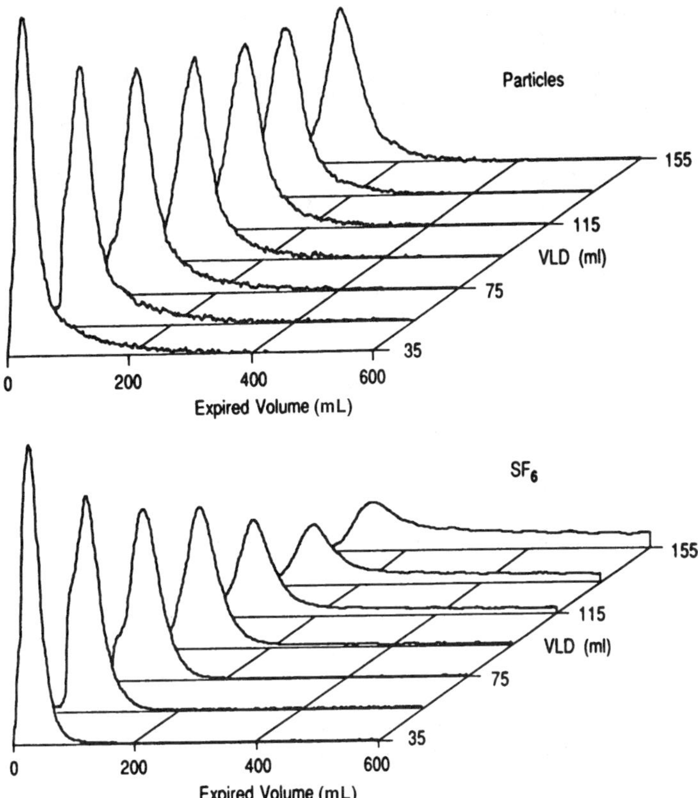

Figure 10 Tracings of expired particles and SF$_6$ boluses inspired to penetration volumes ranging from 35 to 155 mL in beagle lungs. Concentration in relation to expired volume is given in arbitrary units. While aerosol and SF$_6$ expired tracings are similar in the conducting airways (volumetric lung depth, VLD $\equiv$ V$_p$ < 110 mL), SF$_6$ boluses expired deeper in the lung showed a more and more developed tail compared with expired particle boluses because of the importance of diffusion in gas transport in peripheral airways. (From Ref. 38.)

for the largest G level as illustrated in Figure 11 for 1-μm-diameter particles. Several factors contribute to the increase in bolus dispersion. One factor is the cumulative effects of the nonreversibilities of the flow at each airway bifurcation. As discussed in the previous section, because of the branching structure of the bronchial tree, there is a directional asymmetry in the airflow field preventing the bolus from recovering its original shape.

Another factor affecting dispersion is the inhomogeneities in lung ventila-

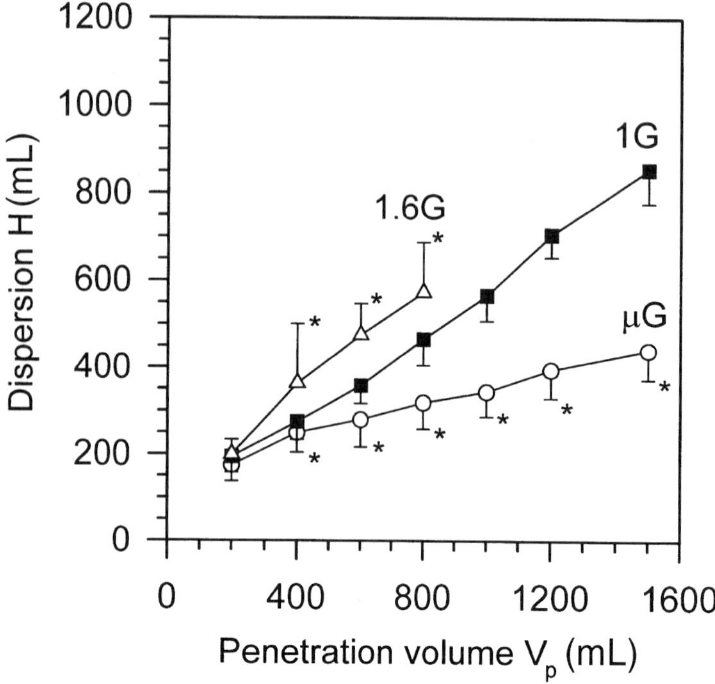

Figure 11 Dispersion of boluses of 1-μm-diameter aerosol inhaled to different lung penetration volumes. Data were collected in 1 G, in μG, and in 1.6 G. * Indicates significantly different from 1 G ($p < 0.05$). Dispersion increased with increasing V_p at each G level. There were significant differences between μG and 1 G and between 1 G and 1.6 G for $V_p \geq 400$ mL. (From Ref. 6.)

tion. As discussed in Chapter 4, inhomogeneities in lung ventilation are dependent on the gravity level. Based on rebreathing tests performed in μG and in 1 G, Verbanck et al. (53) showed that ventilatory inhomogeneities are present in μG, and that they are at least as large as the gravity-dependent inhomogeneity. Guy et al. (54) studied ventilatory inhomogeneity during sustained microgravity using single-breath nitrogen washouts. They also showed the presence of ventilatory inhomogeneity during μG, although ventilatory inhomogeneity was reduced compared with 1 G measurements. Finally, Prisk et al. (29) showed the presence of both gravitational and nongravitational ventilatory inhomogeneities in a study of single-breath washin tests of helium and sulfur hexafluoride in μG. The gravity-independent ventilatory inhomogeneity includes both convective and diffusive effects that are impossible to separate when gases are used. Because particles have negligible diffusive properties compared with gases, the measurement of aerosol bolus dispersion in μG allows an estimate of the extent of convective ventilatory inhomogeneities in the lung that is not gravitational in origin.

The sedimentation of the particles also affects the dispersion of the bolus as illustrated in Figure 12, where the dispersion of 0.5-, 1-, and 2-μm-diameter aerosol are plotted as a function of penetration volume in μG and 1 G (7). In

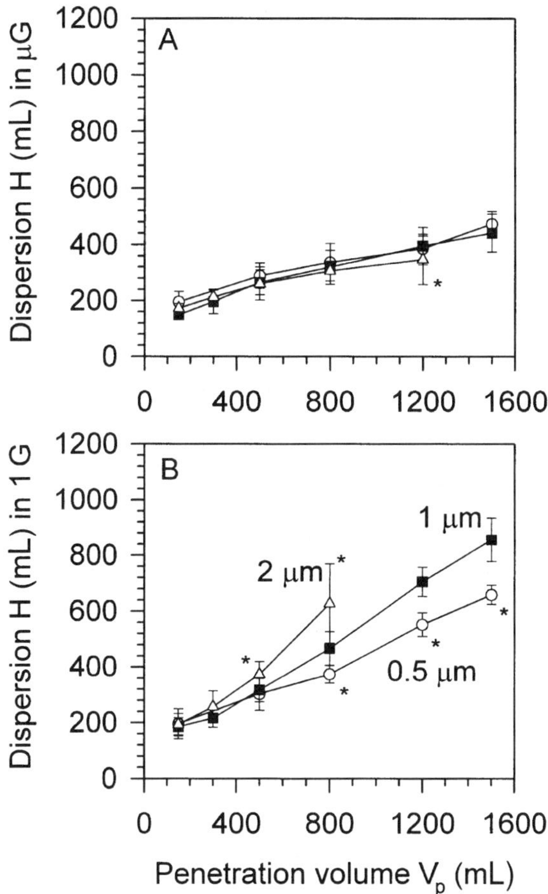

Figure 12 Dispersion of aerosol bolus (mean ± SD) averaged over four subjects and plotted as a function of V_p. ○, d_p = 0.5 μm; ■, d_p = 1 μm; △, d_p = 2 μm. * Indicates significantly different from 1 μm ($p < 0.05$). (A) μG. There was no significant difference between particle size. (B) 1 G. There was significant difference in dispersion for $V_p \geq$ 500 mL between 1- and 2-μm-diameter particles and for $V_p \geq$ 800 mL between 0.5- and 1-μm-diameter particles. (From Ref. 7.)

μG, when sedimentation is absent, there are no significant differences in the bolus dispersion between particle size, except for $V_p = 1200$ mL, where the dispersion of 2-μm aerosol bolus ($H_{2\mu m}$) is significantly smaller than the dispersion of both 0.5- ($H_{0.5\mu m}$) and 1-μm ($H_{1\mu m}$) aerosol boluses (see Fig. 12A). At $V_p = 1200$ mL, particle inertia is negligible (see Table 1). Therefore, the only intrinsic motion that can be responsible for dispersion is diffusion. Diffusion is more effective in the distal part of the lung where gas velocities are low and, therefore, residence time is high. The smaller diffusive properties of 2-μm particles compared with 0.5 and 1 μm (see Table 1) may explain the lower dispersion observed for 2-μm particles at $V_p = 1200$ mL. Even if particle diffusion does play a role in dispersion, the data suggest that its effect is small.

This is in sharp contrast to the situation in 1 G, where dispersion increases with particle size (see Fig. 12B). The difference between dispersion in μG and 1 G reflects both the increase in ventilatory inhomogeneities from μG to 1 G and the effect of sedimentation on aerosol dispersion. This increase can be characterized by the change in the slopes of the regression lines of dispersion as a function of penetration volume between μG and 1 G (Fig. 13). While the slopes are not significantly different between particle sizes in μG, they increase significantly with increasing particle size in 1 G. The increase in the slope in 1 G with increasing particle size reflects the effect of sedimentation on aerosol dispersion. For 0.5-μm particles, in 1 G sedimentation settling is low compared with diffusive motion (see Table 1), and cannot explain the increase of the slope between μG and 1 G. Therefore, for this particle size, the increase in the slope between μG and 1 G most probably indicates the effect of gravitational convective inhomogeneity on aerosol bolus dispersion. This implies that the increase in the 1 G slope between 0.5-μm-diameter particles and the larger particle sizes is due to sedimentation. The data obtained with 0.5-μm particles also suggest that the nongravitational convective inhomogeneity is as large as the gravitational convective inhomogeneity, as the slope almost doubles between μG and 1 G (see Fig. 13).

In the bolus inhalations performed in μG, the lung is in the most uniform state possible, and there are no losses due to sedimentation. Despite this, there is clearly a consistent increase in bolus dispersion as penetration volume increases that is independent of the particle size (see Fig. 12A). This is clear evidence that airway geometries, lung expansion, and the flow patterns that they generate directly result in convective mixing in the human lung.

Several authors (35,55) have also suggested cardiogenic mixing as a mechanism that contributes to the dispersion of the aerosol bolus. This effect can be differentiated from the mixing resulting from the inhaled and exhaled flows by looking at the evolution of aerosol dispersion when various breathholdings are introduced after the inhalation of an aerosol bolus. Figure 14 displays such dispersion of 1-μm particles for penetration volumes of 150 and 500 mL (8). The experiments were the same as those used to calculate the effect of residence time on

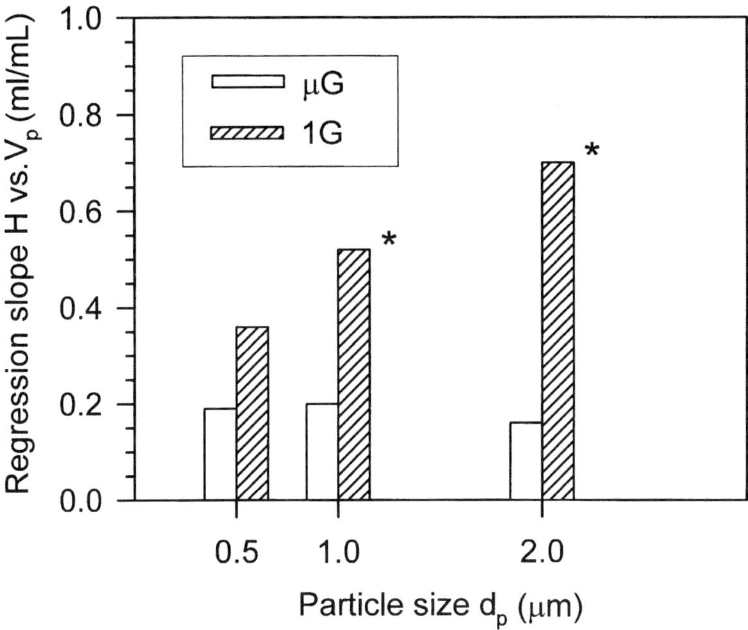

Figure 13 Slope of the regression lines of aerosol bolus dispersion as a function of penetration volumes. The slopes are plotted as a function of particle size both in μG and 1 G. * Indicates significantly different from slope for 0.5 μm. The regression slope did not vary significantly between particle size (d_p) in μG while it increased significantly with d_p in 1 G.

aerosol deposition shown in Figure 9. During the breathholding dispersion is affected by cardiogenic mixing and by the intrinsic motions of the particles (sedimentation and diffusion). The data show that the effect of sedimentation on dispersion is negligible at shallow penetration volumes where dispersion is almost the same in μG and 1 G. This is in sharp contrast with the situation in the alveolar regions of the lung (V_p = 500 mL) where dispersion is gravity dependent (see Fig. 14B).

In μG, when sedimentation is absent, cardiogenic mixing and diffusion are the only mechanisms available to explain the increase in dispersion depicted in Figure 14. Brownian displacements of 1-μm particles are small (see Table 1) and cannot explain the increase in dispersion. Based on Weibel's symmetric model of the lung (56) (for a lung volume of 4 L), the dispersion measured at the mouth after a 5-s breathholding would increase by less than 1 mL for V_p = 150 mL because of axial diffusion and by ~25 mL for V_p = 500 mL (8). In addition,

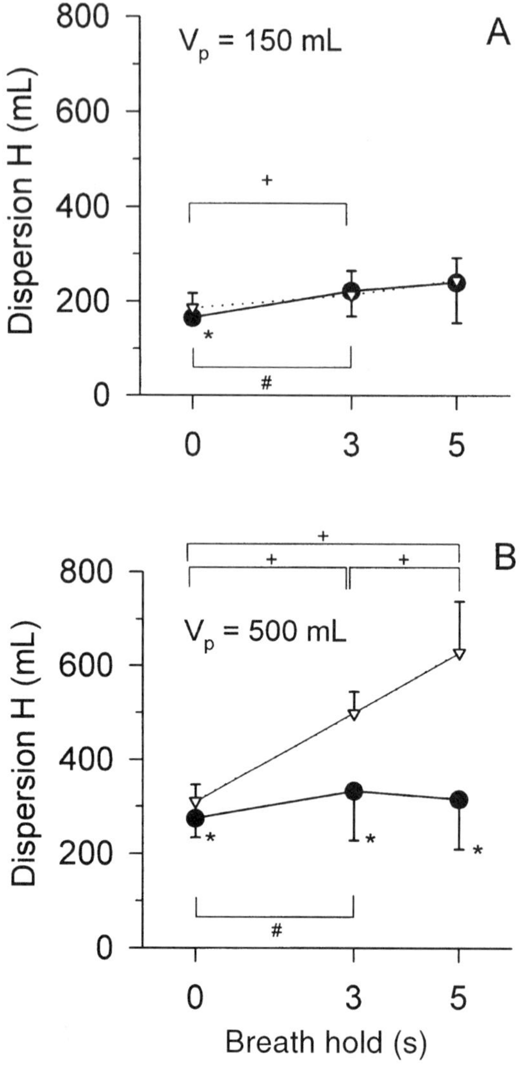

Figure 14 Dispersion of 1-μm-diameter aerosol bolus as a function of breathholding time following the end of the bolus inspiration. ●, μG; ▽, 1 G. * Indicates significantly different (p < 0.05) from 1 G data. #, + Indicate significant differences between breathholding time in μG and 1 G, respectively. (A) V_p = 150 mL. (B) V_p = 500 mL. (From Ref. 8.)

radial diffusion causes a change in the radial position of the particles, which therefore follow different streamlines during inspiration and expiration. This also results in the spreading of the exhaled bolus. This effect is similar to that resulting from the particle sedimentation. However, it is hard to quantify the diffusive effect compared with the gravitational effect, as diffusion acts in all directions while sedimentation occurs only in the direction of the gravity vector. It is, however, unlikely that the increase in measured dispersion with increasing breathholding time can be explained solely by the diffusive properties of the particles.

The increase in dispersion observed in Figure 14 is therefore likely due to cardiogenic mixing. The physical motion of the heart generates an oscillating flow in the lungs, which results in an enhanced gas mixing in the airways (24,57). In a study of cardiac action on gas mixing in dog lungs, Horsfield et al. (58) showed that the flow pulsations resulting from the motion of the heart increased the effective diffusion coefficient in the conducting airways while mixing in the distal part of the lung was only slightly affected. They explained the increase in mixing in the central airways simply by the result of the mechanical action of the heart. Scheuch and Stahlhofen (55) studied the effect of cardiogenic mixing on aerosol bolus behavior. They performed bolus inhalations on a subject at rest and after exercise when the heart rate was increased by more than a factor of 2. They showed that the motion of the heart considerably influences both the aerosol dispersion and deposition and that the effect was more obvious at shallow penetration volumes. The effect of cardiogenic mixing on aerosol dispersion may be characterized by the slope of the regression line of dispersion as a function of breathholding time in μG. In the data shown in Figure 14, the slope was 15.3 mL/s for $V_p = 150$ mL and 9.1 mL/s for $V_p = 500$ mL. The larger slope at small penetration volume suggests a larger effect of cardiogenic mixing on aerosol dispersion in the central airways than in the alveolar regions of the lung, in agreement with previous studies (55).

Mode Shift

The mode shift (MS) is a parameter that provides information on the symmetry and reversibility of ventilation. Mode shift is defined by the difference between the mode of the exhaled bolus and the penetration volume. If filling and emptying of the lung occurred accordingly to the first-in, last-out principle and if there were no intrinsic particle motions, the mode of the exhaled bolus should be identical to the penetration volume. The data show that at small penetration volumes, there are little or no differences between M_{ex} and V_p (6) and MS ≈ 0 (Fig. 15). This indicates that ventilation is quite reversible at shallow depth within the lung. In the distal part of the lung however, larger differences are found between M_{ex} and V_p, suggesting larger ventilation inhomogeneities. The differences result in

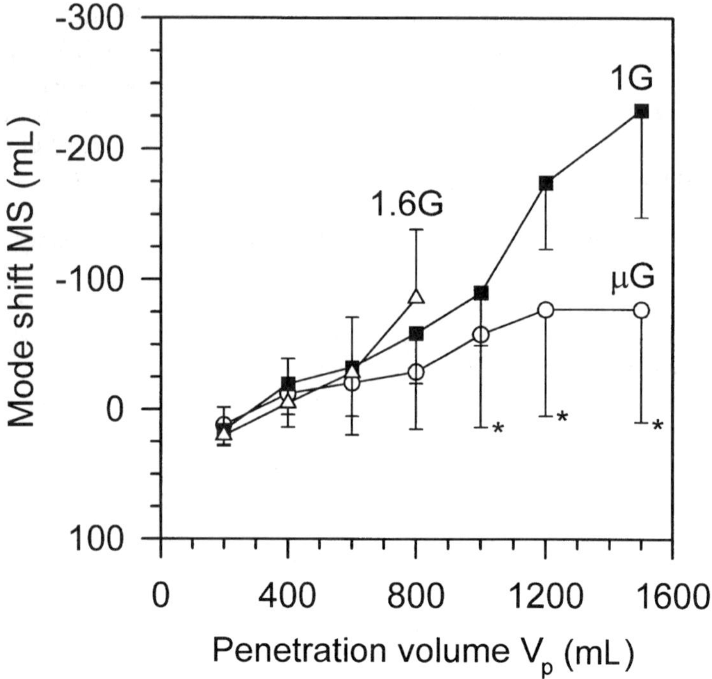

Figure 15 Mode shift of boluses of 1-μm-diameter aerosol inhaled to different lung penetration volumes. Data were collected in 1 G, in μG, and in 1.6 G. * Indicates significantly different from 1 G ($p < 0.05$). Mode shift increased with increasing V_p at each G level with the expired bolus moving toward the mouth. (From Ref. 6.)

a negative mode shift, which indicates that the mode of the exhaled bolus is shifted toward the mouth. Mode shift also becomes more negative with increasing G. This is more likely due to the increase in ventilation inhomogeneities with increasing G. In μG, ventilation inhomogeneities are reduced as is MS.

Another factor that may explain the increase in mode shift with increasing penetration volume is the higher deposition occurring at higher V_p (see Fig. 7). The particles that penetrate deeper in the lung deposit more and erode the tail of the bolus, shifting the mode of the exhaled bolus proximally. This effect is accentuated with increasing G level as deposition by sedimentation is increased.

IV. Summary and Perspectives

To date, all the studies examining the effect of gravity on aerosol behavior in the human lung have been performed during parabolic flights aboard high-powered jet aircraft. During such flights, the effects of sedimentation on the deposition and dispersion of particles can be directly manipulated since sedimentation is a gravitationally driven process. Weightlessness can be maintained for a period of approximately 25 s during parabolic flights. While this limited time has dictated the length of the protocols used in the experiments and while the period of hypergravity preceding the μG phase may have affected the results, useful information can still be obtained from such studies.

Measurements of total deposition (4,5) showed that for 2- and 3-μm-diameter particles, deposition is proportional to G. More interestingly, the study by Darquenne et al. (5) showed that, in μG, the deposition of small particles (0.5 and 1 μm) was higher than that predicted by models. The higher deposition was explained by a larger deposition by diffusion because of the absence of sedimentation and by the nonreversibility of the flow, producing an additional mixing effect. The higher deposition likely occurs in the alveolar regions of the lung. If this additional mixing is active in μG, then it must also be active in 1 G, suggesting that alveolar deposition may, in fact, be higher than previously suspected.

While the absolute magnitude of the increase in total deposition due to enhanced diffusion may be small in comparison with that due to sedimentation, the health implications may be disproportionately large. Much of the sedimentation in 1 G is thought to occur in the small ciliated airways, where the deposited particles may be readily removed from the lung. However, the studies of stretch and fold by Butler and Tsuda (25) suggest that complex mixing may occur in the small airways inside the acinus. If enhanced deposition occurs within the acinus where the mucociliary removal system is not present, the alveolar macrophages must act as the primary defense mechanism. This may exacerbate the deleterious health effects of inhaled toxins, as alveolar macrophages are readily disabled by ingestion of toxic materials. This finding may have relevance to the clinical observation that many lung diseases originate in the small airways (59) and to the consequences of long-term exposure to respirable aerosols in long-duration space flight. This finding may also be beneficial in the development of therapeutic applications using aerosols. The data of Darquenne et al. (5) suggest that, while the increase in overall deposition is quite small in 1 G, deposition in the more-sensitive alveolar zone of the lungs may be twice that predicted.

The bolus studies provide a way to probe aerosol deposition and dispersion at different depths within the lung. The data showed that both deposition and dispersion increased with penetration volume. At shallow penetration volumes, when the aerosol was mostly in the larger airways, both deposition and dispersion were not different between G levels (6,7). In contrast, at larger penetration volumes, when the aerosol reaches the alveolar regions of the lung, deposition and dispersion were strongly gravity dependent.

Aerosol bolus inhalations are also used as a probe for ventilatory convective inhomogeneity. For that purpose, they complement studies using gases that have shown that convective inhomogeneity is present in the lung even in the absence of gravity (29,54,60). However, gases diffuse readily while particles are almost unaffected by diffusion. Studies using gases cannot easily distinguish between the diffusive and convective effects involved in ventilatory inhomogeneities, while aerosols are excellent probes of convective mixing. In particular, when 0.5-μm-diameter particles were used, for which settling due to sedimentation is low compared with diffusive motions, the difference between dispersion in μG and 1 G likely reflects the gravity-dependent ventilatory inhomogeneity.

The experimental data described in this chapter provide a unique set of data to distinguish between gravitational (sedimentation) and nongravitational (inertia and diffusion) deposition mechanisms. These data represent a powerful tool to improve the modeling of aerosol transport and deposition in the human lung. Indeed, the data when compared with numerical models show significant and potentially important discrepancies (see Fig. 5). The effect of gravity on aerosol dispersion can also be modeled based on these experimental data. This will lead to better models and therefore better predictions of aerosol transport in the lung. The models can then be used to simulate aerosol exposure under various conditions and better predict aerosol deposition in the respiratory tract. This is especially important when dealing with pollutants for which experimental data in humans are often limited. Such an application is a good example of how studies performed in μG can directly benefit applications on Earth.

In the future, performing aerosol studies aboard the Space Shuttle and/or the International Space Station would permit use of longer protocols than those used during parabolic flights where the duration of weightlessness is limited to ~25 sec. This would also provide verification that parabolic flights are a reasonable model of sustained weightlessness for aerosol studies. While it has been shown that nitrogen single-breath washouts performed during parabolic flights were fairly accurate predictions of the subsequent results obtained in sustained μG (54), Lauzon et al. (61) showed that helium and sulfur hexafluoride single-breath washouts performed during parabolic flights gave different results from those obtained in sustained μG. These results suggested that changes seen in peripheral gas mixing in sustained μG required more than 25 sec of weightlessness to occur. At present, we have no means to be certain that the results of aerosol deposition and dispersion obtained during short periods of μG are accurate predictions of what would happen in space.

Acknowledgments

This work was supported by NASA grant NAGW 4372. The author is a Parker B. Francis fellow in pulmonary research.

References

1. West JB, Guy HJB, Elliott AR, Prisk GK. Respiratory system in microgravity. In: Fregly MJ, Blatteis CM, eds. The Handbook of Physiology. Sect. 4. Environmental Physiology. New York: Oxford University Press, 1996:675–689.

2. Muir DCF. Influence of gravitational changes on the deposition of aerosols in the lungs of man. Aerosp Med 1967; 38:159–161.

3. Beeckmans JM. Alveolar deposition of aerosols on the moon and in outer space. Nature 1966; 211:208.

4. Hoffman RA, Billingham J. Effect of altered G levels on deposition of particulates in the human respiratory tract. J Appl Physiol 1975; 38:955–960.

5. Darquenne C, Paiva M, West JB, Prisk GK. Effect of microgravity and hypergravity on deposition of 0.5- to 3-μm-diameter aerosol in the human lung. J Appl Physiol 1997; 83:2029–2036.

6. Darquenne C, West JB, Prisk GK. Deposition and dispersion of 1 μm aerosol boluses in the human lung: Effect of micro- and hypergravity. J Appl Physiol 1998; 85: 1252–1259.

7. Darquenne C, West JB, Prisk GK. Dispersion of 0.5–2 μm aerosol in micro- and hypergravity as a probe of convective inhomogeneity in the human lung. J Appl Physiol 1999; 86:1402–1409.

8. Darquenne C, Paiva M, Prisk GK. Effect of gravity on aerosol dispersion and deposition in the human lung after periods of breath-holding. J Appl Physiol 2000; 89: 1787–1792.

9. Henney MR, Kropp KD, Pierson DL. Microbial monitoring of orbiter air. Aviat Space Environ Med 1984; 55:466.

10. James J. Environmental health monitoring results for STS-40/Space Life Sciences (SLS-1). Spacelab Life Sciences-1 180-Day Preliminary Results 1992; 4:4.1.3–4.1.3-34 (abstr).

11. Burrough B. Dragonfly, NASA and the Crisis Aboard Mir. New York: HarperCollins, 1998.

12. Firket J. The cause of the symptoms found in the Meuse Valley during the fog of December 1930. Bull R Acad Med Belgium 1931; 11:683–741.

13. Logan WPD. Mortality in London fog incident. Lancet 1953; 1:336–338.

14. Dockery DW, Pope A. Epidemiology of acute health effects: Summary of time-series studies. In: Wilson R, Spengler J, eds. Particles in Our Air: Concentrations and Health Effects. Cambridge: Harvard University Press, 1996:123–147.

15. Pope A, Dockery DW. Epidemiology of chronic health effects: Cross-sectional studies. In: Wilson R, Spengler J, eds. Particles in Our Air: Concentrations and Health Effects. Cambridge: Harvard University Press, 1996:149–167.

16. Thurston GD, Ito K, Hayes CG, Bates DV, Lippmann M. Respiratory hospital admissions and summertime haze air pollution in Toronto, Ontario: Consideration of the role of acid aerosols. Environ Res 1994; 65:271–290.

17. Environmental Protection Agency. Revised Requirements for Designation of Reference and Equivalent Methods for $PM_{2.5}$ and Ambient Air Quality Surveillance for Particulate Matter (40 CFR Parts 53 and 58). Federal Register 1997; 62:38763–38854.

18. Lawrence ID, Patterson R. Topical respiratory therapy. Adv Intern Med 1990; 35: 27–44.

19. Konig P. Inhaled corticosteroids: Their present and future role in the management of asthma. J Allergy Clin Immunol 1988; 82:297–306.

20. Hiller FC. Therapeutic aerosols: An overview from a clinical perspective. In: Hickey AJ, ed. Pharmaceutical Inhalation Aerosol Technology. New York: Marcel Dekker, 1992:289–306.

21. Darquenne C, Paiva M. One-dimensional simulation of aerosol transport and deposition in the human lung. J Appl Physiol 1994; 77:2889–2898.

22. Haefeli-Bleuer B, Weibel ER. Morphometry of the human pulmonary acinus. Anat Rec 1988; 220:401–414.

23. Schroter RC, Sudlow MF. Flow patterns in models of the human bronchial airways. Respir Physiol 1969; 7:341–355.

24. West JB, Hugh-Jones P. Pulsatile gas flow in bronchi caused by the heart beat. J Appl Physiol 1961; 16:697–702.

25. Butler JP, Tsuda A. Effect of convective stretching and folding on aerosol mixing deep in the lung, assessed by approximate entropy. J Appl Physiol 1998; 83:800–809.

26. Edyvean J, Estenne M, Paiva M, Engel LA. Lung and chest wall mechanics in microgravity. J Appl Physiol 1991; 71:1956–1966.

27. Elliott AR, Prisk GK, Guy HJB, West JB. Lung volumes during sustained microgravity on spacelab SLS-1. J Appl Physiol 1994; 77:2005–2014.

28. Paiva M, Estenne M, Engel LA. Lung volumes, chest wall configuration, and pattern of breathing in microgravity. J Appl Physiol 1989; 67:1542–1550.

29. Prisk GK, Guy HJB, Elliott AR, Paiva M, West JB. Ventilatory inhomogeneity determined from multiple-breath washouts during sustained microgravity on Spacelab SLS-1. J Appl Physiol 1995; 78:597–607.

30. Davies CN, Heyder J, Subba Ramu MC. Breathing of half-micron aerosols. I. Experimental. J Appl Physiol 1972; 32:591–600.

31. Heyder J, Armbruster L, Gebhart J, Grein E, Stahlhofen W. Total deposition of aerosol particles in the human respiratory tract for nose and mouth breathing. J Aerosol Sci 1975; 6:311–328.

32. Altshuler B, Palmes ED, Yarmus L, Nelson N. Intrapulmonary mixing of gases studied with aerosols. J Appl Physiol 1959; 14:321–327.

33. Brand P, Rieger C, Schulz H, Beinert T, Heyder J. Aerosol bolus dispersion in healthy subjects. Eur Respir J 1997; 10:460–467.

34. Brown JS, Gerrity TR, Bennett WD, Kim CS, House DE. Dispersion of aerosol boluses in the human lung: Dependence on lung volume, bolus volume, and gender. J Appl Physiol 1995; 79:1787–1795.

35. Heyder J, Blanchard JD, Feldman HA, Brain JD. Convective mixing in human respiratory tract: Estimates with aerosol boli. J Appl Physiol 1988; 64:1273–1278.

36. Rosenthal FS, Blanchard JD, Anderson PJ. Aerosol bolus dispersion and convective mixing in human and dog lungs and physical models. J Appl Physiol 1992; 73:862–873.

37. Altshuler B. Behaviour of airborne particles in the respiratory tract. In: Wolstenholme GEW, Knight J, eds. Circulatory and Respiratory Mass Transport. London: J & A Churchill, 1969:215–231.

38. Schulz H, Heilmann P, Hillebrecht A, Gebhart J, Meyer M, Piiper J, Heyder J. Convective and diffusive gas transport in canine intrapulmonary airways. J Appl Physiol 1992; 72:1557–1562.

39. Darquenne C, Paiva M. Gas and particle transport in the lung. In: Hlastala MP, Robertson HT, eds. Complexity in structure and function of the lung. New York: Marcel Dekker, 1998:297–323.

40. Kim CS, Hu SC, DeWitt P, Gerrity TR. Assessment of regional deposition of inhaled particles in human lungs by serial bolus delivery method. J Appl Physiol 1996; 81: 2203–2213.

41. Bennett WD, Scheuch G, Zeman KL, Brown JS, Kim CS, Heyder J, Stahlhofen W. Regional deposition and retention of particles in shallow, inhaled boluses: Effect of lung volume. J Appl Physiol 1999; 86:168–173.

42. Bennett WD, Scheuch G, Zeman KL, Brown JS, Kim CS, Heyder J, Stahlhofen W. Bronchial airway deposition and retention of particles in inhaled boluses: Effect of anatomic dead space. J Appl Physiol 1998; 85:685–694.

43. Anderson PJ, Blanchard JD, Brain JD, Feldman HA, McNamara JJ, Heyder J. Effect of cystic fibrosis on inhaled aerosol boluses. Am Rev Respir Dis 1989; 140:1317–1324.

44. Brand P, Tuch T, Manuwald O, Bischof W, Heinrich J, Wichmann HE, Beinert T, Heyder J. Detection of early lung impairment with aerosol bolus dispersion. Eur Respir J 1994; 7:1830–1838.

45. Anderson PJ, Hardy KG, Gann LP, Cole R, Hiller FC. Detection of small airway dysfunction in asymptomatic smokers using aerosol bolus behavior. Am J Respir Crit Care Med 1994; 150:995–1001.

46. Schulz H, Schulz A, Brand P, Tuch T, Von Mutius E, Erdl R, Reinhardt D, Heyder J. Aerosol bolus dispersion and effective airway diameters in mildly asthmatic children. Eur Respir J 1995; 8:566–573.

47. Beinert T, Brand P, Huber A, Stahlhofen W, Heyder J. Aerosol morphometry and aerosol dispersion to study obstructive lung disease. Am Rev Respir Dis 1990; 141(suppl):A236.

48. Blanchard JD. Aerosol bolus dispersion and aerosol-derived airway morphometry: Assessment of lung pathology and response to therapy. Part 1. J Aerosol Med 1996; 9:183–205.

49. Blanchard JD. Aerosol bolus dispersion and aerosol-derived airway morphometry: Assessment of lung pathology and response to therapy. Part 2. J Aerosol Med 1996; 9:453–476.

50. Hickey AJ. Methods of aerosol particle size characterization. In: Hickey AJ, ed. Pharmaceutical Inhalation Aerosol Technology. New York: Marcel Dekker, 1992: 219–253.

51. Ultman JS, Doll BE, Spiegel R, Thomas MW. Longitudinal mixing in pulmonary airways-normal subjects respiring at a constant flow. J Appl Physiol 1978; 44:297–303.

52. Ultman JS, Blatman HS. Longitudinal mixing in pulmonary airways. Analysis of inert gas dispersion in symmetric tube network models. Respir Physiol 1977; 30: 349–367.

53. Verbanck S, Linnarsson D, Prisk GK, Paiva M. Specific ventilation distribution in microgravity. J Appl Physiol 1996; 80:1458–1465.

54. Guy HJB, Prisk GK, Elliott AR, Deutschman RAI, West JB. Inhomogeneity of pulmonary ventilation during sustained microgravity as determined by single-breath washouts. J Appl Physiol 1994; 76:1719–1729.

55. Scheuch G, Stahlhofen W. Effect of heart rate on aerosol recovery and dispersion in human conducting airways after periods of breathholding. Exp Lung Res 1991; 17:763–787.

56. Weibel ER. Morphometry of the Human Lung. New York: Academic, 1963.

57. Engel LA, Menkes H, Wood LDH, Utz G, Joubert J, Macklem PT. Gas mixing during breath holding studied by intrapulmonary gas sampling. J Appl Physiol 1973; 35:9–17.

58. Horsfield K, Gabe I, Mills C, Buckman M, Cumming G. Effect of heart rate and stroke volume on gas mixing in dog lung. J Appl Physiol 1982; 53:1603–1607.

59. Hogg JC, Macklem PT, Thurlbeck WM. The resistance of collateral channels in excised human lungs. J Clin Invest 1969; 48:421–431.

60. Prisk GK, Elliott AR, Guy HJB, Verbanck S, Paiva M, West JB. Multiple-breath washin of helium and sulfur hexafluoride in sustained microgravity. J Appl Physiol 1998; 84:244–252.

61. Lauzon A-M, Prisk GK, Elliott AR, Verbanck S, Paiva M, West JB. Paradoxical helium and sulfur hexafluoride single-breath washouts in short-term vs. sustained microgravity. J Appl Physiol 1997; 82:859–865.

62. Braun JD, Valberg PA. Deposition of aerosol in the respiratory tract. Am Rev Respir Dis 1979; 120:1325–1373.

6

Pulmonary Perfusion
Gravitational Components

G. KIM PRISK

University of California, San Diego
La Jolla, California

I. Introduction

Although it is well known, in any discussion relating to gravitational effects on pulmonary perfusion it is worth reiterating that pulmonary vascular pressures are very low in normal subjects. At rest in a normal adult, mean pulmonary artery pressure is only about 15 mmHg and pulmonary venous pressure about 5 mmHg (1). In units of cmH_2O, the pulmonary arterial pressure is therefore ~ 20 cmH_2O. The pulmonary artery enters the lung at the hilum, and in an average-sized adult, the lung extends approximately 20 cm above this point. As a consequence, the hydrostatic pressure gradient that exists in the pulmonary vasculature is of a comparable magnitude to the vertical size of the human lung. This would lead one to believe that gravity must play an important role in the nature of pulmonary perfusion, at least at rest, in the upright human.

Despite this seemingly obvious a priori evidence for an effect of gravity, there exists today considerable controversy and discussion regarding exactly how much influence gravity actually has on the distribution of pulmonary perfusion. This chapter discusses some of the older work that forms the foundation for the current understanding of the gravitational effects on pulmonary perfusion. Those studies show clearly that in the upright human lung, gravity has an overwhelming influence on the distribution of pulmonary perfusion. However, as Chapter 7 discusses, there is considerable evidence for nongravitational gradients in pulmonary perfusion in animals. At this time, it is probably fair to say that the controversy between the gravitational argument and the nongravitational argument remains unresolved. It may be that studies performed in microgravity may prove useful in the resolution of this in the future.

II. Differences in Pulmonary Perfusion

A. The Zone 1, 2, and 3 Model

The current understanding of the gravitational effects on the vertical distribution of pulmonary perfusion stems from the studies of pulmonary blood flow performed by West et al. (2–4,6) in the late 1950s and early 1960s. West and Dollery (2) used the clearance of inhaled radioactive CO_2 to examine the vertical distribution of pulmonary blood flow in upright humans. Subjects inhaled radioactive CO_2 and the uptake from the lung was measured using externally placed radioactive counters during breathholding following an inhalation from FRC of 1 L of the radioactive gas. The results showed that in normal upright men at rest, there was a ninefold increase in blood flow between the apex and base of the lung. This gradient in blood flow was subject to change. For example, the supine posture resulted in an equalization of this gradient, clear evidence of a gravitational effect. Exercise and other conditions, such as intracardiac shunts, which serve to raise pulmonary blood flow (and thus pressure) reduced (but did not eliminate) the vertical gradient in pulmonary blood flow (3,4).

The studies in intact humans provided excellent evidence for a strong effect of gravity on the distribution of pulmonary perfusion. The actual understanding of the mechanisms responsible for the results came from an extensive series of studies performed in isolated dog lungs (2), and from the ideas of Permutt et al. (5). In these studies, isolated dog lungs were perfused while suspended in a negative pressure box providing the ventilation. Importantly, all the relevant pressures (pulmonary arterial, pulmonary venous, and alveolar) could be independently controlled. The pulmonary blood flow was measured using radioactive xenon (^{133}Xe) that was dissolved in saline and infused into the isolated lung preparation. Since Xe is only minimally soluble in blood, almost all the gas is given off in a single pass through the alveoli, effectively marking the alveolar gas dependent on the blood flow. The counts were then determined from externally placed radiation counters.

The results showed that there was a vertical gradient in pulmonary perfusion in these isolated lungs. Importantly, when pulmonary perfusion pressures were low, regional counts in the highest regions of the lung fell to essentially 0, indicating that these regions of the lung became unperfused. Similarly, increasing the alveolar pressure while holding vascular pressures constant resulted in the development of unperfused regions in the uppermost portions of the lung.

When taken together, the results led to the development of the now famous zone 1, 2, 3 model describing the vertical distribution of pulmonary blood flow in terms of pulmonary arterial, pulmonary venous, and alveolar pressures (Fig. 1). The lung is perfused with a pulmonary arterial pressure (Pa) and venous pressure (Pv), and there is a vertical gradient in these pressures due the hydrostatic

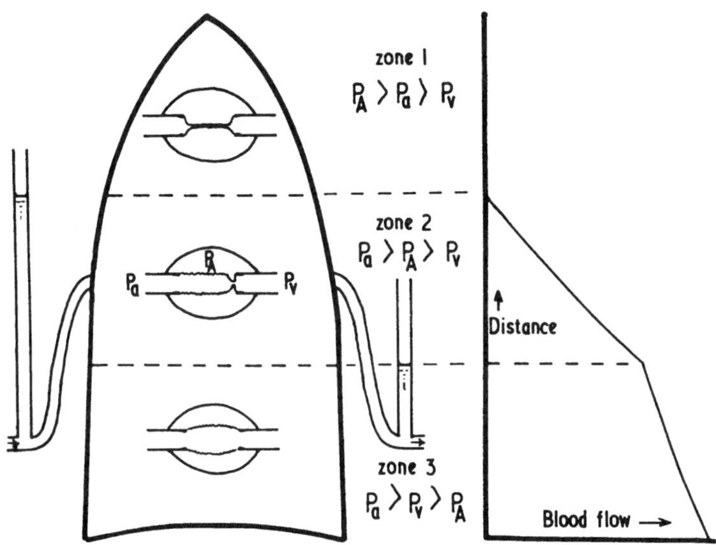

Figure 1 The zone 1, 2, 3 model of pulmonary perfusion. In a lung perfused with a pulmonary arterial pressure (Pa) and venous pressure (Pv), there is a vertical gradient in these pressures due to the hydrostatic effect. Thus, vascular pressures are lowest at the lung apex, and greatest at the lung base. In contrast, alveolar pressure (P_A) is constant throughout the lung. At the apex of the lung (zone 1), alveolar pressure exceeds both venous and arterial pressure, and as a result the collapsible pulmonary capillaries are held shut, eliminating blood flow. In the middle of the lung (zone 2), arterial pressure exceeds alveolar pressure, but venous pressure remains below alveolar pressure. As a consequence, the pulmonary capillaries become constricted at some point, limiting blood flow. However, because this constriction occurs independently of venous pressure, there is an increase in perfusion pressure that depends solely on the hydrostatic increase in pulmonary arterial pressure. Near the base of the lung (zone 3), both arterial and venous pressures exceed alveolar pressure and so the pulmonary capillaries are held open. Since both pressures increase as the distance down the lung increases, the pressure gradient is constant down the lung. The increase in blood flow toward the base results from the increasing caliber of the pulmonary capillaries as the distending pressure increases. (From Ref. 2.)

effect. This is due to the weight of a column of blood in the gravitational field. As a consequence, vascular pressures are lowest at the lung apex, and greatest at the lung base. In contrast, alveolar pressure (P_A) is constant throughout the lung since all airways connect to the atmosphere. Zone 1 occurs at the apex of the lung where alveolar pressure exceeds both venous and arterial pressure, and as a result the collapsible pulmonary capillaries are held shut, eliminating blood flow. Zone 2 occurs in the middle of the lung where arterial pressure exceeds alveolar pressure, but venous pressure remains below alveolar pressure. As a consequence, the pulmonary capillaries become constricted at some point when vascular pressure falls below alveolar pressure, limiting blood flow. However, because this constriction occurs independently of venous pressure, there is an increase in perfusion pressure that depends solely on the hydrostatic increase in pulmonary arterial pressure. The flow is independent of the downstream pressure. This Starling resistor effect results in a large vertical gradient in pulmonary blood flow. Zone 3 is near the base of the lung where both arterial and venous pressures exceed alveolar pressure and so the pulmonary capillaries are held open. Since both pressures increase as the distance down the lung increases, the pressure gradient is constant down the lung. The increase in blood flow toward the base of the lung results from the increasing caliber of the pulmonary capillaries as the distending pressure increases.

B. Zone 4

The initial measurements made of the distribution of pulmonary perfusion in intact humans were all made at lung volumes above FRC. West and Dollery (6) performed their studies at 1 L above FRC, and Ball et al. (7) at TLC. If the measurements are made at TLC, all the alveoli are more or less at a similar lung volume, and so regional blood flow can be measured directly. However, the high lung volumes may perturb the distribution of pulmonary perfusion seen in the lung under normal conditions. Anthonisen and Milic-Emili (8) performed regional perfusion studies at different lung volumes, using a technique to correct for differences in regional lung volume. At FRC, all three zones could be identified. However, at RV, the entire lung had a much more uniform blood flow, a finding at odds with the Zone 1, 2, 3 model (9).

Hughes et al. (10) performed a comprehensive series of measurements over a wide range of lung volumes. They found that at FRC, blood flow decreased down the lung over the lower one-third of the lung, an effect that was accentuated at even lower lung volumes. At high lung volumes, the effect became smaller. At RV, there was virtually no vertical gradient in regional blood flow per alveolus. The zone 1, 2, 3 model is unable to explain these results. The results were attributed to the effect of lung expansion on the extraalveolar vessels. These vessels are attached to the lung parenchyma, and when the lung expands, are

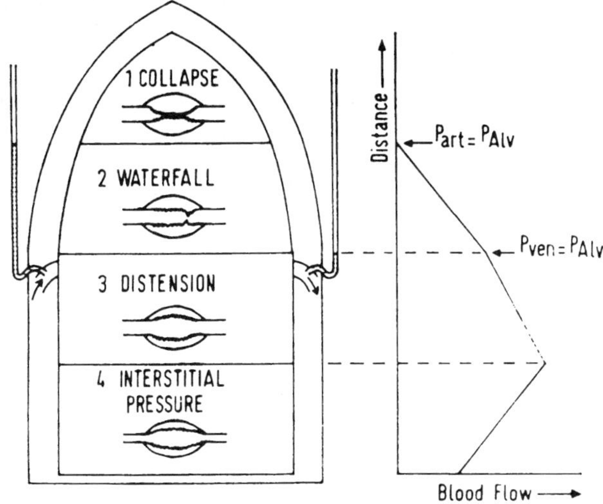

Figure 2   The zone 4 model is the same as that described in Figure 1, except that in the most dependent lung regions, blood flow decreases with distance down the lung. This cannot be explained on the basis of the model in Figure 1. In this so-called zone 4, the relatively poor lung expansion reduces the radial traction on the extraalveolar vessels, reducing their caliber and increasing vascular resistance. The effect is to reduce blood flow with a greater effect being seen at low lung volumes. (From Ref. 10.)

subject to an expansion force as the lung stretches. Thus, at low lung volumes, the lack of an outward pull on these vessels results in a reduction in the caliber of the extraalveolar vessels, reducing regional blood flow. At lung volumes below FRC, the relatively poor expansion of the lower zones leaves vascular resistance high in these areas, reducing blood flow. This region of the lung was termed zone 4 (Fig. 2), and demonstrates the gravitational effect that differences in lung expansion (and hence in pulmonary ventilation) may have on pulmonary perfusion.

C.   Interregional Differences

Apart from the effects of gravity, there are numerous other potential causes of nonuniform pulmonary perfusion. These causes are extensively discussed in chapter 7, and so they will be only briefly discussed here. As well as gravity, regional lung expansion clearly affects pulmonary blood flow and there have been studies that show that regional lung expansion is at times markedly heterogeneous (10–12). Perivascular edema may affect blood flow (13). Hypoxic pulmonary vasoconstriction may have significant effects on regional blood flow (14).

There is also considerable evidence for nonuniformity of blood flow within iso-gravitational planes (15–20).

There are documented differences in regional vascular conductance, at least in dogs. Beck and Rehder (21) used radioactive microspheres and showed that, independently of gravity, regional lung expansion, and hypoxic pulmonary vaso-constriction, there were differences in the pulmonary vascular conductance re-lated to anatomical location. In particular, they showed that in the dog, the dorso-caudal regions of the lung had a higher vascular conductance than did the ventrocephalad regions, and attributed these findings to differences in regional vascular anatomy. Whether these differences are similar in humans is largely unknown.

Hakim et al. (18) used single-positron imaging techniques to infer the pres-ence of a central-to-peripheral gradient in pulmonary perfusion in supine humans. This model provides for the greatest perfusion near the hilum, with decreasing values toward the periphery (Fig. 3). Because of the shape of the upright human lung, the planar images of perfusion look similar to the vertical distributions of

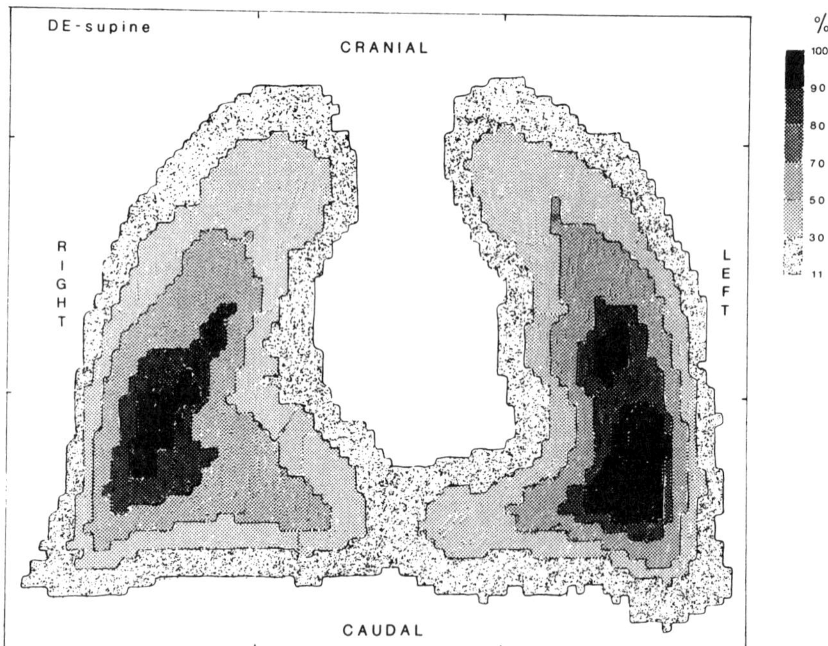

Figure 3 Mapping of isoflow regions in single-positron emission computed tomo-graphic images in humans. Subjects were supine when radioactively labeled microaggre-gated albumin was injected. Note the apparent central-to-peripheral gradient in blood flow. (From Ref. 18.)

blood flow seen in the earlier radioactive counting studies (6). The results also suggest that zone 4 may not be solely a dependent lung region effect, but rather a peripheral lung region effect. However, independent studies performed in dogs failed to confirm the presence of such a gravity-independent distribution in pulmonary perfusion (22).

D. Small-Scale Nontopographical Differences

There is also evidence for inhomogeneity of pulmonary perfusion on a scale much smaller than that addressed by the studies of topographical differences. Data from rat lungs suggest that there is an approximately 4:1 gradient in blood flow per unit of lung volume from the apex to the base of the secondary lobule (23). Such a gradient had previously been inferred in human lung (24). Later West et al. (25) confirmed the presence of a gradient in perfusion along the acinus using fluid-filled lungs. Ewan et al. (26) showed that in humans, there was unevenness in intraregional perfusion.

Warrell et al. (27) showed that in dog lungs, the recruitment of pulmonary capillaries was essentially a stochastic process. Under zone 2 conditions the variation in capillary filling occurred entirely within that part of the capillary bed supplied by single arterioles, as opposed to between areas supplied by different arterioles. In a subsequent modeling study, West et al. (28) suggested that recruitment of pulmonary capillaries was ''patchy'' until quite high vascular pressures were reached. This implies that even in the absence of other effects such as gravity, pulmonary perfusion may remain quite uneven, especially at rest. However, in contrast to this, recent studies in isolated dog lung lobes using fluorescent dye techniques suggest that acinar perfusion is unexpectedly homogeneous (9).

Glenny et al. (29) have made suggestions of a fractal nature in the distribution of pulmonary perfusion. In these studies, in which labeled microspheres were injected at FRC, the results suggested that the dominant factor in the heterogeneity of pulmonary blood flow was its fractal property, and that gravity played a secondary role in these dogs. Other studies by the same group suggest that in supine dogs under zone 3 conditions at FRC, gravity plays only a minor role in the distribution of pulmonary perfusion (30,31). However, there have been objections raised about the lung volumes at the time of injection (32). In addition, the subsequent processing of the lungs, in which they are dried at TLC, may influence the results. Chapter 7 has a more complete discussion of these and other related studies.

III. Studies in Microgravity

Studies in microgravity (μG) have been few to date, and all of them suffer from the same problem, namely, that in humans, only indirect techniques have been available to be used. While these provide some inferential data on the nature of the changes in pulmonary perfusion, direct measurements are preferable. Unfortunately, the technical challenges associated with more direct studies are considerable.

A. Parabolic Flight

As a part of the studies of pulmonary ventilation in transient microgravity in parabolic flight, Michels and West (33) also performed a pilot study of the effects of microgravity on the inhomogeneity of pulmonary perfusion in a single subject. These studies were performed in a Learjet, which provided from 22 to 27 s of μG in each maneuver. Measurements were made in μG, 1 G, and 2 G.

The maneuver was that of a hyperventilation, breathholding type. The subject hyperventilated while breathing air for 5 s, thus lowering the CO_2 concentration throughout the lung, in this case to $\sim 4\%$. He then performed breathholding at TLC for approximately 15 s. During this period, CO_2 was added to the alveolar space at a rate that depends on the local perfusion per unit lung volume. Since the lung was at TLC, alveolar volume was largely uniform, and so the level of CO_2 in the alveoli is a marker for regional lung perfusion. The subject then exhaled in a controlled fashion (0.5 L/s) to RV. During this exhalation, the size of the cardiogenic oscillations in the CO_2 was taken as a marker of the degree of unevenness of blood flow. The maneuver depends on the fact that, as shown by Fukuchi et al. (34), the cardiogenic oscillations arise from the direct mechanical action of the heart on the lung, displacing gas preferentially from lung regions near the heart during diastole. Figure 4 shows an example of the expired tracing in CO_2 seen using this technique.

The studies in parabolic flight (33) clearly showed that the size of the cardiogenic oscillations was strongly dependent on the G level present during the breathholding period (that time during which CO_2 evolves into the alveolar space) (Fig. 5). The studies were performed so that the expiration always occurred at the same G level (thus minimizing the effects of changes in the distribution of ventilation). There was an almost linear increase in the size of the cardiogenic oscillations in both CO_2 and O_2 with the G level during breathholding. Interestingly however, there were persisting cardiogenic oscillations at μG. However, because of the period of hypergravity immediately preceding μG, it was not possible to determine whether or not this was representative of sustained μG, or an artifact resulting from the inability of the lung and/or pulmonary perfusion to respond rapidly to an abrupt change in G.

B. Space Flight

As part of the Spacelab Life Sciences-1 flight, flown in 1991, Prisk et al. (35) repeated the hyperventilation breathhold measurements of Michels and West (33) in sustained μG. Because the period of μG was not constrained as it was in the

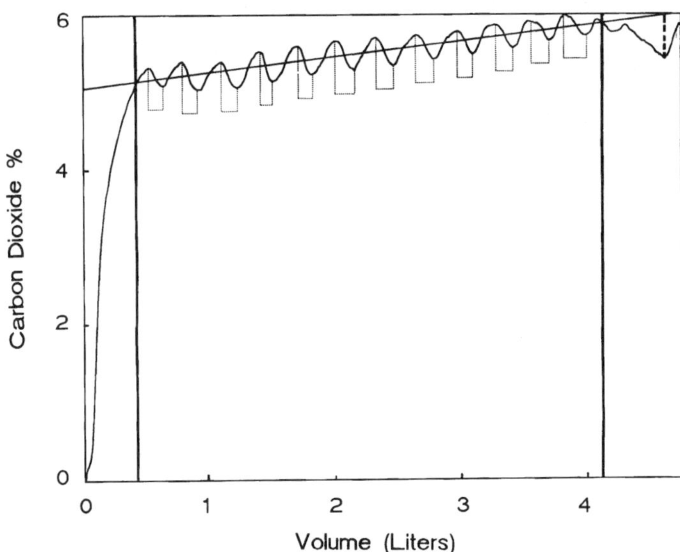

Figure 4 Hyperventilation breathholding measurement of the inequality of pulmonary perfusion. Expired CO_2 concentration is plotted as a function of expired volume over the course of a controlled (0.5 L/s) exhalation from TLC to RV. In this case, the subject had hyperventilated for 20 s and breathheld for 15 s. In the case of the Learjet studies of Michels and West (1978), hyperventilation was for 5 s and the breathholding 15 s. The cardiogenic oscillations are marked as is the onset of airway closure and the height of the subsequent terminal fall in CO_2 concentration. (From Ref. 35.)

Learjet, the period of hyperventilation was extended to 20 s, which reduced the alveolar CO_2 to ~2.5%, but the test was otherwise the same. In addition, data were collected with the subjects in both the upright and supine postures on the ground.

The cardiogenic oscillations seen in μG essentially matched the results seen in parabolic flight (33) (compare Figs. 5 and 6). Importantly, the cardiogenic oscillations persisted even in sustained μG, indicating that there remains considerable inhomogeneity of pulmonary perfusion in the absence of gravity. The scale of this inhomogeneity must be sufficiently large (i.e., the lung units with differing perfusion sufficiently far apart) that path length differences from the alveoli and mixing within the airways were unable to smear out the oscillations.

The other marker of regional inhomogeneity of perfusion in these single-breath studies is the terminal fall in CO_2. In 1 G, this fall occurs when gravitationally dependent lung units close. The flow at the mouth then has a greater contribu-

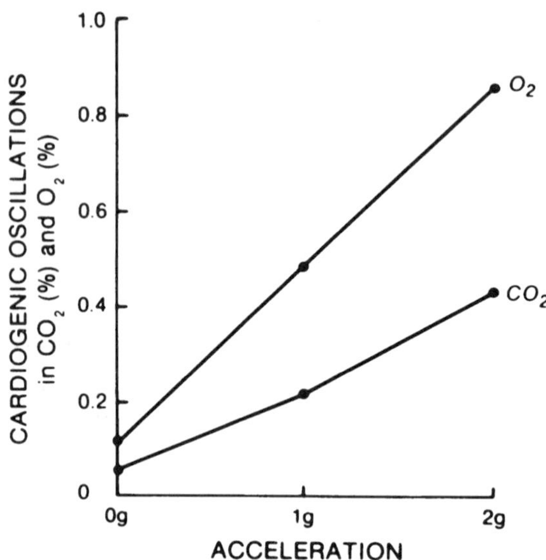

Figure 5 The size of cardiogenic oscillations from the hyperventilation breathholding maneuver performed in the Learjet. Expiration was at the same G level in all cases. The G level during the 15-s breathholding period is indicated. Note that while the size of the cardiogenic oscillations decreased with decreasing G level, they persist in the absence of gravity. (From Ref. 33.)

tion from upper lung regions. Since these have a lower CO_2 due to lower blood flow, CO_2 measured at the mouth falls. The studies of pulmonary ventilation performed in the same subjects (36) showed clearly that airway closure occurs at approximately the same lung volume in μG as it does upright in 1 G. While airway closure still occurs, it seems likely that in μG, instead of it being predominately basal units that close, a pattern of patchy airway closure ensues (37).

In sharp contrast to the cardiogenic oscillations, the height of the terminal fall in CO_2 that occurs after airway closure (see Fig. 4) was absent in μG (see Fig. 6). The absence of a terminal fall in CO_2 in the hyperventilation breathholding test, in the presence of continuing airways closure, is clear evidence that the pulmonary perfusion to units that close near RV and units that remain open must be similar in μG. The situation in μG is markedly different from that in the supine posture. In that case, while there was a reduction in cardiogenic oscillation size, the terminal fall in CO_2 persisted.

A plausible, although not definitive, explanation for the results shown in Fig. 6 is that in μG, the normally present gravitational gradient in pulmonary perfusion is removed. This abolishes the coherent change in CO_2 concentration

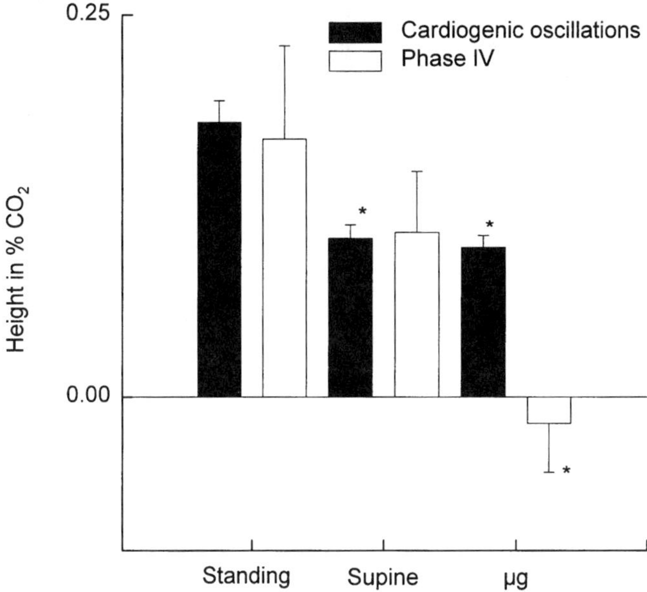

Figure 6 Relative sizes of cardiogenic oscillations in CO_2 and the height of the terminal fall (inverted for the purposes of comparison) standing and supine in 1 G, and in sustained μG. Note that in μG the terminal fall is absent, but cardiogenic oscillations persist. (From Ref. 35.)

that occurs after the onset of airway closure, abolishing the terminal fall. However, the cardiogenic oscillations persist, principally reflecting the residual (and necessarily nongravitational) differences in pulmonary perfusion that exist in the lung in μG. Alternatively, it may be that the pattern of airway closure in the absence of gravity is such that regions that close do so irrespective of the perfusion of the regions they subtend. However, this is perhaps somewhat less likely given the fact that there is a persisting phase IV (terminal rise) seen in the studies of pulmonary ventilation (36). Overall, the results are consistent with the abolition of a top-to-bottom gradient in pulmonary perfusion, but with persisting regional inhomogeneity in μG.

Other data from SLS-1 support the notion that pulmonary perfusion became much more uniform in μG. There was an increase in cardiac output (especially early inflight) (38), which should have served to better perfuse the entire lung. Importantly, diffusing capacity of the lung for carbon monoxide increased by ~25% above that measured in the upright posture, and this increase remained unchanged over the course of the 9-day flight. The increase in diffusing capacity was a consequence of similar increases in both the membrane diffusing capacity and the pulmonary capillary blood volume. This was in contrast to the supine posture, where there was an increase in pulmonary capillary blood volume, but membrane diffusing capacity remained unchanged. The results were consistent with the lung becoming evenly perfused with blood in μG, and the abolition of the zone 1, 2, 3 pattern normally seen upright in 1 G. In contrast, in the supine posture, the gravitational gradient in pulmonary perfusion persists. Thus, the upper portions of the lung remain, relatively speaking, poorly perfused with some pulmonary capillaries remaining unperfused, and this is reflected in the lack of increase in the membrane diffusing capacity. Similar sustained changes in diffusing capacity were seen in data collected during the German spacelab mission D-2 (39).

C. Time Course of Changes in Pulmonary Perfusion

Little is known about the time course of the changes in pulmonary perfusion that occur in response to a change in gravity. The data from parabolic flight (33) show a substantial reduction in the size of cardiogenic oscillations when breathholding occurred during a $\sim$25-s period of μG. The data from space flight (35) show no change in pulmonary perfusion over the course of a 9-day flight, with the first data being taken $\sim$24 hr after the onset of μG. However, the slight differences in protocol between these two studies preclude direct comparison. In addition, it may well be that the 25 s of μG available in parabolic flight is insufficient time for circulatory changes to be complete.

References

1. West JB. Respiratory Physiology: The Essentials. Baltimore: Williams & Wilkins, 2000.
2. West JB, Dollery CT, Naimark A. Distribution of bloodflow in isolated lung: Relation to vascular and alveolar pressures. J Appl Physiol 1964; 19:713–724.
3. Dollery CT, West JB, Wilcken DEL, Goodwin JF, Hugh-Jones P. Regional pulmonary blood flow in patients with circulatory shunts. Br Heart J 1961; 23:225–235.
4. Dollery CT, West JB. Regional uptake of radioactive oxygen, carbon monoxide and carbon dioxide in the lungs of patients with mitral stenosis. Circ Res 1960; 8:765–771.
5. Permutt S, Bromberger-Barnea B, Bane HN. Alveolar pressure, pulmonary venous pressure and the vascular waterfall. Med Thorac 1962; 19:239–260.
6. West JB, Dollery CT. Distribution of blood flow and ventilation-perfusion ratio in the lung, measured with radioactive CO_2. J Appl Physiol 1960; 15:405–410.
7. Ball WC Jr, Stewart PB, Newsham LGS, Bates DV. Regional pulmonary function studied with xenon-133. J Clin Invest 1962; 41:519–531.
8. Anthonisen NR, Milic-Emili J. Distribution of pulmonary perfusion in erect man. J Appl Physiol 1966; 21:760–766.

9.   Tanabe N, Todoran TM, Zenk GM, Bunton BR, Wagner WW Jr, Presson RG Jr. Perfusion heterogeneity in the pulmonary acinus. J Appl Physiol 1998; 84:933–938.

10. Hughes JMB, Glazier JB, Maloney JE, West JB. Effect of lung volume on the distribution of pulmonary blood flow in man. Respir Physiol 1968; 4:58–72.

11. Olson LE, Rodarte JR. Regional differences in expansion in excised dog lung lobes. J Appl Physiol 1984; 57:1710–1714.

12. Hughes JMB, Glazier JB, Maloney JE, West JB. Effect of extra-alveolar vessels on distribution of blood flow in the dog lung. J Appl Physiol 1968; 25:701–712.

13. West JB, Dollery CT, Heard BE. Increased pulmonary vascular resistance in the dependent zone of the isolated dog lung caused by perivascular edema. Circ Res 1965; 17:191–206.

14. Prefaut C, Engel LA. Vertical distribution of perfusion and inspired gas in supine man. Respir Physiol 1981; 43:209–219.

15. Amis TC, Jones HA, Hughes JMB. Effect of posture on inter-regional distribution of pulmonary perfusion and VA/Q ratios in man. Respir Physiol 1984; 56:169–182.

16. Bryan AC, Bentivoglio LG, Beerel F, MacLeish H, Zidulka A, Bates DV. Factors affecting regional distribution of ventilation and perfusion in the lung. J Appl Physiol 1964; 19:395–402.

17. Engel LA, Prefaut C. Cranio-caudal distribution of inspired gas and perfusion in supine man. Respir Physiol 1981; 45:43–53.

18. Hakim TS, Lisbona R, Dean GW. Gravity-independent inequality of pulmonary blood flow in humans. J Appl Physiol 1987; 63:1114–1121.

19. Hogg JC, Holst P, Corry P, Ruff F, Housley E, Morris E. Effect of regional lung expansion and body position on pulmonary perfusion in dogs. J Appl Physiol 1971; 31:97–101.

20. Reed JH Jr, Wood EH. Effect of body position on vertical distribution of pulmonary blood flow. J Appl Physiol 1970; 28:303–311.

21. Beck KC, Rehder K. Differences in regional vascular conductances in isolated dog lung. J Appl Physiol 1986; 61:530–538.

22. Nicolaysen G, Shepard J, Orizukal M, Tanita T, Hattner RS, Staub NC. No gravity-independent gradient of blood flow distribution in dog lung. J Appl Physiol 1987; 63:540–545.

23. Wagner PD, McRae J, Read J. Stratified distribution of blood flow in secondary lobule of the rat lung. J Appl Physiol 1967; 22:1115–1123.

24. Read J. Stratification of ventilation and blood flow in the normal lung. J Appl Physiol 1966; 21:1521–1531.

25. West JB, Maloney JE, Castle BL. Effect of stratified inequality of blood flow on gas exchange in liquid-filling lungs. J Appl Physiol 1972; 32:357–361.

26. Ewan PW, Jones HA, Nosil J, Obdrzalek J, Hughes JMB. Uneven perfusion and ventilation within lung regions studied with nitrogen-13. Respir Physiol 1978; 34:45–59.

27. Warrell DA, Evans JW, Clarke RO, Kingaby GP, West JB. Patterns of filling in the pulmonary capillary bed. J Appl Physiol 1972; 32:346–356.

28. West JB, Schneider AM, Mitchell MM. Recruitment in networks of pulmonary capillaries. J Appl Physiol 1975; 39:976–984.

29. Glenny RW, Robertson HT. Fractal properties of pulmonary blood flow: Characterization of spatial heterogeneity. J Appl Physiol 1990; 69:532–545.

30. Glenny RW, Lamm WJE, Albert RK, Robertson HT. Gravity is a minor determinant of pulmonary blood flow distribution. J Appl Physiol 1991; 72:620–629.
31. Glenny RW, Polissar L, Robertson HT. Relative contribution of gravity to pulmonary perfusion heterogeneity. J Appl Physiol 1991; 71:2449–2452.
32. West JB. Gravity and pulmonary blood flow distribution. J Appl Physiol 1992; 73:2201–2202.
33. Michels DB, West JB. Distribution of pulmonary ventilation and perfusion during short periods of weightlessness. J Appl Physiol 1978; 45:987–998.
34. Fukuchi Y, Cosio M, Kelly S, Engel LA. Influence of pericardial fluid on cardiogenic gas mixing in the lung. J Appl Physiol 1977; 42:5–12.
35. Prisk GK, Guy HJB, Elliott AR, West JB. Inhomogeneity of pulmonary perfusion during sustained microgravity on SLS-1. J Appl Physiol 1994; 76:1730–1738.
36. Guy HJB, Prisk GK, Elliott AR, Deutschman RA III, West JB. Inhomogeneity of pulmonary ventilation during sustained microgravity as determined by single-breath washouts. J Appl Physiol 1994; 76:1719–1729.
37. Engel LA, Grassino A, Anthonisen NR. Demonstration of airway closure in man. J Appl Physiol 1975; 38:1117–1125.
38. Prisk GK, Guy HJB, Elliott AR, Deutschman RA, West JB. Pulmonary diffusing capacity, capillary blood volume and cardiac output during sustained microgravity. J Appl Physiol 1993; 75:15–26.
39. Verbanck S, Larsson H, Linnarsson D, Prisk GK, West JB, Paiva M. Pulmonary tissue volume, cardiac output, and diffusing capacity in sustained microgravity. J Appl Physiol 1997; 83:810–816.

7

Pulmonary Perfusion Distribution

Nongravitational Factors

**MICHAEL P. HLASTALA, H. THOMAS ROBERTSON,
and ROBB W. GLENNY**

University of Washington
Seattle, Washington

I. Resolution of Methods

The distribution of ventilation and perfusion and, hence, the gas exchange function of the lung are governed by both gravitational and nongravitational factors. While the gravitational model provided a persuasive hypothesis to explain measurements made in the 1960s of the distribution of regional blood flow, recent observations using higher-resolution techniques have revealed factors that influence perfusion and ventilation independent of the gravitational vector. This chapter describes the nature and significance of these nongravitational factors.

Insight into the mechanisms of pulmonary blood flow distribution has progressed as new techniques have developed. With the application of inert radioactive gases and scintillation counters to lung physiology (1), researchers observed a vertical gradient of pulmonary blood flow distribution and attributed this gradient to intravascular hydrostatic pressure differences. The improved resolution was achieved with a counter that averaged blood flow within horizontal rod-shaped regions of the intact (2) or isolated lung (3). These studies have been reviewed in Chapters 1 and 6 of this book.

Reed and Wood (4), utilizing the injection of radionuclide-labeled microspheres to measure regional perfusion, were the first investigators to demonstrate marked heterogeneity of pulmonary blood flow within isogravitational planes. They sampled cylinders of tissue across the lung with a 2-inch by 2-inch area, and noted that perfusion was not uniform within isogravitational planes, a finding not explained by the simple gravitational explanation for flow heterogeneity. They concluded that ''the interrelationships of determinants of regional pulmonary blood flow in the intact animal are sufficiently complex so that achievement of an adequate description by a relatively simple model is not possible at this time.'' They evoked regional hypoxic pulmonary vasoconstriction as a possible explanation. Greenleaf and associates confirmed the observation of heterogeneous blood flow within isogravitational planes using methods that assessed a cylinder of tissue of 1 cm in diameter across the lung—less than $^1/_{25}$ of the piece size used by Reed and Wood (5). The large heterogeneity of isogravitational pulmonary blood flow has since been confirmed in a variety of laboratory animals (6–11). In all of these studies, radiolabeled blood flow markers were used and the inflated lungs were systematically sampled with piece volumes ranging from 1- to 2-cm^3 pieces. In those studies, isogravitational perfusion heterogeneity was almost as large as the perfusion heterogeneity across the entire lung (8,12).

A dorsocaudal dominance of pulmonary blood flow, independent of posture, was first identified by Beck and Rehder (6). An illustration from that paper (Fig. 1) demonstrates high-flow regions (darker shading) in the dorsocaudal regions of the dog lung independent of posture or inflation volumes. Another high-resolution study seeking radial gradients of flow in dogs was performed by Nico-

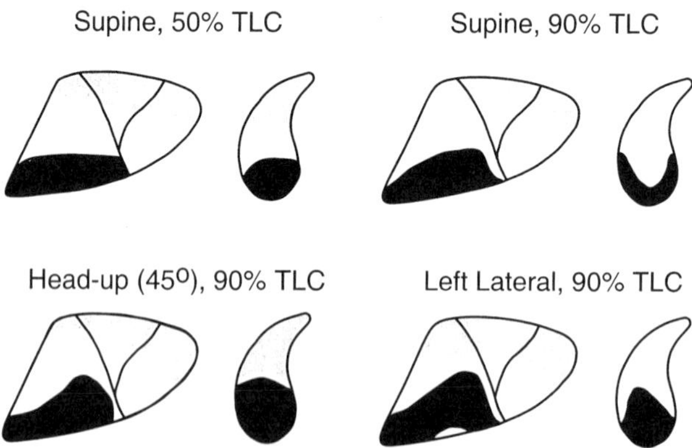

Figure 1 Diagrammatic representation of anatomical locations of lung regions occupied by samples with the 25% highest (dark area) and 25% lowest (lightly shaded area) vascular conductances in three different postures and two different lung volumes showing dorsal dominance of blood flow. (From Ref. 6.)

laysen et al. (13). The lungs were labeled with Tc-macroaggregates, cut in concentric rings, and separately counted. The authors were unable to demonstrate any radial gradient by these direct measurements contrary to what had been proposed from earlier in vivo images obtained with clinical single-photon emission computed tomography (SPECT) scanners (14,15).

A. Nonrandom Distribution Within Isogravitational Planes

The distribution of regional pulmonary perfusion of a supine dog within a coronal (isogravitational) plane was determined by intravenous injection of radionuclide-labeled 15-μm microspheres (12). Individual 2-cm^3 lung pieces were counted (Fig. 2). Note that high-flow pieces tend to be near other regions of high flow, and low-flow regions are more likely to be near other areas of low flow. This is a pattern consistent with the flow being determined by a branching vascular tree in which neighboring regions share parent branches and, hence, have similar flows.

When examined at the resolution of approximately 2 cm^3, the overall heterogeneity of perfusion varies according to experimental conditions and the species studied. Previous studies have observed the coefficient of variation of perfusion heterogeneity of 41 to 52% in anesthetized dogs (10,12), 31% in unanesthetized

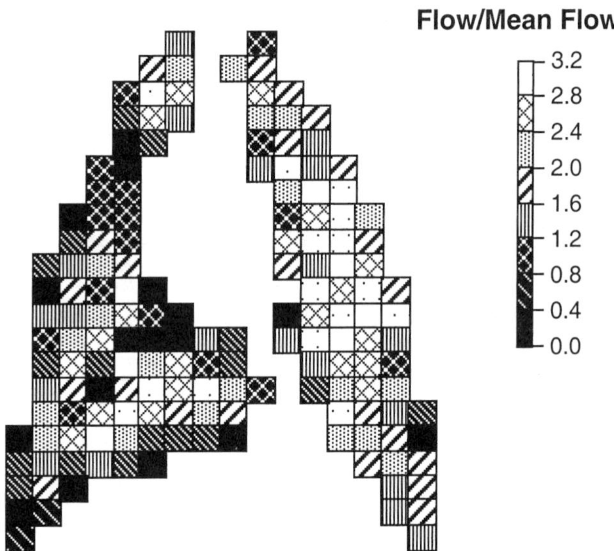

Figure 2 Distribution of regional perfusion within a coronal (isogravitational) plane in a prone dog.

horses (16,17), 30% in unanesthetized sheep (18), 38% in awake goats (8), 73% (supine) and 68% (prone) in anesthetized pigs (19,20), 38% and 45% in unanesthetized pigs (21,22), and 65% in upright baboons (23).

B.   Local Correlation

In all of the blood flow distribution studies described above, the flow heterogeneity was not random. A more quantitative approach to characterizing the heterogeneity was proposed by Glenny (24), where correlations were calculated for all lung pieces as a function of the distance between pieces (Fig. 3). Even though a significant heterogeneity exists across isogravitational planes in the mammalian lung, there is a strong correlation among adjacent pieces (24) and a negative correlation among pieces that are very far apart. High-flow pieces are next to high-flow pieces and low-flow pieces neighbor on low-flow pieces. This spatial

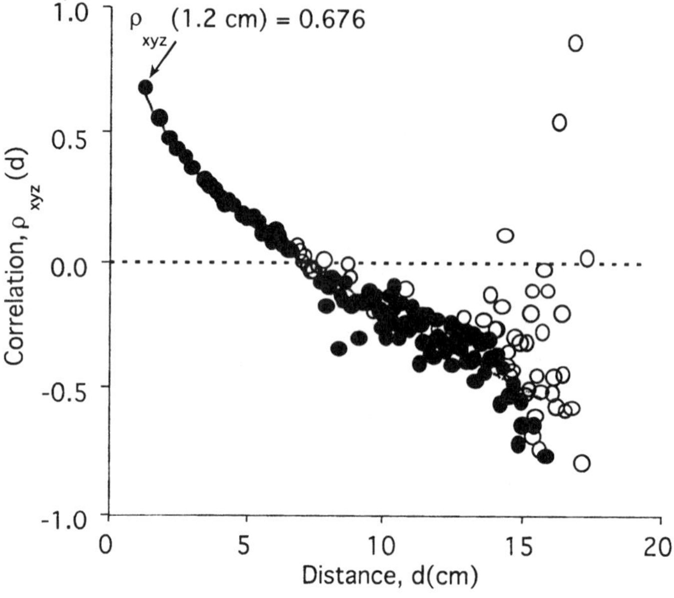

Figure 3   Correlogram of regional blood flow as a function of distance between the pieces. Pieces separated by 1.2 cm have similar flows ($\rho = 0.676$, where ρ is the correlation coefficient), and flows become less similar and eventually negatively correlated at larger distances. Solid circles indicate points with spatial correlation significantly different from 0.0 ($p < 0.05$).

pattern of pulmonary blood flow is not random and remains very stable over days (24).

C. Erect Animals

Laboratory animals have been used for recent high-resolution studies that require destructive sampling of the lung. Although humans are the only true mammalian biped, baboons spend most of their time in the upright posture and have pulmonary structures and physiology remarkably similar to those of humans. The baboon pulmonary vascular tree parallels the human system from its gross anatomy to the degree of muscularization at the arteriolar and venular level (25). Gas exchange and hemodynamic response to hypoxia in the baboon are also similar to humans (26). In the baboon, changes in gas exchange with postural changes are identical to those in humans, and they tolerate the upright posture well.

Pulmonary blood flow in anesthetized baboons has been measured by injected fluorescent microspheres in the supine, prone, upright, and head-down postures (23). The vertical distribution of pulmonary blood flow in the erect baboon is shown in Figure 4. Perfusion heterogeneity was greatest in the upright posture and least when prone. Using multiple-stepwise regression, it was found that in the upright lung, vertical height up the lung contributed 25% of the total variance. When the animals were in either the supine or prone posture, the variance due to vertical height up the lung was only 7 and 5%, respectively. Although important, gravity is not the predominant determinant of pulmonary perfusion heterogeneity in upright primates. Because of anatomical similarities, the same may be true for humans.

The data for the erect baboon can be compared with what would be found using the original low-resolution external scintillation counter methods by averaging the blood flow within isogravitational planes as shown in Figure 5. As the only large-scale feature of perfusion heterogeneity, the gravitational gradient appears to be the only relevant mechanism at low image resolution (Fig. 5) compared with high resolution (see Fig. 4).

The authors estimated that 25% of the observed perfusion heterogeneity in the upright primate can be attributable to the gravity-induced hydrostatic gradient down the lung. The residual perfusion heterogeneity observed in isogravitational planes cannot be explained by hydrostatic gradients. This is an overestimate because the experimental design attributes changes in perfusion observed with postural changes to an effect of hydrostatic pressure differences. Mediastinal structures, abdominal contents, and the diaphragm likely shift position in the head-down compared with the upright posture. Any alterations in regional perfusion induced by these structural changes are attributed to a gravitational effect. Any

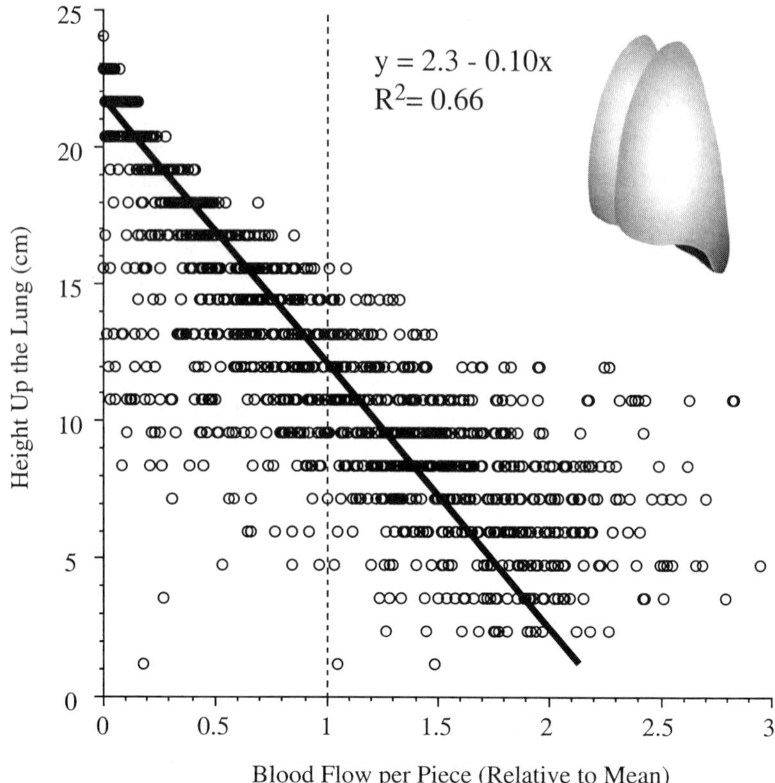

Figure 4 Vertical distributions of blood flow in the baboon in the upright posture. Independent and dependent axes have been interchanged so that the formats are similar to previously published plots. (From Ref. 23.)

change in regional ventilation that alters local alveolar O_2 pressures may invoke local changes in perfusion through hypoxic pulmonary vasoconstriction. Hence, the studies may overestimate the effect of gravity on perfusion heterogeneity in a single posture.

The important findings of this baboon study are (1) pulmonary blood flow is heterogeneously distributed in the upright primate model, (2) the relative contribution of gravity to pulmonary perfusion heterogeneity is similar in supine and prone primates compared with other laboratory animals, (3) gravity plays a greater role in perfusion heterogeneity when animals are upright, and (4) although important, gravity remains a secondary determinant of regional pulmonary blood flow heterogeneity in the upright primate.

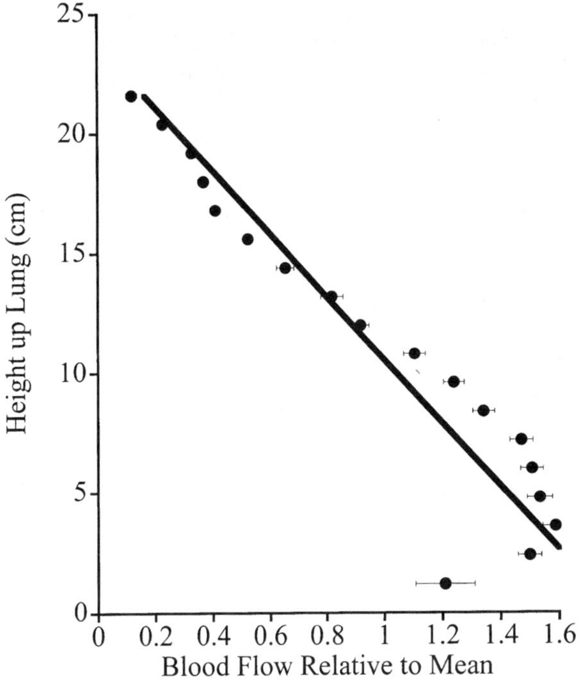

Figure 5 Height up the lung (cm) vs. blood flow relative to mean. Data are taken from Figure 4 and averaged within isogravitational planes.

To examine the hypothesis that vasoactivity constitutes a significant mechanism for explaining the marked isogravitational heterogeneity in the erect baboon, a study was performed to determine if the pulmonary blood flow heterogeneity would be altered by infusion of prostacycline (PGI2), a potent pulmonary vasodilator. Glenny et al. (27) found that despite heterogeneous perfusion, active regulation of regional blood flow is not required for efficient gas exchange in the normal lung when breathing air. However, with pathology or hypoxic environments, hypoxic pulmonary vasoconstriction must serve an important function in the normalization of lung function.

II. Vascular Structure Dependence

The variation in flow across the experimental perturbations can be assessed to determine the relative role of structure vs. the perturbation variable (such as posture, acceleration, or exercise level) in influencing pulmonary blood flow distribution. The component of variation due to structure can be thought of as variation arising from a biological pattern that endured across the experimental alterations. The overall variance of pulmonary blood flow across the individual regions and the experimental states is divided into the variance due to (1) a component due to spatial location of the piece within the lung and (2) a component due to experimental state, variation over time, and methodological noise. The component of variation due to spatial location can be thought of as variation arising from a biological pattern common to the states being considered. The second component represents the changeable portion of flow across the experimental states.

Regional blood flow remains relatively unchanged when the animal is exposed to change or stress of various types. The considerable stability under these conditions is demonstrated by examining the variation in flow and partitioning this variation among the various sources. During exercise in thoroughbred horses, the variance due to vascular structure contributed 70% of the total variance among all regions over all four states from rest to gallop, while the similarity among trot, canter, and gallop contributed 81.2% (16). Thus, exercise and methodological noise accounted for less than 30% of the redistribution in pulmonary blood flow.

Pulmonary blood flow variance has also been studied in dogs with unilateral alveolar hypoxia (28) by alteration of left lung inspired $F_{I_{O_2}}$ over values of 1.0, 0.09, and 0.03. The coefficient of variation (CV) of blood flow distribution increased in the hypoxic lung and was unchanged in the hyperoxic lung. Most of the variance in flow (91 to 94%) in the hyperoxic lung was attributable to structure compared with 74 to 79% in the hypoxic lung. Hypoxic vasoconstriction alters the regional distribution of blood flow in the hypoxic lung.

A method of stressing the role of structure in determining pulmonary blood flow distribution is to apply a strong acceleration stress. If structure has a minimal role, it would be expected that acceleration would have a profound influence on the variance of pulmonary blood distribution (see Chapter 2). Pigs have been exposed to $-1\,G_X$, $-2\,G_X$, and $-3\,G_X$ (dorsal to ventral direction) and pulmonary blood flow spatial distribution was measured with fluorescent microspheres at each G level (21). Analysis of variance showed that structure accounted for 81% of the total variance, indicating that acceleration had minimal effect on regional perfusion variance. A related study of pigs exposed to $+G_Z$ direction (cranial to caudal direction) demonstrated that 60.6% of the variance resulted from acceleration (22). The overall acceleration stress was much greater (up to $+9\,G_Z$) in the second study and the vector was cranial to caudal, resulting in a lesser structure dependence than the $-G_X$ study. It is also interesting to note that there was only a minimal effect of gravity on pulmonary blood flow distribution when the chest wall and abdomen configuration were held constant with an anti-G suit. The apparent shift of blood flow with anti-G–suit inflation is due to minimizing the effect of acceleration on the configuration of the chest wall, diaphragm, and abdominal contents.

A. Stability over Time

Regional pulmonary blood flow in awake dogs remains stable across days. High-flow pieces are always high flow, and low-flow pieces are always low flow. The r^2 between regional flow measurements obtained daily over five consecutive days averages 0.865. Hence, 87% of the variability in piece flow is explained by the regional flow on another day, suggesting that the mechanisms determining regional blood flow are relatively fixed over time in lung free of pathology. Whereas regional perfusion can vary about a local mean, the displacement from this mean is limited. Correlation of regional flows is, therefore, very strong, regardless of the time interval between observations.

Despite the long time interval between microsphere injections, the temporal variations in regional perfusion are not random. Although the largest component of perfusion heterogeneity is invariant over time, temporal changes in regional flow do exist, but even this variability demonstrates order. Utilizing cluster analysis, Glenny et al. (29) demonstrated that the temporal variability of perfusion showed strong local correlation. That is, adjacent regions of lung gained or lost fractional perfusion in concert with each other. These observations were made based on perfusion labels injected into dogs at 20-min intervals.

B. Fractal Analysis

A fractal structure or fractal process can be defined as having a characteristic form that remains constant over a wide range of scales. This is the quality of self-similarity, also termed scale independence.

A schematic representation of a fractal bifurcating vascular tree is shown in Figure 6. The asymmetry is maintained through each generation. The relative distribution of flow at any bifurcation is similar to the distribution at subsequent bifurcations. Note that with this constant asymmetry, the heterogeneity of flow distribution increases with each generation.

Heterogeneity of regional blood flow in an organ can be characterized by measuring the coefficient of variation (CV = SD/mean) of the regional flows when the organ is divided into a number of pieces. While lung pieces cut by rectilinear coordinates cannot exactly isolate vascular bifurcations diagrammed in Figure 6, they do represent an estimate of heterogeneity that will be determined by piece size.

When the heterogeneity of organ blood flow is characterized by this means, the calculated spatial CV (CV_S) is dependent on the size of the sampled pieces. If the blood flow in each of four pieces of the lung is measured, one can obtain

                                                          Hlastala et al.

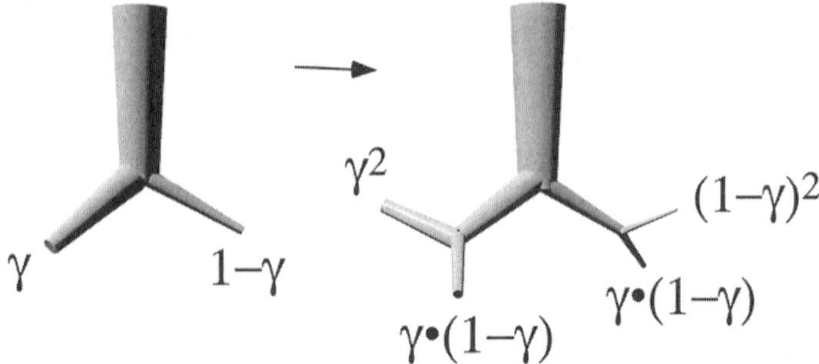

Figure 6   Fractal bifurcation pattern of a vascular tree. At each bifurcation, the daughter branches have relative flows of γ and $1 - \gamma$. The next generation is similarly bifurcated.

the mean, SD, and hence CV of flow in the lung. If these same pieces are progressively subdivided, then for 8, 16, 32, 64, or more regions, the mean remains constant but the estimate of SD and CV increases. Even after appropriate corrections for experimental error, the largest estimate of CV_S will be obtained from the finest subdivisions of the lung.

The studies discussed above were obtained using lung pieces of approximately 2 cm^3. The magnitude of the heterogeneity is dependent on the scale of resolution used to examine the local blood flow. As progressively smaller lung regions are examined, the observed variability in perfusion increases. Over the range of observations that have been made, this scale-dependent heterogeneity has been well characterized by a fractal model. As the volume of observation decreases, the measured perfusion heterogeneity increases in a linear fashion on a log–log plot (30,31). In large and medium-sized mammals, this fractal relationship remains linear down to lung pieces at the $\sim$2-cm^3 level of scale.

Glenny and Robertson (31) showed that regional distribution of pulmonary blood flow in dogs can be characterized by a fractal dimension. Described first by Mandelbrot (32), a fractal dimension is measured by a log–log plot of the coefficient of variation as a function of increasing piece size (Fig. 7). If a system is fractal, this plot will have a constant slope. Fractal dimension (D_S) is equal to $1 -$ slope of this plot. The more random the relationship among adjacent pieces, the higher the fractal dimension value. With this approach to data analysis, fractal dimension approaches 1 as regional perfusion becomes more spatially correlated (i.e., neighboring pieces will have more similar flows), whereas fractal dimension of 1.5 defines a totally random system. Unlike the CV, the fractal dimension does not change with changes in spatial resolution (i.e., fractal dimension is a scale-independent measure of scale-dependent heterogeneity). Values of fractal dimension have been determined for pulmonary blood flow distribution in several spe-

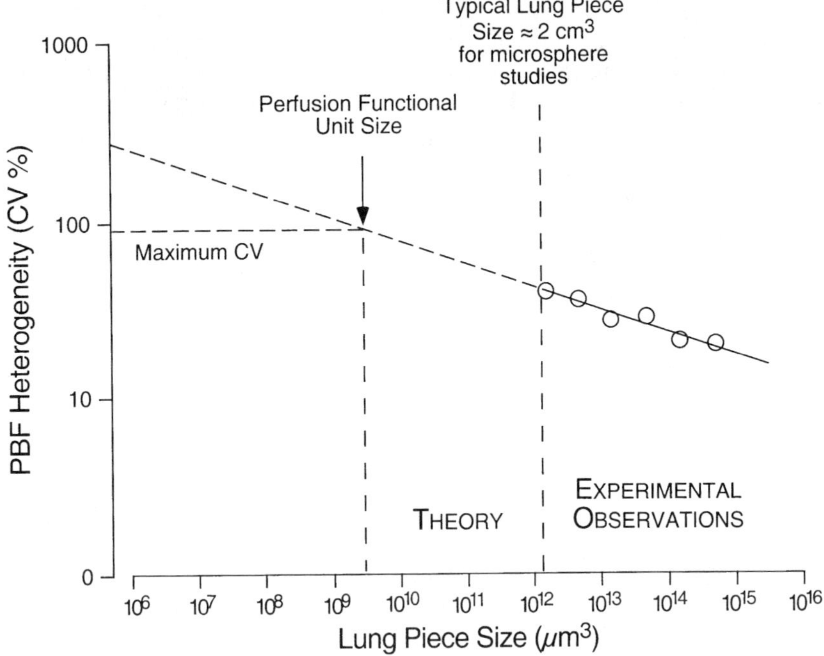

Figure 7 Coefficient of variation (CV) of pulmonary blood flow vs. lung piece size. The fractal dimension (D_S) is equal to 1 − slope of the regression line through the data points.

cies, e.g., dogs 1.09 ± 0.02 (31), 1.22 ± 0.08 (33), or 1.15 to 1.20 (34), and sheep 1.14 ± 0.09 (35). The fractal dimension remained constant with increasing pulmonary blood flow in an in situ, pump-perfused sheep model, despite a decrease in CV at high flows (35). Additionally, no change in fractal dimension was observed from rest to exercise in dogs (10). However, changes in fractal dimension have been observed by Deem et al. (36) with hemodilution. In thoroughbred horses, Sinclair et al. (37) found a mean resting fractal dimension of 1.18, but fractal dimension decreased from 1.216 at trot to 1.173 at gallop.

Even higher resolution methods for measuring regional perfusion could demonstrate one of two possibilities. Either perfusion heterogeneity will continue to increase in a linear fashion down to alveolar capillary level, or it will reach a level of scale where it becomes uniform within regions (38). At a piece size of ~2 cm³, perfusion heterogeneity, as measured by the coefficient of variation is 0.45 in dogs. If the fractal relationship is extrapolated to an acinar volume of ~1 mm³ (39), the observed perfusion heterogeneity would increase by a factor of 3. This large variability in blood flow has important implications for gas exchange and the mechanisms matching ventilation and perfusion.

Using a specially designed imaging cryomicrotome, Glenny et al. (40) studied the rat lung with a sampling size of 0.53 mm^3 and found that pulmonary blood flow heterogeneity (CV = 0.310 ± 0.043) increases past the volume of an acinar unit in the rat and that the spatial distribution of pulmonary perfusion is similar between the rat and other species. Another study of rats using injected microspheres (41) and a sampling volume of 200 mm^3 found a CV of 0.113. Using the mean fractal dimension of 1.1 measured in rats by Glenny et al., the predicted CV at a 200 mm^3 volume would be 0.095, a figure reasonably close to that found by Kuwahira et al. (41).

Despite the much smaller scale of observation, blood flow heterogeneity in these rats is similar to that previously observed in larger awake laboratory animals. Using piece sizes of ~2 cm^3 (4000 times larger than in the rats), CVs of between 0.259 and 0.356 have been observed in larger animals (17,18,42). If the fractal relationship remains linear in these larger species, the CV of perfusion will approach 1.15 at the smallest sampling volume. Although this degree of variability in regional perfusion will strain the mechanisms matching ventilation and perfusion, they must manage, given the efficient gas exchange seen in these animals. Larger ventilatory units will compensate somewhat for the greater perfusion heterogeneity in larger animals.

Wagner et al. (43) recently demonstrated that flow patterns within neighboring alveolar regions vary independently of each other. Pulmonary capillary perfusion was analyzed from videomicroscopic recordings to determine flow switching characteristics among capillary segments in isolated, blood-perfused dog lungs. Within each alveolus, the rapid switching pattern was repetitive and was nonrandom (with fractal dimensions of 1.12 ± 0.04, mean ± SD). This self-similarity over time was unexpected in a network widely considered to be passive. Among adjacent alveoli, the relationship among the switching patterns was even more surprising, for there was virtually no relationship between the perfusion patterns. These findings demonstrate that the perfusion patterns in individual alveolar walls were independent of their adjacent neighbors. The lack of dependence among neighboring networks suggests an interesting characteristic: the failure of one alveolar-capillary bed would leave its neighbors relatively unaffected. Their study suggests that, whereas high-flow regions are near other high-flow regions, blood flow at the capillary level varies independently within alveoli. One potential explanation is that the geometry of larger feeder vessels determines the spatial correlation of blood flow, whereas other mechanisms determined temporal fluctuations in blood flow at the alveolar level.

The fractal characterization of blood flow, at a minimum, provides us with a significant tool: a model or means of describing the contribution of sample piece size to the calculated heterogeneity of flow. Thus, the results of different studies of flow heterogeneity that use different sample piece sizes can be compared with each other using the estimate of the fractal D_S. Fractal dimension therefore affords a measure of the variability in pulmonary blood flow that can be used for comparison among interventions, species, and laboratories. It is a scale-independent measure of spatial organization of pulmonary blood flow.

West et al. (44) have used teleologic arguments and mathematical models to provide support for fractal vascular systems. They noted that fractal vascular trees are conserved across a wide range of species and deduced that such systems must offer significant advantages to the organisms. They argue that fractal vascular systems afford three important physiological advantages to an organism. First, fractal vascular trees are space filling and provide the maximum surface area for gas exchange. Through successive branchings, the pulmonary artery transforms into a sheet of capillaries covering 90% of the alveolar surface area. Second, fractal structures minimize the energy needed to deliver oxygen to tissues and to build the vascular tree. Third, because a fractal tree fills a space with the least amount of volume, fractal vascular systems minimize the amount of blood needed to deliver nutrients or exchange gases.

Another advantage of fractal systems is that they can be constructed with recursive algorithms. By repeatedly applying the same branching rule, a complicated and robust vascular tree can be generated from a single simple recursive rule. If this rule is encoded in the DNA, the vascular tree can be constructed from a small section of DNA. Hence, genetic code does not need every alveolar–capillary unit, thereby conserving DNA.

Along with these advantages, a fractal vascular system incurs one significant problem. Because of the asymmetries in branching, perfusion distributions become heterogeneous. The heterogeneous distributions of blood flow must somehow be matched to function so that gases are exchanged efficiently. While feedback mechanisms must play an important role in matching ventilation and perfusion in the pathological lung, feedback mechanisms have not been shown to play an important role in gas exchange in the normal lung during normoxia (27). This suggests that the fractal vascular and airway trees are anatomically matched (45).

III. Microgravity

All prior studies of pulmonary blood flow distribution using high-resolution methods have been performed in a setting of gravity. The optimal approach to determine the relative contributions of structure of gravity is to remove the influence of gravity. To quantitatively assess pulmonary blood flow heterogeneity with fluorescent microspheres in microgravity, anesthetized pigs were flown on the National Aeronautics and Space Administration (NASA) KC-135 aircraft

(46). A series of parabolas created alternating weightlessness and 1.8 G conditions. Fluorescent microspheres of varying colors were injected into the pulmonary circulation to mark regional blood flow during different postural and gravitational conditions. Perfusion heterogeneity did not change significantly during weightlessness compared with normal and increased gravitational forces. Regional blood flow to each lung piece changed little, despite alterations in posture and gravitational forces. Both gravity and the geometry of the pulmonary vascular tree influence regional pulmonary blood flow. However, the structure of the vascular tree was the primary determinant of regional perfusion in these animals.

One potential concern with transient microgravity studies is whether the unsteady-state nature of the microgravity (over 25 s) allows for stability of blood flow distribution. The temporal response of the pulmonary capillary bed to rapid changes in flow has been studied by Jaryszak et al. (47) using a pump-perfused lung preparation. These investigators showed that pulmonary capillary recruitment reached a new steady state within 4 s of a step change in flow. The microspheres were injected after the 4-s time for stability in the microgravity study (46).

In microgravity, there is no relationship between pulmonary blood flow and the ventral-to-dorsal direction. Considerable heterogeneity is apparent within any plane perpendicular to the dorsal-to-ventral axis (Fig. 8). The coefficient of variation of perfusion was similar under all gravitational conditions, varying between an average of 0.67, 0.64, and 0.66 in the prone posture during 0, 1, and

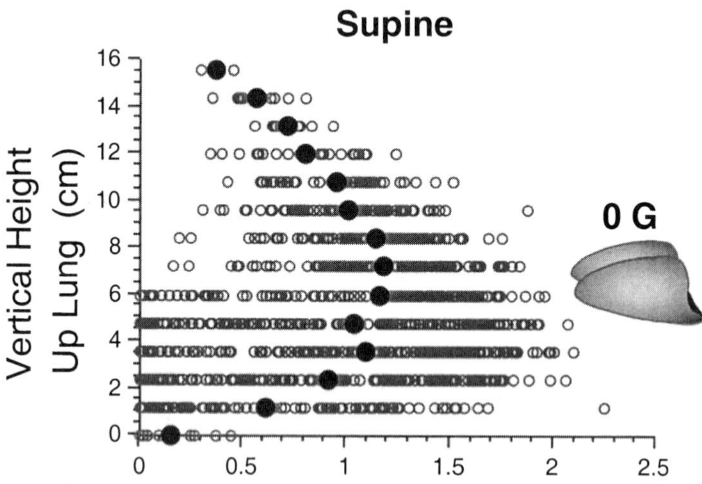

Figure 8 Vertical distribution of blood flow during 0 G conditions in one animal in the supine posture.

1.8 G, respectively, and 0.71, 0.71, and 0.77 in the supine posture during 0, 1, and 1.8 G conditions. During weightlessness, regional blood flows between supine and prone postures were highly correlated, with an r^2 value of 0.84.

When the spatial resolution of these data are averaged into isogravitational planes comparable to those of previous studies (1), they appear quite similar (compare Figs. 8 and 9). Without the increased resolution of these measurements, it is not possible to observe the underlying pattern of perfusion heterogeneity. Indirect data obtained from humans in prolonged weightlessness by Prisk et al. (48) have demonstrated that the majority of perfusion heterogeneity remains in humans in microgravity as well.

When averaged within isogravitational planes, blood flow increases and then decreases down the lung (in the ventral-to-dorsal direction) (see Fig. 9). This pattern persists regardless of posture or gravitational force. Traditionally, the increase in perfusion has been attributed to the hydrostatic gradient down the lung that distends dependent vessels and that increased interstitial pressures cause vessels to collapse in dependent regions (49). The observation that this pattern of increasing and then decreasing perfusion persists during weightlessness sug-

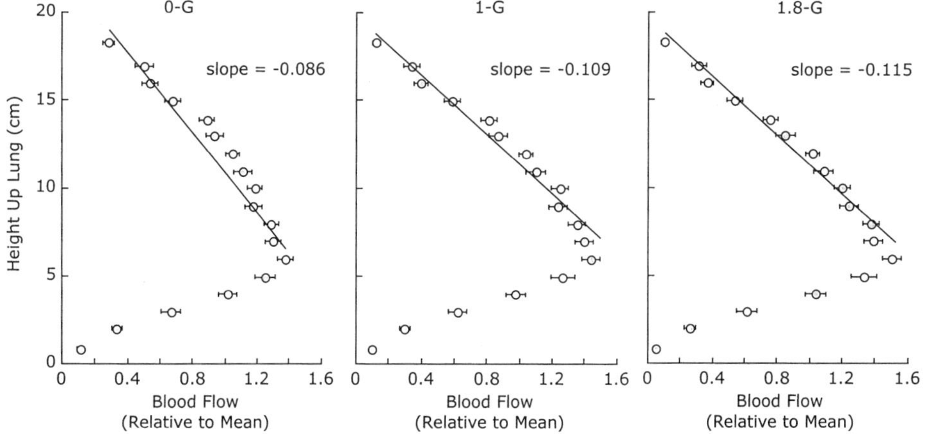

Figure 9 Vertical distribution of blood flow during 0, 1, and 1.8 G conditions in one animal (different from the animal shown in Fig. 8) in the supine posture. Slopes were determined from a least-squares linear fit to blood flow as a function of height up the lung. Traditional zone 4 regions were excluded and only points with increasing blood flow down the lung were fit. Dependent- and independent-variable axes have been interchanged so that the formats are similar to previously published plots. Vertical gradient of perfusion was steeper at higher gravitational forces. Note that the general pattern of blood flow increasing and then decreasing down the lung (zones 3 and 4) persists during weightlessness. (From Ref. 46.)

gests that vertical pressure gradients cannot explain this phenomenon in normal gravitational conditions. The relatively large extent of zone 4 conditions we observed (see Fig. 9) probably related to the thin dorsocaudal lung in the pig, which is consistently compressed in the supine position under anesthesia. In these experiments, the lungs were hyperinflated before each microsphere injection. It may be that this lung region remained underinflated relative to its state in awake prone animals. However, the similarity of the distribution of pulmonary blood flow at 0.0 G, +1.0 G, and +1.8 G is remarkable.

IV. Summary

The single most important factor determining perfusion heterogeneity in the normal lung is the scale-dependent heterogeneity of flow that arises as a consequence of the progressive branching of the pulmonary vasculature. This property of increasing measured heterogeneity with increasing resolution is characteristic of all progressively branching distribution systems, and permits description of this branching characteristic by a fractal dimension. Recent high-resolution studies of perfusion heterogeneity in the lung during transient microgravity have demonstrated that overall perfusion heterogeneity is nearly unchanged by the transition from 1 G to microgravity, although significant macroscopic alterations can be demonstrated at higher G forces. While both temporal changes and exercise-associated changes in pulmonary perfusion are demonstrable, high-flow regions remain high, and low-flow regions remain low in many different experimental interventions. Hence, the conclusion remains that pulmonary blood flow heterogeneity is determined primarily by anatomical features of the pulmonary vasculature, with a minimal influence from both gravity and vasomotion.

References

1. West J, Dollery C. Distribution of blood flow and ventilation-perfusion ratio in the lung, measured with radioactive CO_2. J Appl Physiol 1960; 15:405–410.
2. West JB, Dollery CT. Distribution of blood flow and the pressure-flow relations of the whole lung. J Appl Physiol 1964; 20:175–183.
3. West JB, Dollery CT, Naimark A. Distribution of blood flow in isolated lung: Relation to vascular and alveolar pressure. J Appl Physiol 1964; 19:713–724.
4. Reed J, Wood E. Effect of body position on vertical distribution of pulmonary blood flow. J Appl Physiol 1970; 28:303–311.
5. Greenleaf JF, Ritman EL, Sass DJ, Wood EH. Spatial distribution of pulmonary blood flow in dogs in left decubitus position. Am J Physiol 1974; 227:230–244.
6. Beck KC, Rehder K. Differences in regional vascular conductances in isolated dog lungs. J Appl Physiol 1986; 61:530–538.

7. Beck KC, Vettermann J, Rehder K. Gas exchange in dogs in the prone and supine positions. J Appl Physiol 1992; 72:2292–2297.

8. Melsom MN, Flatebø T, Kramer-Johansen J, et al. Both gravity and non-gravity dependent factors determine regional blood flow within the goat lung. Acta Physiol Scand 1995; 153:343–353.

9. Melsom M, Kramer-Johansen J, Flatebø T, Müller C, Nicolaysen G. Distribution of pulmonary ventilation and perfusion measured simultaneously in awake goats. Acta Physiol Scand 1997; 159:199–208.

10. Parker J, Ardell J, Hamm C, Barman S, Coker P. Regional pulmonary blood flow during rest, tilt, and exercise in unanesthetized dogs. J Appl Physiol 1995; 78:838–845.

11. Glenny RW, Lamm WJE, Albert RK, Robertson HT. Gravity is a minor determinant of pulmonary blood flow distribution. J Appl Physiol 1991; 71:620–629.

12. Glenny RW, Polissar L, Robertson HT. Relative contribution of gravity to pulmonary perfusion heterogeneity. J Appl Physiol 1991; 71:2449–2452.

13. Nicolaysen G, Shepard J, Onizuka M, Tanita T, Hattner RS, Staub NC. No gravity-independent gradient of blood flow distribution in dog lung. J Appl Physiol 1987; 63:540–545.

14. Hakim TS, Lisbona R, Dean GW. Gravity-independent inequality in pulmonary blood flow in humans. J Appl Physiol 1987; 63:1114–1121.

15. Hakim TS, Dean GW, Lisbona R. Effect of body posture on spatial distribution of pulmonary blood flow. J Appl Physiol 1988; 64:1160–1170.

16. Bernard S, Glenny R, Erickson H, et al. Minimal redistribution of pulmonary blood flow with exercise in racehorses. J Appl Physiol 1996; 81:1062–1070.

17. Hlastala M, Bernard S, Erickson H, et al. Pulmonary blood flow distribution in standing horses is not dominated by gravity. J Appl Physiol 1996; 81:1051–1061.

18. Walther S, Domino K, Glenny R, Polissar N, Hlastala M. Pulmonary blood flow distribution in sheep: Effects of anesthesia, mechanical ventilation and change in posture. Anesthesiology 1997; 87:335–342.

19. Altemeier W, Robertson H, Glenny R. Pulmonary gas-exchange analysis by using simultaneous deposition of aerosolized and injected microspheres. J Appl Physiol 1998; 85:2344–2351.

20. Mure M, Domino K, Lindahl S, Hlastala M, Altemeier W, Glenny R. Regional ventilation-perfusion distribution is more uniform in the prone position. J Appl Physiol 2000; 88:1076–1083.

21. Hlastala M, Chornuk M, Self D, et al. Pulmonary blood flow redistribution by increased gravitational force. J Appl Physiol 1998; 84:1278–1288.

22. Chornuk M, Bernard S, Burns J, et al. The effects of inertial load and countermeasures on the distribution of pulmonary blood flow. J Appl Physiol 2000; 89:445–457.

23. Glenny R, Bernard S, Robertson H, Hlastala M. Gravity is an important but secondary determinant of regional pulmonary blood flow in upright primates. J Appl Physiol 1999; 86:623–632.

24. Glenny RW. Spatial correlation of regional pulmonary perfusion. J Appl Physiol 1992; 72:2378–2386.

25. Kay M. Comparative morphologic features of the pulmonary vasculature in mammals. Am Rev Respir Dis 1983; 128:S52–S57.

26. Herman D, Homer L, Horwitz D. Effects of sedation and posture on pulmonary gas exchange in the baboon. J Appl Physiol 1971; 30:498–501.

27. Glenny R, Bernard S, Robertson H, Hlastala M. Vasomotor regulation plays a minor role in perfusion heterogeneity and gas exchange in normal lungs during normoxia. J Appl Physiol 2000; 89:2263–2267.

28. Mann CM, Domino KB, Walther SM, Glenny RW, Polissar NL, Hlastala MP. Redistribution of pulmonary blood flow during unilateral hypoxia in prone and supine dogs. Anesthesiology 1998.

29. Glenny RW, Polissar NL, McKinney S, Robertson HT. Temporal heterogeneity of regional pulmonary perfusion is spatially clustered. J Appl Physiol 1995; 79:986–1001.

30. Bassingthwaighte J, King R, Roger S. Fractal nature of regional myocardial blood flow heterogeneity. Circ Res 1989; 65:578–590.

31. Glenny RW, Robertson HT. Fractal properties of pulmonary blood flow: Characterization of spatial heterogeneity. J Appl Physiol 1990; 69:532–545.

32. Mandelbrot B. The Fractal Geometry of Nature. New York: Freeman, 1983.

33. Barman S, McCloud L, Catravas J, Ehrhart I. Measurement of pulmonary blood flow by fractal analysis of flow heterogeneity in isolated canine lungs. J Appl Physiol 1996; 81:2039–2045.

34. Parker JC, Cave CB, Ardell JL, Hamm CR, Williams SG. Vascular tree structure affects lung blood flow heterogeneity simulated in three dimensions. J Appl Physiol 1997; 83:1370–1382.

35. Caruthers S, Harris T. Effects of pulmonary blood flow on the fractal nature of flow heterogeneity in sheep lungs. J Appl Physiol 1994; 77:1474–1479.

36. Deem S, Hedges R, McKinney S, Polissar N, Alberts M, Swenson E. Mechanisms of improvement in pulmonary gas exchange during isovolemic hemodilution. J Appl Physiol 1999; 87:132–141.

37. Sinclair S, McKinney S, Glenny R, Bernard S, Hlastala M. Exercise alters fractal dimension and spatial correlation of pulmonary blood flow in the horse. J Appl Physiol 2000; 88:2269–2278.

38. Bassingthwaighte J. Physiologic heterogeneity: Fractals link determinism and randomness in structures and functions. News Physiol Sci 1988; 3:5–10.

39. Mercer R, Crapo J. Three-dimensional reconstruction of the rat acinus. J Appl Physiol 1987; 63.

40. Glenny R, Bernard S, Robertson H. Pulmonary blood flow remains fractal down to the level of gas exchange. J Appl Physiol 2000; 89:742–748.

41. Kuwahira I, Moue Y, Ohta Y, Mori H, Gonzalez N. Distribution of pulmonary blood flow in conscious resting rats. Respir Physiol 1994; 97:309–321.

42. Glenny R, McKinney S, Robertson H. Spatial pattern of pulmonary blood flow distribution is stable over days. J Appl Physiol 1997; 82:902–907.

43. Wagner WW Jr, Todoran T, Tanabe N, et al. Pulmonary capillary perfusion: Intraalveolar fractal patterns and interalveolar independence. J Appl Physiol 1999; 86:825–831.

44. West G, Brown J. The fourth dimension of life: Fractal geometry and allometric scaling of organisms. Science 1999; 284:1677–1679.
45. Weibel E. Fractal geometry: A design principle for living organisms. Am J Physiol 1991; 261 (Lung Cell Mol Physiol 5):L361–L369.
46. Glenny R, Lamm W, Bernard S, et al. Redistribution of pulmonary perfusion during weightlessness and increased gravity. J Appl Physiol 2000; 89:1239–1248.
47. Jaryszak E, Baumgartner WJ, Peterson A, Presson RJ, Glenny R, Wagner WJ. Measuring the response time of pulmonary capillary recruitment to sudden flow changes. J Appl Physiol 2000; 89:1233–1238.
48. Prisk GK, Guy HJ, Elliott AR, West JB. Inhomogeneity of pulmonary perfusion during sustained microgravity on SLS-1. J Appl Physiol 1994; 76:1730–1738.
49. Hughes JMB, Glazier JB, Maloney JE, West JB. Effect of lung volume on the distribution of pulmonary blood flow in man. Respir Physiol 1968; 4:58–72.

8

Pulmonary Gas Exchange

JOHN B. WEST

University of California, San Diego
La Jolla, California

I. Normal Gravity

A. Early Measurements

Some of the early studies of the effects of gravity on pulmonary gas exchange were alluded to in Chapter 1. Much of the initial information came from sampling gas in different regions of the lung and this work is briefly reviewed here.

According to Bjorkman (1), the first attempt to measure the function of one part of the lung was made in 1871 by Wolffberg. He used a catheter with two concentric tubes, a design suggested by Pfluger. The outer tube had a thin wall at its distal end so that it could be inflated to occlude a bronchus, and gas could then be aspirated from the lung through the inner tube. The catheter was used to study the alveolar gas composition in dogs. Loewy and Schrotter (2) have the distinction of being the first to attempt measurements of regional lung function in humans. They used a modified Pfluger catheter and were also mainly concerned with the composition of alveolar gas.

A major step forward was the introduction of the double lumen broncho-scope by Jacobaeus and his colleagues in 1932. Various improvements in the catheter were made and the most satisfactory was that described by Carlens (3). The catheter consisted of two relatively thin-walled rubber tubes mounted side by side. The upper part of the catheter was curved to fit the pharynx and trachea, and one tube was extended in a curve to fit the left main bronchus. There were inflatable rubber cuffs for both the trachea and left main bronchus. A number of studies were made with this catheter showing that gravity had an influence on pulmonary gas exchange. For example, several investigators measured the function of one lung when the subject was first in the right and then in the left lateral position, and they reported that the oxygen uptake increased when the lung was dependent (1,4). Additional studies were made of hypoxic pulmonary vasoconstriction. For example, Rahn and Bahnson (5) in dogs, and Himmelstein et al. (6) in humans, demonstrated a reduced perfusion in the lung that was made hypoxic by lowering the inspired oxygen concentration.

A disadvantage of bronchospirometry is that large regions of the lung are pooled together because only one lung can be compared with another. It is true that Mattson and Carlens (7) described a triple-lumen catheter that enabled the ventilation and oxygen uptake of the right upper lobe to be recorded independently of the remainder of the right lung, but in practice this technique proved very difficult to carry out. Therefore, it was a substantial advance when investigators began to use fine catheters that were introduced into various lobes of the lung.

The first measurements were made by Martin et al. (8) when they aspirated alveolar gas through fine radiopaque catheters that had been guided into the bronchi of the upper and lower lobes of the lung under fluoroscopic control. The subjects were normal volunteers and patients undergoing treatment for pulmonary tuberculosis in a sanatorium. The results showed that the respiratory exchange ratios of the gas samples from the upper-lobe bronchi were significantly higher than those of the lower-lobe samples when the subject was in the erect position, but there was not a systematic difference with the subject in the supine position. A second study (9) showed that the differences in respiratory exchange ratios were not correlated with changes in resting lung volumes, and the authors concluded that the distribution of ventilation and/or perfusion must have been affected by alterations in body position.

A similar technique was employed by Rahn et al. (10) in dogs. The animals were anesthetized and polyethylene tubes of 1.25-mm outside diameter were placed in various lobar bronchi under direct vision by threading the catheters through 2-mm-diameter brass tubes that were withdrawn afterward. Simultaneous gas samples were aspirated and analyzed. The results showed that the respiratory exchange ratio was almost invariably higher in the upper lobe than the lower for both lungs in the supine dog, and that the differences were exaggerated by tilting the animal into the erect position. The most likely explanation was a change in the distribution of pulmonary blood flow.

Additional information about the effects of gravity on pulmonary gas exchange come from measurements of physiological dead space by Riley et al.

(11) and Bjurstedt et al. (12). Both groups demonstrated appreciable increases in physiological dead space in humans in the upright position, particularly after some minutes with the subject standing still. Gerst et al. (13) showed that physiological dead space increased by some 60% in anesthetized dogs who were bled, whereas venous mixtures fell slightly compared with the control values. Barr (14) reported that a large alveolar dead space develops in humans during exposure to increased acceleration with the vector passing from the head to the feet. All of the above results can be explained by the effects of gravity on the distribution of pulmonary blood flow.

B. Distribution of Ventilation–Perfusion Ratios in the Upright Human Lung

As discussed in Chapters 4 and 6, both the distribution of ventilation and blood flow in the normal upright human lung are influenced by gravity, and both increase from apex to base of the lung. However, the changes in the distribution of blood flow are greater than those for ventilation, and therefore the ventilation–perfusion ratio is relatively high at the apex of the upright lung and low at the base (Fig. 1). Since the ventilation–perfusion ratio determines the gas exchange

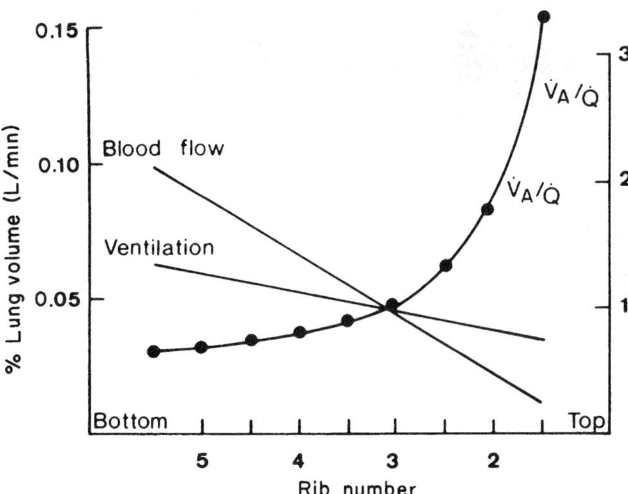

Figure 1 Change in ventilation–perfusion ratio ($\dot{V}_A/\dot{Q}$) from the base to apex of the normal upright human lung. The topographical differences of blood flow and ventilation as measured with radioactive gases, and from which the ventilation–perfusion ratio was derived, are also shown. (Modified from Ref. 15.)

in any region of the lung, the inevitable result is that there are topographical differences of pulmonary gas exchange.

Figure 2 shows how the regional differences of ventilation and blood flow, and therefore ventilation–perfusion ratio, can be used to determine the topographical pattern of gas exchange in the upright human lung (15). The ventilation and blood flow in each of nine imaginary slices of the lung were determined from radioactive gas measurements. The nine circles in Figure 2 show the amounts of ventilation and blood flow at each level, and the regional gas exchange can be read off a ventilation–perfusion line joining a normal mixed venous point (P_{O_2} 40, P_{CO_2} 45 mmHg) and inspired gas point (P_{O_2} 149, P_{CO_2} 0 mmHg).

Figure 3 shows the results of this analysis. Although the gas exchange was calculated for each of the nine imaginary slices, only the values for the uppermost and lowermost slices are shown to avoid undue complexity. The first column of

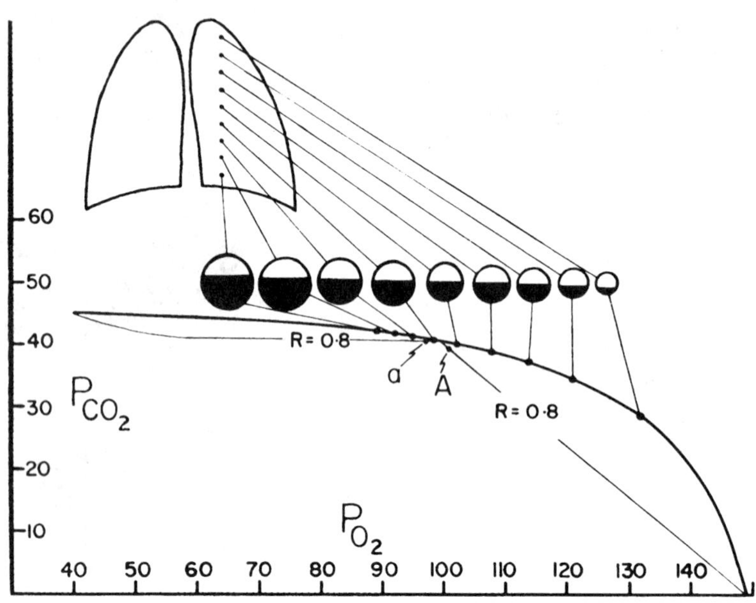

Figure 2 Oxygen–carbon dioxide diagram showing a $\dot{V}_A/\dot{Q}$ line (curving down to right-hand lower corner), which gives all the possible O_2 and CO_2 partial pressures in a lung inspiring air and having mixed venous O_2 and CO_2 partial pressures of 40 and 45 mmHg, respectively. The position of each of the nine slices on this line is shown. Areas within the large circles indicate the relative ventilations (white) and blood flows (black), and therefore ventilation–perfusion ratio of each slice. Mixed alveolar (A) and mixed arterial (a) points can be seen on the respective R = 0.8 lines for gas and blood. (From Ref. 15.)

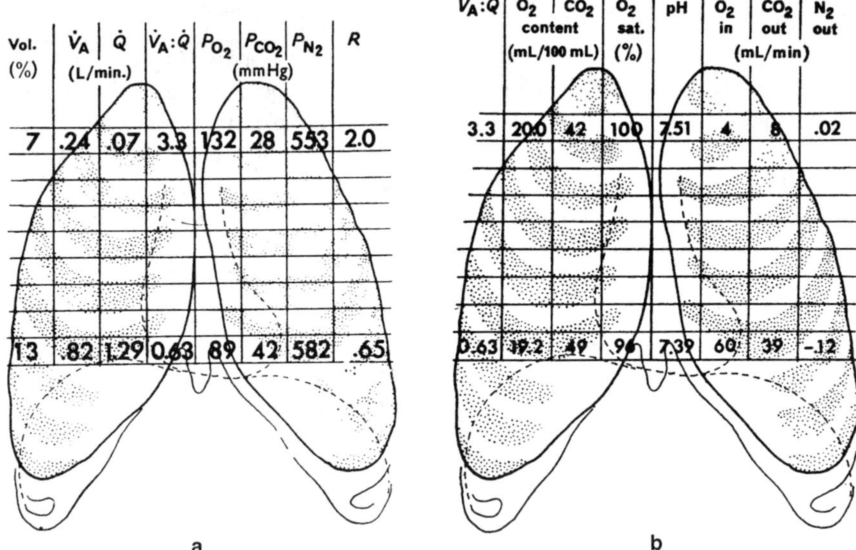

Figure 3   In this figure, the lung is divided into nine imaginary horizontal slices, but only the figures for the uppermost and lowermost slices are shown. (a) Table shows relative lung volume (percentage of total examined), ventilation ($\dot{V}_A$), blood flow ($\dot{Q}$), ventilation–perfusion ratio ($\dot{V}_A/\dot{Q}$), gas partial pressures (P_{O_2}, P_{CO_2}, P_{N_2}), and respiratory exchange ratio (R) in each slice. (b) Table shows ventilation–perfusion ratio ($\dot{V}_A/\dot{Q}$), end-capillary blood values (concentrations of O_2 and CO_2, O_2 saturation, pH), and local gas exchange (O_2, CO_2, and N_2). (From Ref. 54.)

Figure 3a shows that the volume of the uppermost slice is less than that of the lowermost, simply because of the geometry of the lung. The next two columns show that both ventilation and blood flow increase down the upright lung, but because the differences of blood flow are much greater than those for ventilation, the ventilation–perfusion ratio (column 4) is much higher at the top than the bottom of the lung. (The normal ventilation–perfusion ratio is about 0.8.) These differences in ventilation–perfusion ratio determine the gas exchange in the remaining columns of Figure 3a, and all the columns of Figure 3b. Note that the P_{O_2} is more than 40 mmHg higher at the top of the lung than the bottom, and there are also appreciable differences in alveolar P_{CO_2}. The variation in P_{N2} can be thought of as by default, since the sum of the alveolar gas partial pressures must be equal to the total barometric pressure. The alveolar partial pressures of O_2, CO_2, and N_2 can be used to calculate the regional respiratory exchange ratio

(R), and it can be seen that this is high at the top of the lung and low at the bottom.

Figure 3b extends these calculations to the blood phase (15). Here we are assuming that the partial pressures in the capillary blood draining from different regions of the lung are in equilibrium with the partial pressures in the gas phase. This is a reasonable assumption in the normal lung. Note that because the P_{O_2} is high at the top of the lung, the oxygen concentration of the effluent blood is greater at the top of the lung than the bottom. The reverse is the case for the CO_2 concentration. The oxygen saturation reflects the differences in oxygen concentration. The differences in pH are particularly interesting. Because the P_{CO_2} is so low at the top of the lung (because the ventilation–perfusion ratio is so high), and the base excess of the blood is the same in all regions of the lung, the calculated pH of the blood is much higher at the top of the lung. The next column shows that amounts of oxygen being taken in by each imaginary slice. The oxygen uptake at the top of the lung is very low, because the blood flow is so small. The arterial-venous difference for oxygen concentration is actually greater at the apex of the lung than the base, but the oxygen uptake is dominated by the blood flow term. The differences of CO_2 output are not as marked. Incidentally, the differences in O_2 uptake and CO_2 output reflect the fact that O_2 uptake is closely related to blood flow, and CO_2 output more closely related to ventilation. The last column in Figure 3b shows an interesting nitrogen cycle in the lung. Because of the much lower P_{N_2} at the top of the lung than the bottom, as shown in Figure 3a, nitrogen is eliminated from the mixed venous blood at the apex of the lung and absorbed at the bottom (15). Of course, the net nitrogen exchange is 0. Incidentally, undue reliance should not be placed on the exact values shown in Figure 3; the intent of this analysis is to show the general pattern.

C. Implications for Disease

Apical Localization of Pulmonary Tuberculosis

We saw in Chapter 1 that the German pathologist, Johannes Orth wrote a treatise on "Etiological and Anatomical Considerations of Phthisis" in which he argued the weight of the column of blood in the lung might cause anemia at the apex and thus contribute to the predilection of adult tuberculosis for that region. He added that the known high incidence of pulmonary tuberculosis in patients with pulmonary artery stenosis, which causes a reduction in pulmonary artery pressure, supported this hypothesis.

There are a number of clinical observations that favor the hypothesis that a region of lung with a low blood flow is particularly susceptible to tuberculosis infection. For example, the disease classically is more common in people of asthenic build who have long chests and therefore a particularly low blood flow at the lung apex. A correlation has been described between the degree of topographical inequality of blood flow and chest height (16), this conclusion being reached by an analysis of cardiogenic oscillations in expired gas. An increased incidence of the disease in patients with pulmonary stenosis has been remarked on by many pathologists over the years and has been particularly well documented by Auerbach and Stemmerman (17), Norris (18), and Abbott (19).

The exact opposite pattern is seen in patients with mitral stenosis who have an increased pulmonary artery pressure, and who have been shown to be relatively protected from pulmonary tuberculosis. For example, in a series of 300 cases of mitral stenosis, only 1 case (0.3%) of pulmonary tuberculosis was found, and in another series of 20,000 cases of pulmonary tuberculosis, there was only 1 case of mitral stenosis (20). These figures are consistent with a much lower prevalence of pulmonary tuberculosis in patients with mitral stenosis than in the general population at the time the series was collected (i.e., 1919). In this context, it is also of interest that several earlier studies, including that of Dock (21), showed that the incidence of pulmonary tuberculosis at the right apex of the lung exceeded that at the left apex. Remarkably, Laennec was aware of this as early as 1826 (22, p. 552)! The reduced incidence on the left side was attributed to the higher blood flow to the left apex. Dock argued that this could be explained by the leftward inclination of the main pulmonary artery. In fact, measurements of the clearance rate of radioactive carbon dioxide confirm that there is more blood flow to the left lung apex than the right on the average in normal upright subjects (23). Dock's explanation is probably correct.

It is probable that it is not the low blood flow per se that is responsible for the localization of tuberculosis, but the high P_{O_2} as shown in Figure 3a. Evidence to support this comes from studies in which part of the lung is supplied with large amounts of arterial blood via an appropriate anastomosis. These regions are then prone to develop tuberculosis infection in spite of their high blood flow (24). The importance of the local P_{O_2} for the growth of the mycobacterium bacillus was demonstrated in vitro by Kempner (25), who found a progressive fall in metabolism of the bacillus when the P_{O_2} was reduced below 100 mmHg. There is also evidence that a reduced P_{CO_2}, as is found at the apex of the upright lung (see Fig. 3a), favors the growth of tubercle bacilli (26).

Another explanation has been advanced for the increased localization of tuberculosis at the apex of the lung, this being the smaller lymph flow and therefore impairment of tissue clearance mechanisms (27). The authors point out that this explanation would be consistent with the tendency of other pulmonary diseases to attack the upper regions of the lung, including chronic pulmonary histoplasmosis and progressive massive fibrosis. However, against this theory is the fact, mentioned above, that in experimental studies where the lung is supplied with large amounts of arterial blood, there is an increased risk of tuberculosis.

High pH at the Lung Apex

As Figure 3a shows, the pH at the apex of the upright lung is remarkably high because of the reduced P_{CO_2} in that region. It is possible that this is a factor in the localization of metastatic calcification of the lung. This is an uncommon condition, associated with a variety of diseases, including destructive bone disease, chronic renal disease, and parathyroid tumors. Kaltreider et al. (28) and Mulligan (29) suggested that metastatic calcification of the lung had a preponderance for the upper lobes. Virchow (30) pointed out long ago that calcification occurs chiefly in those organs, such as the lung, stomach, and kidney, that excrete acid and are therefore likely to have alkaline conditions in their tissues. Holmes (31) has emphasized that the processes of extraosseous tissue uptake of calcium are very complicated and poorly understood, but it may be that the most alkaline regions of the lung will be most vulnerable. Figure 4 shows a striking example of metastatic calcification in a 27-year-old woman who had acute leukemia with

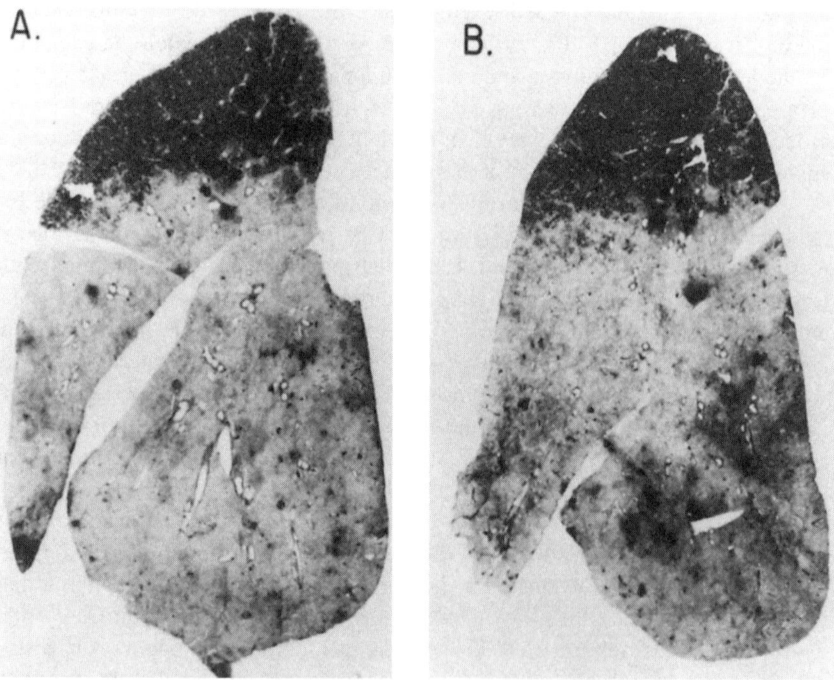

Figure 4 Thin lung sections from a 27-year-old woman with pulmonary metastatic calcification: (A) right lung, (B) left lung. Note the clear preponderance for the upper lobe, which might be explained by the relatively high pH there.

extensive bone destruction. The remarkable localization of the disease in the upper lobes is clearly seen.

II. Increased Acceleration

Pulmonary gas exchange during exposure to increased gravitational loading was discussed extensively in Chapter 2. As indicated there, the marked changes in the distribution of ventilation and blood flow cause severe impairment of gas exchange. In particular, acceleration in the head-to-foot direction causes compression of lung tissue and cessation of ventilation in the dependent regions of the lung, with the result that any blood flow is shunted through unventilated lung causing severe arterial hypoxemia. In effect, most of the pulmonary blood flow goes to the lower regions of the lung because of hydrostatic effects, but most of the ventilation goes to upper regions because of airway closure in the dependent zones, and only the middle section of the lung has any effective gas exchange! A particularly interesting situation occurs when the subject is exposed to acceleration while breathing a high concentration of oxygen. The ensuing atelectasis of the lower regions of the lung means that gas exchange remains impaired when the acceleration ceases (32). These topics have been considered at length in Chapter 2 and are not further reviewed here.

III. Microgravity

A. Distribution of Ventilation and Blood Flow

The efficiency of gas exchange in the lung depends on the matching of ventilation and blood flow. Therefore, it is pertinent first to summarize the effects of microgravity on the normal distributions of ventilation and blood flow in the lung.

As described in Chapter 4, the inhomogeneity of ventilation in the lung has been measured by both single-breath and multibreath nitrogen washouts (33–35). When the single-breath washout measurements were made after a vital capacity breath of oxygen, with an argon bolus inhaled at residual volume, a marked decrease in ventilatory inhomogeneity was found in microgravity, as evidenced by the significant reductions in cardiogenic oscillations, the slope of phase III, and the height of phase IV for both nitrogen and argon. This result was consistent with the known effect of gravity on the distribution of ventilation because of distortion of the lung by its weight. However, considerable ventilatory inhomogeneity remained, showing that nongravitational factors, such as diffusion and convection inhomogeneity in small lung units, play an important role.

By contrast, when multibreath nitrogen washouts were carried out at tidal volumes of approximately 700 mL, that is, near the normal tidal volume, it was found that the degree of ventilatory inhomogeneity was not significantly different in microgravity, compared with that observed in normal gravity with the subject standing. However, in the supine position during normal gravity there was a significant increase in ventilatory inhomogeneity, confirming the sensitivity of the measuring technique. The somewhat surprising conclusion from this study was that the primary determinants of ventilatory inhomogeneity during tidal breathing in the upright posture are not gravitational in origin.

The single-breath nitrogen washouts also allowed the closing volume of the lung to be measured. In normal gravity, this is generally believed to be the volume of the lung at which the airways near the base of the lung close because of distortion of the lung by its weight (36). It was therefore surprising to find that, in microgravity, the closing volume measured from the argon tracing was not significantly different from that in normal gravity, with the subject standing (33), although there was considerable variability among subjects. The inevitable conclusion seems to be that some airways close at lung volumes above residual volume in the absence of gravity, and that presumably means that this process is determined by the distribution of mechanical properties of the airways and parenchyma. Although this finding runs counter to much contemporary thinking, it is consistent with earlier measurements showing some airway closure in excised lungs (37). Closing capacity, that is, the total volume of gas in the lung when airway closure was first identified, was reduced consistent with the reduction in residual volume.

The effects of microgravity on the distribution of blood flow in the lung were discussed in Chapter 6. In brief, the results of an indirect measurement of the degree of blood flow inhomogeneity using the magnitude of cardiogenic oscillations in expired gas, showed that the distribution of pulmonary blood flow was more uniform in microgravity than in normal gravity. For example, the size of the cardiogenic oscillations in phase III of the test expiration was reduced to approximately 60% of the preflight standing values, and the height of phase IV was almost abolished. These results are consistent with the disappearance of gravity-dependent topographical inequality of blood flow.

However, there was considerable residual inhomogeneity of pulmonary perfusion, as evidenced by the fact that the height of the cardiogenic oscillations was essentially the same in microgravity as in the supine position in normal gravity. The small height of phase IV referred to above indicated that the differences in perfusion were probably in lung units that were not far apart. On the other hand, the inhomogeneity was probably not within the same acinus, because otherwise the cardiogenic oscillations would have been eliminated by diffusion or mixing. The fact that some inhomogeneity of pulmonary blood flow remained in the absence of gravity is consistent with a number of previous studies suggesting intrinsic mechanisms for uneven blood flow in the lung. Possible mechanisms include regional differences in conductance (38), a gradient of blood flow along the acinus (39,40), and stochastic or fractal inequalities of blood flow (41,42).

B. Ventilation–Perfusion Inequality

Unfortunately, the measurement of ventilation–perfusion inequality is technically very difficult. One traditional method is to measure the partial pressures of oxygen and carbon dioxide in arterial blood and expired gas, and calculate the physiological dead space and physiological shunt (43). The most informative method at the present time is the multiple inert gas elimination technique (MIGET) in which a saline solution of six inert gases is infused into a peripheral vein and, after a steady state of gas exchange has occurred, the concentrations of the inert gases in arterial blood and expired gas are determined (44). However, neither of these two methods is feasible during space flight at the present time. Arterial punctures, in particular, are considered too invasive and hazardous. The continuous infusion of a solution of inert gases until a steady state was obtained would be difficult and the method also requires sampling arterial blood. It is therefore necessary to turn to a noninvasive, indirect method of measuring ventilation–perfusion inequality.

The method chosen was to measure the change in respiratory exchange ratio from expired gas during a long expiration (45). This method depends on the fact that there is a close relationship between the expired respiratory exchange ratio and the ventilation–perfusion ratio of any lung unit for a given composition of inspired gas and mixed venous blood. During the expiration, it is assumed that the composition of mixed venous blood remains constant. The method only gives useful information if regions of the lung having different ventilation–perfusion ratios empty at different times during an expiration. In practice, units that are poorly ventilated, and therefore have relatively low ventilation–perfusion ratios, tend to empty late in an expiration. Thus, the change in respiratory exchange ratio over the course of a single expiration can give useful information about the amount of ventilation–perfusion inequality in the lung (45,46). In practice, the concentrations of oxygen and carbon dioxide are measured continuously with a respiratory mass spectrometer and the respiratory exchange ratio is calculated. The analysis is somewhat complicated and has been described in detail elsewhere (47,48).

Figure 5 shows the results of using this technique during studies of sustained microgravity on Spacelabs SLS-1 and SLS-2 (48). The data were obtained from slow vital capacity expirations, and the analysis separately shows the range of ventilation–perfusion ratios for both phase III and phase IV. The distinction between these two phases is shown by the onset of airway closure, which is indicated by a sharp upward deflection in the tracing for alveolar P_{O_2}, and down-

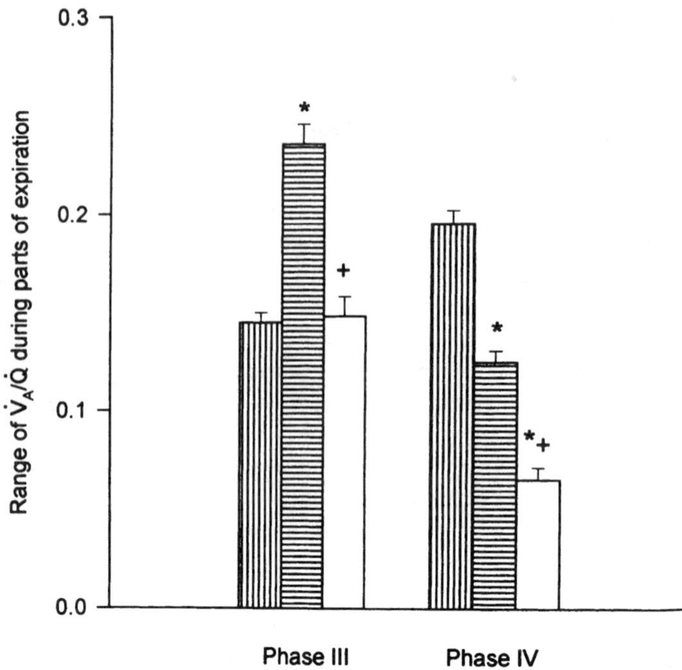

Figure 5 Range of ventilation–perfusion ratios ($\dot{V}_A/\dot{Q}$) seen over phase III and phase IV of vital capacity exhalations in eight subjects studied in the standing posture at 1 G (vertically lined bars), supine posture at 1 G (horizontally lined bars), and in microgravity (open bars). Values are means ± SE. Significantly different ($p < 0.05$) compared with: * standing, + supine. (From Ref. 48.)

ward deflection for P_{CO_2}. It can be seen that during phase III, which comprises most of the vital capacity expiration, the range of ventilation–perfusion ratios in microgravity was the same as in 1 G in the standing posture. However, the range of ventilation–perfusion ratios was greater in the supine position. In phase IV near the end of expiration, the range of ventilation–perfusion ratios was substantially less in microgravity than in both the standing and supine postures in 1 G, but particularly in the former.

   The fact that the range of ventilation–perfusion ratios in phase III in microgravity is essentially the same as that seen in 1 G in the standing posture is perhaps surprising in view of the well-documented reductions in the inequality of both ventilation and blood flow in microgravity. However, as pointed out earlier, although microgravity reduces the degree of inequality of both ventilation and perfusion, there is residual inequality of both presumably resulting from the intrinsic properties of the lung. The results are consistent with the fact that the principal determinants of ventilation–perfusion inequality in the normal upright lung are not gravitational in origin but have to do with the intrinsic variability within the lung itself. The fact that the reduction of the gravitationally determined ventilation and blood flow did not affect the range of ventilation–perfusion ratios may mean that the change in both ventilation and blood flow are matched in the transition from 1 G to microgravity. Indeed, this is not particularly surprising because we know that in the normal upright lung, both ventilation and blood flow increase down the lung, and therefore the changes in these gravitationally determined variables as a result of microgravity will be in the same direction.

Figure 5 also shows that in the phase IV part of the expiration, there was a clear reduction in the range of ventilation–perfusion ratios in microgravity compared with both the standing and supine postures in 1 G. This is consistent with the known reduction in the topographical inequality of ventilation and blood flow in the lung in microgravity because, as airway closure is progressing, the range of ventilation–perfusion ratios in the upright or supine positions in 1 G will be determined by the difference between the upper and lower regions of the lung. We can expect that these differences will be reduced in microgravity.

Additional information about the degree of ventilation–perfusion inequality in the lung can be obtained by looking at the cardiogenic oscillations and the height of phase IV for carbon dioxide in a long expiration following normal air breathing. Just as the ventilation–perfusion ratio of a lung unit determines its respiratory exchange ratio, so it will also determine its P_{CO_2}. During a long expiration, the movement of the heart within the chest will cause preferential emptying of parts of the lung, and insofar as there is ventilation–perfusion inequality between these parts and other regions of the lung, there will be fluctuations in expired P_{CO_2}.

Figure 6 shows the results obtained in Spacelabs SLS-1 and SLS-2 (48). It can be seen that the message from these results is apparently somewhat different from that obtained from the measurements of respiratory exchange ratio shown in Figure 5. The height of the cardiogenic oscillations for carbon dioxide was less in microgravity compared with the standing posture in 1 G, but essentially the same as the supine 1 G measurements. Therefore, these measurements suggest that the degree of ventilation–perfusion inequality between those regions of the lung whose emptying had been enhanced by the movement of the heart compared with other regions of the lung, was reduced in microgravity compared with the standing posture in 1 G.

These results are not necessarily inconsistent with those shown in Figure 5. The factors causing the late emptying of low ventilation–perfusion ratio units during a single expiration are not the same as those that cause differential emptying of different lung units caused by the movement of the heart. Figure 6 also shows that there was a striking reduction in the height of the cardiogenic oscilla-

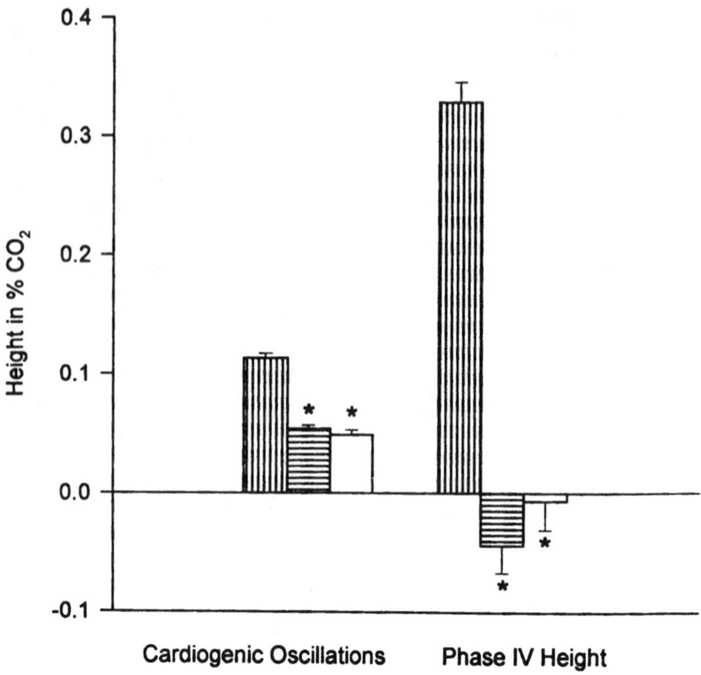

Figure 6 Height of cardiogenic oscillations and height of phase IV for carbon dioxide in eight subjects in the standing posture at 1 G (vertically lined bars), supine posture at 1 G (horizontally lined bars), and in microgravity (open bars). The height of phase IV, which is normally a negative number when standing, has been inverted for ease of comparison. Values are means $\pm$ SE. * Indicates significantly different compared with standing 1 G ($p < 0.05$). (From Ref. 48.)

tions for carbon dioxide in microgravity in phase IV of the expiration. This is consistent with the results shown in Figure 5, although the magnitude of the reduction was apparently considerably greater.

Additional information about the interpretation of these results came from a study of the phase relationships of the cardiogenic oscillations for carbon dioxide and nitrogen during vital capacity expirations in microgravity (49). The analysis also included other inspired inert gases, including helium, argon, and sulfur hexafluoride. When the cross-correlation between the phasing of the cardiogenic oscillations of carbon dioxide, nitrogen, and helium was examined, the conclusion was that in microgravity, areas of high ventilation are associated with high ventilation–perfusion ratios. The analysis helped to explain the apparent contradiction of how a more homogeneous distribution of both ventilation and perfusion in microgravity can occur in spite of an overall unaltered range of ventilation–perfusion ratios in phase III of the expiration as shown in Figure 5.

To summarize these rather complicated measurements of ventilation–perfusion inequality, the results of both of the techniques shown in Figures 5 and 6 agree that the amount of ventilation–perfusion inequality in those regions of the lung emptying in phase IV is reduced in microgravity. This is consistent with the gravitationally determined topographical differences of gas exchange in the normal upright lung where the inequality of ventilation and blood flow cause differences of ventilation–perfusion ratio between the apex and base (see Fig. 3). However, the conclusions about the degree of ventilation–perfusion inequality revealed by most of the single expiration, that is phase III, are somewhat different in Figures 5 and 6. Figure 5 suggests that the principal determinants of ventilation–perfusion inequality are not gravitational in origin in normal subjects, but Figure 6, which is based on cardiogenic oscillations, suggests that there is some reduction in the degree of ventilation–perfusion inequality in microgravity. Possibly the cardiogenic oscillations are caused to some extent by apex–base differences.

Of course, it would be extremely interesting to measure blood gases on crew members exposed to microgravity but, as mentioned above, this has not yet become feasible. However, there is an interesting preliminary report of measurements of P_{O_2} in arterialized blood obtained from cosmonauts on the Russian Mir space station (50). The measurements were made using a hand-held polarograph shaped like a pistol. Arterialized capillary blood was obtained by a stab incision of a hyperemic fingertip. The P_{O_2} values were reported from three cosmonauts between the 83rd and the 108th days in space. These showed reductions in P_{O_2} of the arterialized blood in all three crew members. In one subject, the P_{O_2} fell to 61 mmHg on day 171 compared with a preflight value of 86 mmHg. In the same subject, the P_{O_2} further decreased to 56 mmHg during lower body negative pressure on the spacecraft. These tantalizing though preliminary results suggest that there may be some impairment of pulmonary gas exchange in long periods of exposure to microgravity.

C. Gas Exchange and Alveolar Gas Composition

Measurements of resting pulmonary gas exchange were made during sustained microgravity on Spacelabs SLS-1 and SLS-2 (48). The subjects breathed through a mouthpiece for either 30 or 60 s after a steady state had been established, and from continuous measurements of expired gas concentrations with a respiratory mass spectrometer, and measurements of inspired and expired flow rates, oxygen uptake, carbon dioxide output, end-tidal P_{O_2} and end-tidal P_{CO_2} were obtained. Oxygen consumption averaged 300 ± 6 mL/min in the standing posture in 1 G, and was not significantly altered in either the supine position 1 G or in microgravity (Fig. 7). Carbon dioxide output averaged 250 ± 5 mL/min in the standing

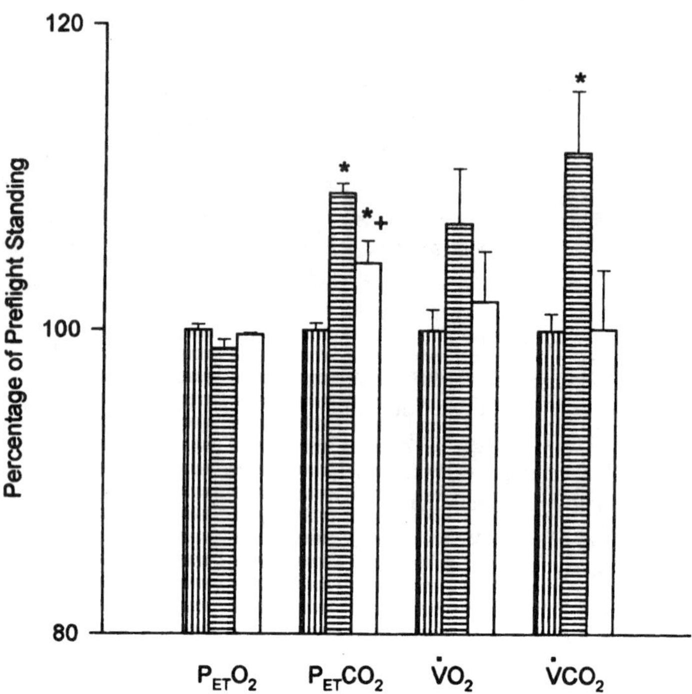

Figure 7   End-tidal P_{O_2} ($P_{ET_{O_2}}$), end-tidal P_{CO_2}($P_{ET_{CO_2}}$), O_2 uptake ($\dot{V}_{O_2}$), and CO_2 output ($\dot{V}_{CO_2}$) standing (vertically lined bars), supine (horizontally lined bars), and in microgravity (open bars) in eight subjects. The increase in $P_{ET_{CO_2}}$ in microgravity was entirely as a result of the SLS-2 flight in which environmental CO_2 was elevated (see text for details). Values are means $\pm$ SE. Significantly different ($p < 0.05$) compared with: * standing, + supine. (From Ref. 48.)

posture in 1 G, and was essentially identical in microgravity. However, there was a slight elevation of 11% or 23 mL/min in 1 G in the supine position. This rise in carbon dioxide output was probably caused by the increase in cardiac output that is observed in the supine position (51) and the fact that insufficient time was allowed in the transition from the seated to supine posture for a completely steady state to be established.

The end-tidal P_{O_2} averaged 114 $\pm$ 1 mmHg in the standing posture in 1 G, and was not significantly altered by changing the posture to supine, or by exposure to microgravity (see Fig. 7). The average inspired oxygen concentration was slightly decreased on Spacelab SLS-1 by 0.30% or 2.3 mmHg, and slightly elevated on Spacelab SLS-2 by 0.29% or 2.1 mmHg so that when the results for the two flights were pooled, there was no overall change. It was found that the small changes in end-tidal P_{O_2} tended to track those of the inspired P_{O_2}. The reasons for the differences in inspired oxygen concentration were small variations in the functioning of the environmental control system.

The end-tidal P_{CO_2} averaged 34.3 ± 0.2 mmHg in the standing posture in 1 G and was significantly increased by 2.0 mmHg in microgravity when the results from the two Spacelabs were pooled. However, this increase was entirely due to the changes observed on Spacelab SLS-2 where the end-tidal P_{CO_2} increased by a mean of 4.5 mmHg. By contrast, there was no change in the end-tidal P_{CO_2} on Spacelab SLS-1 compared with 1 G standing. This increase in end-tidal P_{CO_2} in SLS-2 was associated with an increase in environmental carbon dioxide in the spacecraft because of the inability of the environmental control systems to remove the metabolic carbon dioxide completely. The increase in inspired carbon dioxide was 1 to 3 mmHg, although on the face of it, this seems insufficient to increase the end-tidal P_{CO_2} so much. Note that the end-tidal P_{CO_2} was significantly raised in the supine posture at 1 G. Again, as in the case of the increase in carbon dioxide output, this was presumably related to the increase in cardiac output and the consequent faster washout of carbon dioxide from body tissues as a result of assuming the supine position, and the failure to establish a completely steady state.

D. Other Indices of Pulmonary Gas Exchange

Ventilation

Measurements made on Spacelabs SLS-1 and SLS-2 showed that tidal volume was decreased on the average in microgravity by about 15% compared with the standing posture in 1 G (48,52). In the supine posture, tidal volume was intermediate between the values in the standing posture and in 1 G and significantly different from both (Fig. 8).

In contrast to this reduction in tidal volume in microgravity, there was an increase in respiratory frequency averaging 9% above that observed in the standing posture in 1 G. The resultant effect of these two changes was to decrease total ventilation both in microgravity and the supine posture at 1 G compared with the standing posture 1 G. However, alveolar ventilation calculated as (tidal volume − anatomical dead space) × respiratory frequency was unchanged in microgravity compared with the standing posture in 1 G, and it was slightly reduced in the supine position 1 G (see Fig. 8). The changes in anatomical dead space are discussed below.

The changes in respiratory frequency were analyzed by looking at the proportion of time devoted to inspiration and expiration. Most of the increase in respiratory frequency observed in microgravity was caused by a significant reduction in expiratory time averaging 10%, and a nonsignificant reduction in inspiratory time of about 4%. The result was that the fraction of the total ventilatory

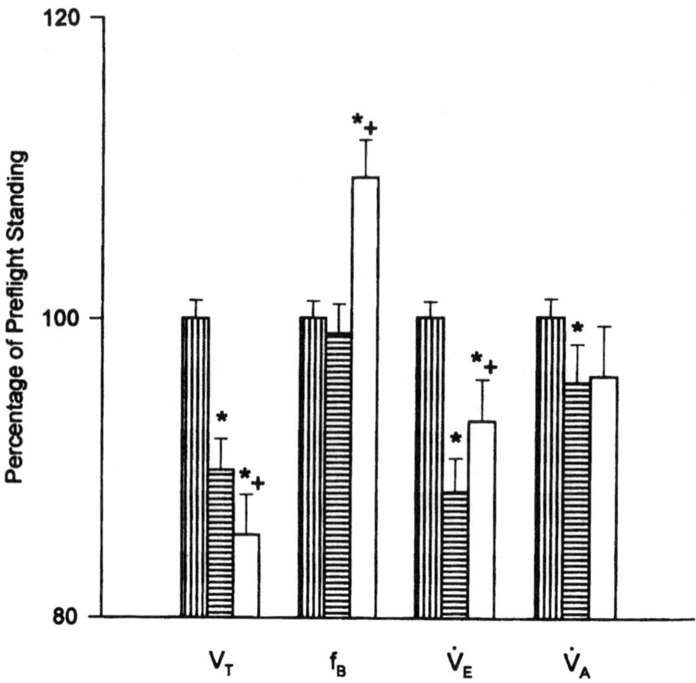

Figure 8 Tidal volume (V_T), respiratory rate (f_b), total ventilation ($\dot{V}_E$), and alveolar ventilation ($\dot{V}_A$) expressed as percentage of preflight standing values for standing (vertically lined bars), supine (horizontally lined bars), and microgravity (open bars) in eight subjects. Values are means $\pm$ SE. Note that in microgravity a compensatory increase in f_b cased $\dot{V}_A$ to remain unchanged although V_T was decreased by 15%. Significantly different ($p < 0.05$) compared with: * standing, + supine. (From Ref. 48.)

cycle occupied by inspiration was increased in microgravity by about 3% compared with both standing and supine measurements at 1 G (Fig. 9). The average inspiratory flow rate calculated by dividing the tidal volume by the inspiratory time was reduced by an average of 10% in microgravity, and reduced even further by 12% in the supine posture in 1 G (see Fig. 9).

Dead Space

Anatomical dead space was measured during Spacelabs SLS-1 and SLS-2 from the expired carbon dioxide tracing using the graphical analysis of Fowler (53). During resting breathing, the anatomical dead space averaged 147 $\pm$ 2 mL in our eight subjects in the standing posture at 1 G, and was significantly decreased

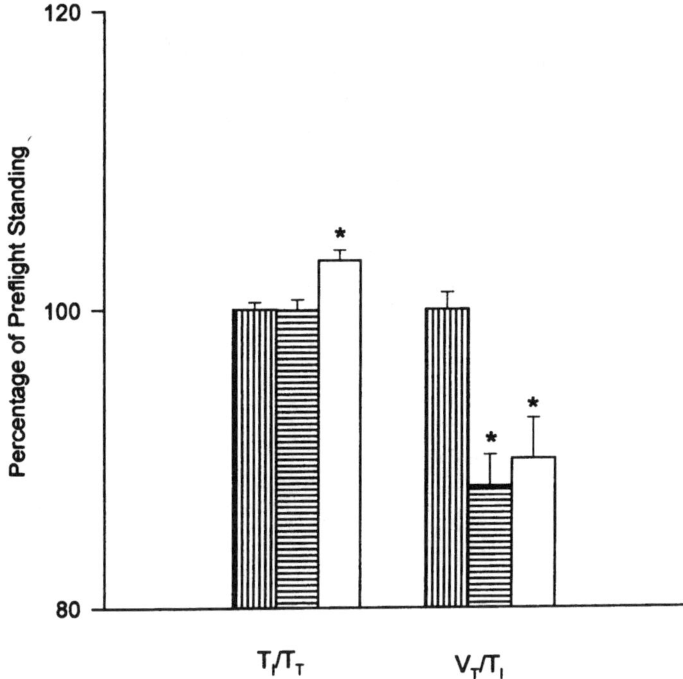

Figure 9 Inspiratory time as a function of total breath time (T_I/T_T) and average inspiratory flow rate (V_T/T_I) for standing (vertically lined bars), supine (horizontally lined bars), and in microgravity (open bars) in eight subjects. Values are means ± SE. Significantly different compared with standing ($p < 0.05$). (From Ref. 48.)

in both microgravity (115 ± 3 mL) and in the supine position at 1 G (98 ± 2 mL) (Fig. 10).

On the same flights, the physiological dead space was calculated from the Bohr equation by using end-tidal and mean expired carbon dioxide concentrations (43). The physiological dead space was largest in the standing position at 1 G (181 ± 3 mL) and was significantly decreased in both microgravity (137 ± 3 mL) and in the supine position at 1 G (126 ± 3 mL). These changes in physiological dead space can be explained by the reduced blood flow to the lung apex in the upright posture, and the more uniform distribution of blood flow in the supine posture and in microgravity. However, the difference in physiological dead space between that measured in the supine position at 1 G and in microgravity was not significant.

Alveolar dead space was calculated by subtracting the anatomical dead space from the physiological dead space. Alveolar dead space was significantly

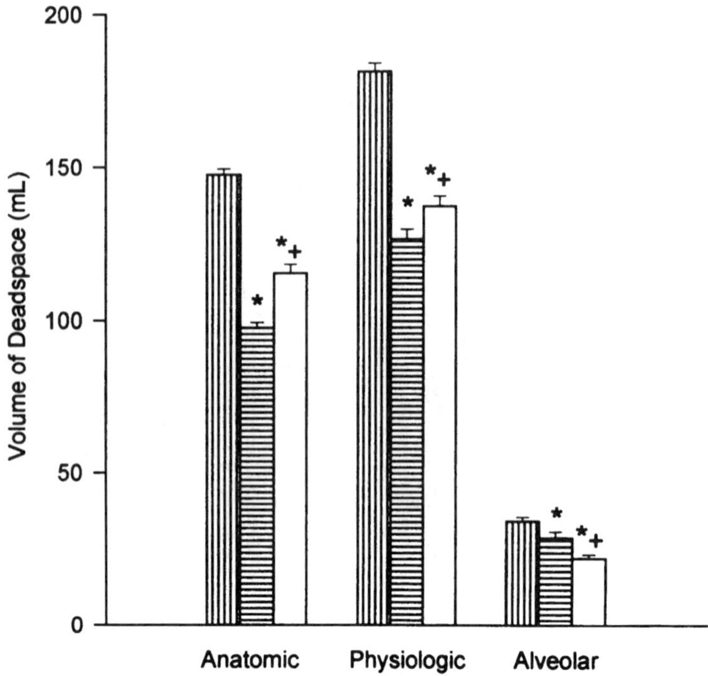

Figure 10 Anatomical physiological, and alveolar dead space measured in the standing posture (vertically lined bars), supine (horizontally lined bars), and in microgravity (open bars) in eight subjects. Note that there is a reduction in alveolar dead space in microgravity that can be explained by the reduction of areas of high ventilation–perfusion ratio in the lung. Values are means $\pm$ SE. Significantly different ($p < 0.05$) compared with: * standing, + supine. (From Ref. 48.)

reduced from 34 ± 1 ml in the standing posture at 1 G to 29 ± 2 mL in the supine posture at 1 G. It was further and significantly reduced to 22 ± 1 mL in microgravity. Again, these changes fit with the expected alterations in the distribution of pulmonary blood flow. The lower dead space in microgravity compared with the supine posture can be explained by the slightly reduced blood flow to the uppermost part of the lung in the latter situation.

The various dead space values were also calculated as a fraction of the tidal volume. There was only a very slight decrease in the ratio of physiological dead space to tidal volume in microgravity, and no change in the ratio of anatomical dead space to tidal volume in microgravity. However, both these measurements were reduced by approximately 0.05 in the supine position at 1 G.

E. Conclusion

A general conclusion from our studies is that although microgravity has profound effects on the function of the lung including the distribution of ventilation and blood flow and lung volumes, overall gas exchange is well preserved. However, this does not mean that it is unnecessary to monitor pulmonary function in crew members who spend extended periods in space, for example, in the International Space Station. The lung, with its gas-exchanging area of 50 to 100 m², presents by far the largest surface in the body to the environment. The atmosphere of a spacecraft can easily change in ways that threaten the lung. One of the most extreme examples was the fire that occurred on the Soviet space station Mir. But apart from such dramatic examples as this, the atmosphere can be altered by failure of the environmental control system, by outgassing from materials within the spacecraft, and especially by aerosol contamination. The absence of gravity means that the normal settling out of particles does not occur. A prudent medical program for space habitation should definitely include the monitoring of pulmonary function.

References

1. Bjorkman S. Bronchospirometrie, eine klinische Methode die Funktion der menschlichen Lungen getrennt und gleichzeitig zu untersuchen. Acta Med Scand (suppl) 1934; 56:1–212.
2. Loewy A, von Schrotter H. Untersuchungen uber die Blutcirculation beim Menschen. Z Exp Pathol Ther 1905; 1:197–310.
3. Carlens E. A new flexible double lumen catheter for bronchospirometry. J Thorac Surg 1949; 18:742–746.
4. Vaccarezza RF, Bence A, Lanari A, Labourt F, Segura RG. The study of the two lungs separately in practical and research work. Dis Chest 1943; 9:95–114.
5. Rahn H, Bahnson HT. Effect of unilateral hypoxia on gas exchange and calculated pulmonary blood flow in each lung. J Appl Physiol 1953; 6:105–112.
6. Himmelstein A, Harris P, Fritts HW, Cournand A. Effect of severe unilateral hypoxia on the partition of pulmonary blood flow in man. J Thorac Surg 1958; 36:369–381.
7. Mattson SB, Carlens E. Lobar ventilation and oxygen uptake in man: influence of body position. J Thorac Surg 1955; 30:676–682.
8. Martin CJ, Cline F, Marshall H. Lobar alveolar gas concentrations: Effect of body position. J Clin Invest 1953; 32:617–621.
9. Martin CJ, Marshall H, Cline F Jr. Lobar alveolar gas concentrations: Effect of reduced lung volumes. J Appl Physiol 1953; 6:209–212.
10. Rahn H, Sadoul P, Farhi LE, Shapiro J. Distribution of ventilation and perfusion in the lobes of the dog's lung in the supine and erect position. J Appl Physiol 1956; 8:417–426.

11. Riley RL, Permutt S, Said S, Godfrey M, Cheng TO, Howell JBL, Shepard RH. Effect of posture on pulmonary dead space in man. J Appl Physiol 1959; 14:339–344.

12. Bjurstedt H, Hesser CM, Liljestrand G, Matell G. Effects of posture on alveolar-arterial CO_2 and O_2 differences and on alveolar dead space in man. Acta Physiol Scand 1962; 54:65–82.

13. Gerst PH, Rattenborg C, Holiday DA. The effects of haemorrhage on pulmonary circulation and respiratory gas exchange. J Clin Invest 1959; 38:524–538.

14. Barr P-O. Pulmonary gas exchange in man as affected by prolonged gravitational stress. Acta Physiol Scand (Suppl) 1963; 207:1–46.

15. West JB. Regional differences in gas exchange in the lung of erect man. J Appl Physiol 1962; 17:893–898.

16. Farber JP. Study of Regional Emptying of the Lung with Soluble Inert Gas. Ph.D. dissertation, State University of New York at Buffalo, 1969.

17. Auerbach O, Stemmerman MG. The development of pulmonary tuberculosis in congenital heart disease. Am J Med Sci 1944; 207:219–230.

18. Norris GW. Tuberculosis and heart disease. Am J Med Sci 1904; 128:649–668.

19. Abbott ME. Atlas of Congenital Cardiac Disease. New York: American Heart Association, 1936.

20. White PD. Heart Disease. 4th ed. New York: Macmillan, 1951.

21. Dock W. Apical localization of phthisis. Am Rev Tuberc 1946; 53:297–305.

22. Laennec R-TH. Traité de L'Auscultation Médiate et des Maladies des Poumons et du Coeur. Paris: J-S Chaudé, Libraire-Éditeur, 1826.

23. Dollery CT, West JB, Wilcken DEL, Hugh-Jones P. A comparison of the pulmonary blood flow between left and right lungs in normal subjects and patients with congenital heart disease. Circulation 1961; 24:617–625.

24. Hanlon CR, Scott Jr. HW, Olson BJ. Experimental tuberculosis: Effects of anastomosis between systemic and pulmonary arteries on tuberculosis in monkeys. Surgery 1950; 28:209–224.

25. Kempner W. Oxygen tension and the tubercle bacillus. Am Rev Tuberc 1939; 39:157–168.

26. Davies R. Effect of carbon dioxide on the growth of tubercle bacillus. Br J Exp Pathol 1940; 21:243–253.

27. Goodwin RA, Des Prez RM. Apical localization of pulmonary tuberculosis, chronic pulmonary histoplasmosis, and progressive massive fibrosis of the lung. Chest 1983; 83:801–805.

28. Kaltreider HB, Baum GL, Bogaty G, McCoy MD, Tucker M. So-called ''metastatic'' calcification of the lung. Am J Med 1969; 46:189–196.

29. Mulligan RM. Metastatic calcification. Arch Path 1947; 43:177–230.

30. Virchow R. Kalk-Metastasen. Virchows Arch [A] 1855; 8:103–113.

31. Holmes RA. Diffuse interstitial pulmonary calcification. J Am Med Assoc 1974; 230:1018–1019.

32. Glaister DH. Effect of acceleration. In: West JB, ed. Regional differences in the lung. New York: Academic, 1977:323–379.

33. Guy HJ, Prisk GK, Elliott AR, Deutschman RA III, West JB. Inhomogeneity of

pulmonary ventilation during sustained microgravity as determined by single-breath washouts. J Appl Physiol 1994; 76:1719–1729.

34. Prisk GK, Guy HJ, Elliott AR, Paiva M, West JB. Ventilatory inhomogeneity determined from multiple-breath washouts during sustained microgravity on Spacelab SLS-1. J Appl Physiol 1995b; 78:597–607.

35. Prisk GK, Elliott AR, Guy HJB, Verbanck S, Paiva M, West JB. Multiple breath washin of helium and sulfur hexafluoride in sustained microgravity. J Appl Physiol 1998; 84:244–252.

36. Milic-Emili J, Henderson AM, Dolovich MB, Trop D, Kaneko K. Regional distribution of inspired gas in the lung. J Appl Physiol 1966; 21:749–759.

37. Glaister DH, Schroter RC, Sudlow MF, Milic-Emili J. Transpulmonary pressure gradient and ventilation distribution in excised lungs. Respir Physiol 1973; 17:365–385.

38. Beck KC, Rehder K. Differences in regional vascular conductances in isolated dog lungs. J Appl Physiol 1986; 61:530–538.

39. Wagner PD, McRae J, Read J. Stratified distribution of blood flow in secondary lobule of the rat lung. J Appl Physiol 1967; 22:1115–1123.

40. West JB, Maloney JE, Castle BL. Effect of stratified inequality of bloodflow on gas exchange in liquid-filling lungs. J Appl Physiol 1972; 32:357–361.

41. Warrell DA, Evans JW, Clarke RO, Kingaby GP, West JB. Patterns of filling in the pulmonary capillary bed. J Appl Physiol 1972; 32:346–356.

42. Glenny RW, Robertson HT. Fractal properties of pulmonary blood flow: Characterization of spatial heterogeneity. J Appl Physiol 1990; 69:532–545.

43. Riley RL, Cournand A. Analysis of factors affecting partial pressures of oxygen and carbon dioxide in gas and blood of lungs: Theory. J Appl Physiol 1951; 4:77–101.

44. Wagner PD, Laravuso RB, Uhl RR, West JB. Continuous distributions of ventilation-perfusion ratios in normal subjects breathing air and 100% O_2. J Clin Invest 1974; 54:54–68.

45. West JB, Fowler KT, Hugh-Jones P, O'Donnell TV. The measurement of the inequality of ventilation and of perfusion in the lung by the analysis of a single expirate. Clin Sci 1957;16:529–547.

46. Reed JW, Guy HJB, Hammond MD, Prisk GK. Reflex compensation of ventilation-perfusion inequality: Comparison of inert gas elimination and intrabreath respiratory exchange ratio (abstr). Physiologist 1986; 29:93.

47. Guy HJB, Gaines RA, Hill PM, Wagner PD, West JB. Computerized, noninvasive tests of lung function. A flexible approach using mass spectrometry. Am Rev Respir Dis 1976; 113:737–744.

48. Prisk GK, Elliott AR, Guy HJ, Kosonen J, West JB. Pulmonary gas exchange and its determinants during sustained microgravity on Spacelab SLS-1 and SLS-2. J Appl Physiol 1995a; 79:1290–1298.

49. Lauzon A-M, Elliott AR, Paiva M, West JB, Prisk GK. Cardiogenic oscillation phase relationships during single-breath tests performed in microgravity. J Appl Physiol 1998; 84:661–668.

50. Haase H, Baranov VM, Asyamolova NM, Polyakov VV, Avak y, Dannenberg R, Jarsumbeck B, König J. First results of P_{O_2} examinations in the capillary blood of

cosmonauts during a long-term flight in the space station ''MIR.'' Proceeding of the 41st Congress of the International Astronautical Federation, Dresden, Germany, 1990:1–4.

51. Prisk GK, Guy HJ, Elliott AR, Deutschman RA III, West JB. Pulmonary diffusing capacity, capillary blood volume and cardiac output during sustained microgravity. J Appl Physiol 1993; 75:15–26.

52. Elliott AR, Prisk GK, Guy HJ, West JB. Lung volumes during sustained microgravity on Spacelab SLS-1. J Appl Physiol 1994; 77:2005–2014.

53. Fowler WS. Lung function studies. II. The respiratory dead space. Am J Physiol 1948; 154:405–416.

54. West JB. Distributions of gas and blood in the normal lungs. Br Med Bull 1963; 19:53–58.

9

Exercise and Gas Exchange

DAG LINNARSSON

Karolinska Institutet
Stockholm, Sweden

I. Introduction

The gas exchange between the environment and the cellular respiration comprises a chain of serially operating processes, including the mechanics of breathing movements, the gas mixing in the air spaces of the lungs, the diffusion through the alveolar–capillary interface, pulmonary perfusion, the central and peripheral systemic circulations, the diffusive gas transport between capillaries and mitochondria, and the energetics of cellular respiration. All these processes may be influenced to a greater or lesser extent by the direction and magnitude of gravitational forces, but it cannot a priori be assumed that normal or increased gravity has a negative influence on each of these processes. For instance, the breathing movements include displacement of the chest against gravity in the upright human and of the abdominal organs against gravity in supine position. If all the potential energy gained during inspiration would be dissipated as useful expiratory work, there would not be any metabolic cost of gravity, but if this is not the case, the work of breathing will increase with the force of gravity.

For an efficient gas exchange within the lungs, the distributions of the two processes, ventilation and perfusion, must be topographically synchronized. To the extent that it is the normal gravity that causes this synchronization, a less efficient gas exchange may be found in microgravity, and, to the extent that the system is optimized for operation at normal gravity, it will be less efficient also in hypergravity. However, if gravity instead is the cause of the normally seen imperfections of the synchronization between the distributions of ventilation and perfusion, then a more optimal synchronization may be found in microgravity than in normal gravity.

The circulatory transport between the lungs and the peripheral tissues may be influenced in two opposite ways by gravity and the resulting hydrostatic pressure gradients in the systemic circulation. On the venous side, gravity will promote peripheral venous pooling and jeopardize venous return to the heart. On the other hand, the hydrostatic pressure gradients on the arterial side may promote perfusion of dependent tissues by distending resistance vessels, and also under some conditions, add to the effective perfusion pressure.

The flux of oxygen and carbon dioxide through this serially operating transport system is markedly increased during exercise. Dynamic exercise with large muscle groups may cause the overall metabolic rate to increase by a factor of 10 or more. Thus, any limitations or facilitations resulting from gravitational changes in any of the serially operating links are likely to become more evident in exercise than during resting conditions. Also, the energy requirement for a given external power output may be different from normal gravity both in hypergravity and microgravity. In hypergravity, the postural work, the work of breathing, and the movement of limbs in the gravity field may be associated with increased energy costs. In microgravity, the body weight can no longer serve as the counterforce to the movements of the pedals in leg exercise, and additional muscular effort from the upper body may be required to produce that counterforce.

II. Gas Transport in the Lungs

A. Hypergravity

The most comprehensive analysis of gas transport in human lungs during exercise in hypergravity has been presented by Rosenhamer (1,2) and Bjurstedt et al. (3). They studied subjects during sitting dynamic leg exercise in a human centrifuge with up to 3 G acting in the head-to-foot direction. Arterial blood gases were determined continuously by arterial sampling and in-line analysis of P_{O_2}, P_{CO_2}, pH and O_2 saturation. Cardiac output was determined by remote-controlled venous injection and arterial analysis of indocyanine green. Standard ventilatory and gas exchange determinations were performed with a remote-controlled Douglas-bag technique. Results are summarized in Figures 1 and 2. Both at rest and during exercise at about 50- and 100-W intensities (300 and 600 kpm/min)

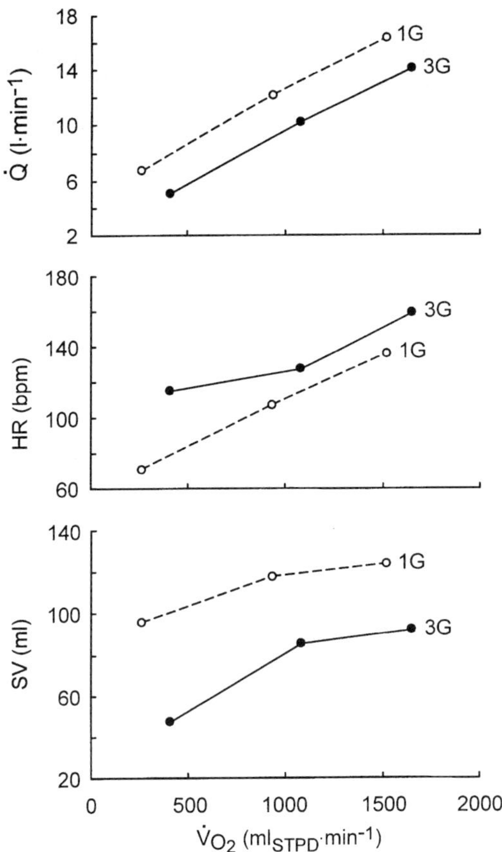

Figure 1 Cardiac output ($\dot{Q}$), heart rate (HR), and stroke volume (SV) as functions of oxygen uptake during rest and dynamic leg exercise at normal (1 G) and 3 times increased gravity (3 G) in the head-to-foot direction. N = 8. (Data from Ref 1.)

pulmonary blood flow ($\dot{Q}_L$) for a given level of $\dot{V}_{O_2}$ was found to be reduced by 2.5 to 3 L/min at 3 G compared with normal G control (see Fig. 1). At the same time, pulmonary ventilation ($\dot{V}_E$) was markedly increased (see Fig. 2). In terms of overall ventilation–perfusion ratio ($\dot{V}_E/\dot{Q}_L$), the efficiency of the lungs as gas exchangers appears to be reduced to about 50% of normal at 3 G during rest and to about 60% during exercise. At least three mechanisms act in concert to increase $\dot{V}_E/\dot{Q}_L$ during high-G exposure: increased respiratory drive, decreased alveolar ventilation/perfusion ($\dot{V}_A/\dot{Q}_L$) homogeneity, and the overall reduction of $\dot{Q}_L$.

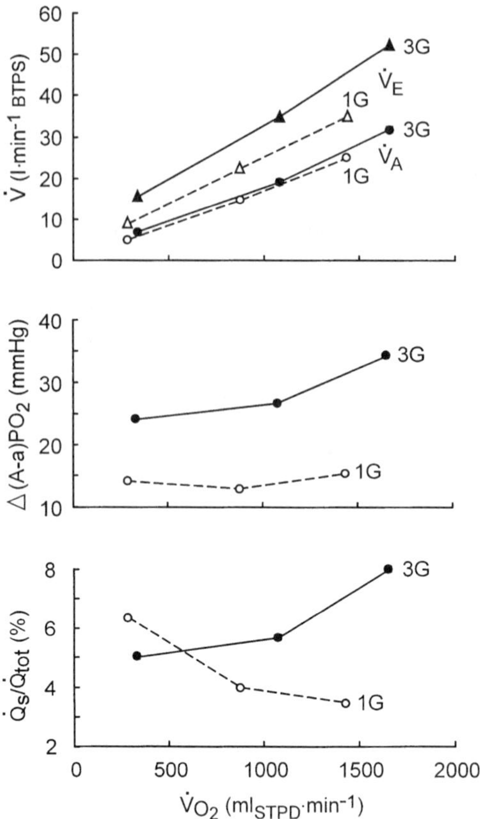

Figure 2   Total ventilation ($\dot{V}_E$), alveolar ventilation ($\dot{V}_A$), alveolar–arterial P_{O_2} difference [$\Delta(A\text{-}a)\, P_{O_2}$], and pulmonary shunt fraction ($\dot{Q}_s/\dot{Q}_{tot}$) as functions of $\dot{V}_{O_2}$ in same experiments as Figure 1. (Data from Ref. 1.)

Respiratory Drive

Figure 3 shows $\dot{V}_E$ as a function of a Pa_{CO2} and exercise intensity, where $\dot{V}_E$ is markedly increased at 3 G_z compared with 1 G. This demonstrates the existence of one or several powerful additional respiratory stimuli apart from Pa_{CO_2}. At the same time, Rosenhamer (1) determined that arterial pH levels did not differ between G conditions. Thus the arterial pH per se could not be that additional stimulus, and Rosenhamer (1) proposed that it rather was the presence of a relative arterial hypotension at the level of the carotid sinus (4,5) that comprised the additional ventilatory stimulus. However, with increasing work intensity, lactacidosis may contribute as a ventilatory stimulus in high-G exercise by displacing

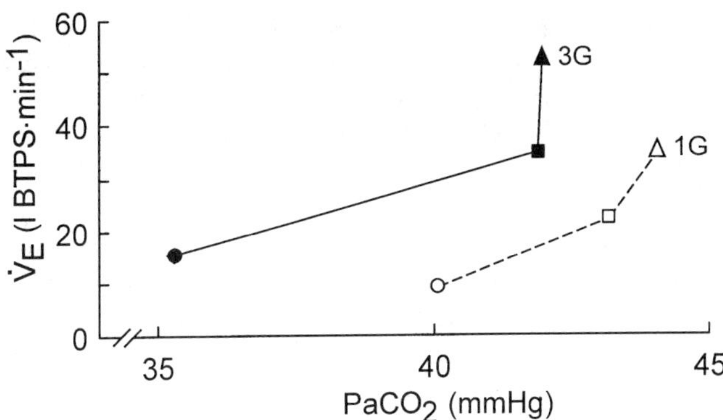

Figure 3 Total ventilation as a function of arterial P_{CO_2} in same experiments as Figure 1. ○ ●, rest; □ ■ 50 W; △ ▲, 100 W. For a given Pa_{CO_2}, $\dot{V}_E$ is markedly elevated at 3 G. (Data from Ref. 1.)

CO_2 from tissue and blood stores, thereby increasing the pulmonary CO_2 load (6) more than in proportion to the increased metabolic cost of exercising in hypergravity (see below, Sec. IV). Bjurstedt et al. (3) demonstrated a gradually increasing lactacidosis with increasing exercise intensity at 3 G; at 150 W, lactate concentrations in arterialized capillary blood averaged 4 to 5 mmol/L, and in two subjects values approached 7 mmol/L.

In summary, the markedly increased $\dot{V}_E$ at 3 G for a given Pa_{CO_2} and metabolic rate both at rest and during exercise appears to be caused by a combination of mechano- and chemoreceptor inputs, and results in a modest respiratory alkalosis at rest, and an essentially compensated metabolic acidosis during exercise.

Ventilation–Perfusion Relationship Within the Lungs

Figure 2 shows effective alveolar ventilation (computed from the ratio $\dot{V}_{CO_2}/Pa_{CO_2}$) as a function of O_2 uptake in resting and exercising men (1). The figure demonstrates that the additional ventilatory effort at 3 G compared with 1 G does not represent an increase of alveolar ventilation; rather $\dot{V}_A$ levels for a given $\dot{V}_{O_2}$ appear identical at 1 G and 3 G. During both rest and exercise, alveolar-to-arterial P_{O_2} differences [Δ(A-a) P_{O_2}] are about twice as large at 3 G than at 1 G (see Fig. 2).

From the data of arterial O_2 saturation obtained at the same time, the shunt fraction could be computed for a theoretical lung model with two compartments, one compartment shunting mixed venous blood through the lungs without any additional oxygenation, and the other providing fully oxygenated pulmonary end-capillary blood. Figure 2 illustrates that the normal pattern of decreasing shunt fraction with transition from rest to exercise is reversed at 3 G; the shunt fraction increases with transition to exercise and increased exercise intensity. Data on the arterial O_2 deficit can be compared with arteriovenous O_2 differences in a quantitative analysis on which level in the O_2 transport chain that G-induced impairments take place. In Rosenhamer's data (1) from 100 W (600 kpm/min) exercise, the mixed venous O_2 deficit (relative to 100% O_2 saturation) averaged 99 and 127 mL/L at 1 G and 3 G, respectively. Out of this reduction of mixed venous O_2 content relative to full saturation, about one-third could be ascribed to the arterial O_2 deficit and about two-thirds to a widening of the arteriovenous O_2 difference. Thus, it appears from a quantitative standpoint that both the distribution and the overall rate of pulmonary perfusion are of importance, but the latter factor is the more critical.

It should be recognized that the Δ(A-a) P_{O_2} data of Rosenhamer (1) were computed from the alveolar gas equation (7) and the alveolar ventilation equation (8). Thus an ideal, representative $P_{A_{O_2}}$ was computed assuming equality of Pa_{CO_2} and a representative $P_{A_{CO_2}}$. This assumption may not be valid at high G, and for each mmHg that an Δ(a-A) P_{CO_2} is underestimated, there is a roughly equivalent underestimation of Δ(A-a) P_{O_2}. Thus, the data presented in Figure 2 represent conservative estimates of lung function impairments in terms of Δ(A-a) P_{O_2} at rest and during exercise in hypergravity.

Further studies from the same laboratory lend support to the notion of a G-induced limitation of exercise capacity. Thus, Bjurstedt et al. (3) showed that out of eight healthy male subjects, one could not complete 6 min of 150-W exercise at 3 G, and in two more the $\dot{V}_{O_2}$ increase with increasing external work load tended to level off at the transition from 100 to 150 W. This study, however, was not designed to allow conclusions about the site (pulmonary, circulatory, etc.) of the O_2 transport limitation. Nunneley et al. (9) performed similar studies on exercising, sitting men including continuous mass spectrometric determinations of intrabreath gas composition. Although they provide no quantitative data, they present a qualitative analysis of cardiogenic oscillations of expired P_{CO_2}. Heat-synchronous oscillations of expired CO_2 can only occur if there is a combination of P_{CO_2} differences between lung regions and time-varying contributions of the expired flow contribution from these regions to the total expired flow (10,11). Nunneley et al. (9) reported that at both normal and high G, the amplitude of cardiogenic P_{CO_2} oscillations were reduced with the transition from rest to exercise but were always larger at 3 G than at normal G. These observations suggest that larger-than-normal intraregional P_{CO_2} differences remain in the lung during exercise in hypergravity, despite the exercise-induced increase of pulmonary blood flow. This lends further support to the findings by Rosenhamer (1) of impaired gas exchange between the lung gas and the pulmonary–capillary blood in sitting, exercising men at 3 times normal gravity.

B. Microgravity

A mere extrapolation of data from hypergravity would predict that $\dot{V}_E/\dot{Q}_L$ be decreased during exercise in microgravity compared with normal gravity, suggesting a more efficient gas exchange in the lungs and a reduced respiratory drive for a given metabolic load. In contrast, early predictions of pulmonary function in microgravity included fears of pulmonary congestion and diffusion limitations due to interstitial edema (12). Indeed, such changes would result in severe impairments of gas transport in the lungs during exercise, when the O_2 flux through the alveolar–capillary membrane is increased. However, throughout 40 years of manned space flight, signs of impaired pulmonary gas exchange have never been reported. During the first systematic studies of exercise gas exchange in long-term microgravity on Skylab (13), three subjects showed essentially normal ventilatory responses to moderately heavy exercise, compared to preflight and postflight test results. Shykoff et al. (14) later studied the relationship between $\dot{V}_E$ and $\dot{V}_{O_2}$ during submaximal exercise in microgravity and found it to be similar to that of supine exercise at 1 G. Also Stegemann et al. (15), studying sitting exercise at 80-W intensity, found similar ventilatory responses in microgravity and on the ground. As previously reviewed in Chapter 6, the diffusion capacity for CO during rest is improved in sustained microgravity (16,17) and the pulmonary tissue volume has been found to be reduced (17). These findings speak very strongly against interstitial edema in the pulmonary tissue in resting humans during weightlessness. Corresponding data for exercise have so far not been published. However, as discussed below (Sec. III), maximum exercise capacity appears to be preserved during sustained (9 to 14 days) microgravity (18), and that observation speaks against any limitations of pulmonary gas exchange specific for the microgravity environment. Also, from ground-based studies at 1 G, it would be expected that gravity-related influences on lung function would become relatively less important with increasing exercise intensity, when flow-dependent pressures in the pulmonary vasculature and flow-dependent airway pressure variations exceed interregional hydrostatic pressure differences (19).

III. Circulatory Transport

A. Hypergravity

Hydrostatic pressure gradients within the systemic circulation will increase with gravity. In the sitting or standing resting human, these gradients tend to curtail systemic blood circulation by two mechanisms (20,21). First, diastolic cardiac filling is reduced because of peripheral pooling of blood. Even at 1 G, the central blood volume in a standing, resting human is reduced by approximately 0.7 L compared with supine position (22). Rosenhamer (1) estimated central blood volume from curve parameters of dye dilution recordings and demonstrated a further decrease of 0.7 L at 3 G compared with 1 G in sitting, resting men. Second, arterial pressures above a hydrostatic indifference point will be reduced with associated reductions of the effective tissue perfusion pressure. A hydrostatic indifference point is defined as a point along the long axis of the body where the intravascular pressure is the same regardless of posture (23). In the upright human, any G-induced reduction of perfusion pressure will be most marked in the head. To a limited extent, the brain will be protected by a counteracting hydrostatic pressure drop in the veins surrounded by bone (24). The perfusion in the eye is especially sensitive to reductions of arterial pressure since its tissue pressure is elevated by up to 20 mmHg.

Dynamic leg exercise results in a dramatically different situation because of the pumping action of the rhythmically contracting muscles on the deep veins of the legs. It has been proposed that up to 30% of the overall pump work required to drive blood through the vascular system during exercise is performed by the muscle pump (25). Both in an animal model (26) and in humans (27), the efficiency of the muscle pump in the lower limbs is greatly enhanced by a hydrostatic pressure gradient in the craniocaudal direction. This is so because venous valves prevent backflow of blood into the muscles resulting in very low—if not subatmospheric—venous pressures between rhythmic contractions. Thus, the effective perfusion pressure in the lower limb of an upright human performing dynamic leg exercise is the sum of the dynamic pressure generated by the cardiac pump and the hydrostatic pressure gradient between the heart and the muscle (21).

The antigravity effects of exercise have been demonstrated by Rosenhamer (2), who compared responses to a 13-min exposure to 3 G during sitting at rest and seated exercise. Out of eight healthy men only three could tolerate the full exposure time at rest, and in three of the subjects, the exposure to 3 G had to be terminated as early as after 5 to 6 min. Six of the subjects experienced dimming of the peripheral vision in the first 4 to 5 min. In the five cases where 3 G exposure was terminated prematurely, this was necessiated by rapidly increased dimming of the vision and impending loss of consciousness. By contrast, when leg exercise at 100 W was performed for the last 12 of the 13 min, the 3 G exposure was perceived as less exhausting than at rest and there were no symptoms to suggest impaired blood perfusion in the eyes or the brain. Figure 4 illustrates the time course of arterial blood pressure at the level of the heart from an experiment performed by Linnarsson and Rosenhamer (4) in which subjects first sat at rest at 1 G and 3 G and then started to exercise after 6 min. There was a clear trend for both mean arterial pressure and pulse pressure to fall toward the end of the resting period, followed by a marked and rapid recovery at the onset of leg exercise.

Gravity-induced alterations of central blood volume differ between rest and exercise. Thus central blood volume, as estimated from parameters of dye dilution

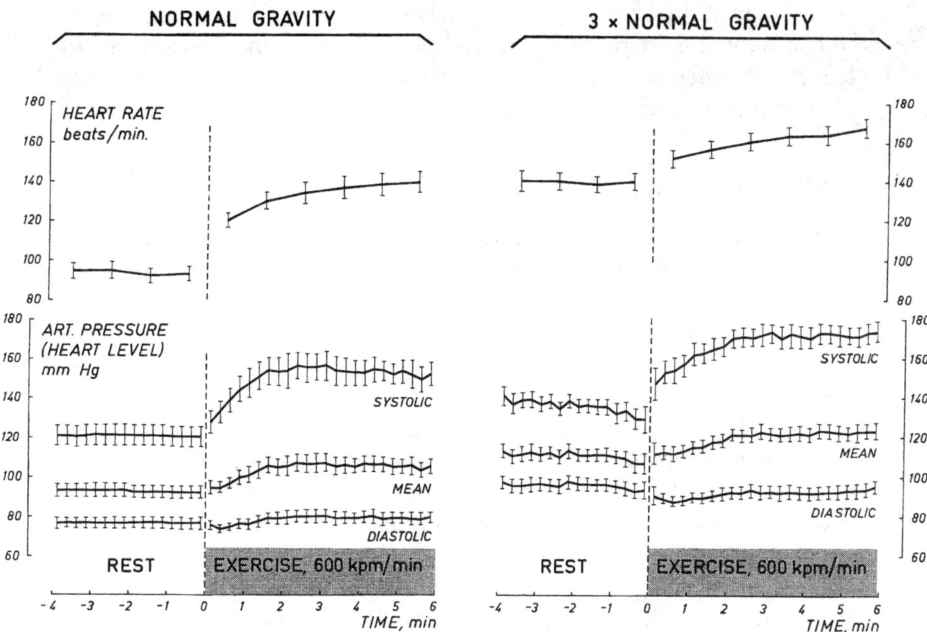

Figure 4 Time courses of heart rate and arterial blood pressure before and after the onset of exercise. Data from sitting men at 1 and 3 G, N = 8. (From Ref. 4.)

curves, do not differ between 1 G and 3 G in sitting exercising men (1). Despite this index of normal intrathoracic blood volume at 3 G, exercise stroke volumes are reduced (see Fig. 1) by 25 to 30%. This may in part be due to an increased cardiac afterload at 3 G since mean arterial blood pressure at heart level is increased at 3 G (4). It may also be speculated that despite an apparently normal central blood volume, some of that blood is functionally sequestered in a congested, dependent part of the pulmonary circulation, so that the normal central blood volume during 3 G exercise does not necessarily reflect a normal preload. No data appear to be available on end-diastolic ventricular dimensions and filling pressures in exercising men during hypergravity conditions.

The dramatic increase of stroke volume in the transition from rest to exercise (see Fig. 1), however, may to some extent be an exaggerated representation of the exercise-induced increase of venous return. Part of the cause of the very low resting stroke volume at 3 G rest (i.e., less than 50% of that at 1 G rest) may be a result of the very marked tachycardia, which does not appear to be effective in maintaining cardiac output in the face of reduced venous return (21). The relative tachycardia at 3 G is less pronounced during exercise (see Fig. 1).

Most likely, the tachycardia is caused by a hypotensive stimulus perceived at the carotid sinus, where arterial pressures are reduced, due to the increased hydrostatic pressure difference between the heart and the carotid sinus, despite an elevated arterial pressure at the level of the heart (4,28).

B. Microgravity

If cardiovascular responses to exercise in microgravity were to be extrapolated from corresponding findings in normal gravity and hypergravity, one would predict that cardiac output for a given submaximal metabolic load would be increased and that exercise capacity would be higher than at 1 G. Actual space flight results do not concur with such extrapolations. Michel et al. (13) compiled data on heart rate and blood pressure in nine astronauts who performed exercise at 25, 50, and 75% of maximal aerobic capacity on Skylab during the 1970s. Inflight measurements did not include cardiac output, but there was nothing in the heart rate and blood pressure data to suggest that hemodynamic responses were different from preflight. Later, not only the submaximal exercise performance (14,15) but also the maximum exercise performance (18) in microgravity was determined; During the two Shuttle flights SLS-1 and SLS-2, six subjects were studied before, during, and after 9 to 14 days of space flight. Breath-by-breath gas exchange was determined with a mass spectrometric technique and pulmonary blood flow was estimated by CO_2 rebreathing (14,18). Subjects performed dynamic leg exercise at intensities corresponding to 30 and 60% of preflight maximum $\dot{V}_{O_2}$ (14). On the ground, experiments were performed in both upright and supine postures. In a comparison between preflight and inflight tests, systemic circulatory responses to submaximal exercise did not demonstrate any striking impairments in weightlessness, but neither did the absence of hydrostatic pressure differences in the circulatory system appear to present any advantage in terms of circulatory gas transport. Heart rate and blood pressure responses to exercise were very much the same as in supine control experiments. However, cardiac output increased much less with increasing $\dot{V}_{O_2}$ in microgravity compared with both supine and upright exercise at 1 G (Fig. 5). For any subject, stroke volumes at rest in microgravity were higher than those either supine or upright; at 30%, exercise was comparable to or lower than supine, and at 60%, exercise was lower than supine and lower or comparable to erect. Assuming that it is the O_2 delivery ($\dot{Q} \times O_2$ content) that is controlled rather than $\dot{Q}$ during exercise, Shykoff et al. (14) speculate that the inflight plasma volume loss and the associated increase in hemoglobin concentration allowed their subjects to maintain O_2 delivery while allowing a lower $\dot{Q}$. Alternative theories include sequestration of part of the effective circulatory blood volume in a more homogeneously filled pulmonary circulation. Further alternative theories include a reduced efficiency of the muscle pump compared with upright exercise at 1 G, in the view of this

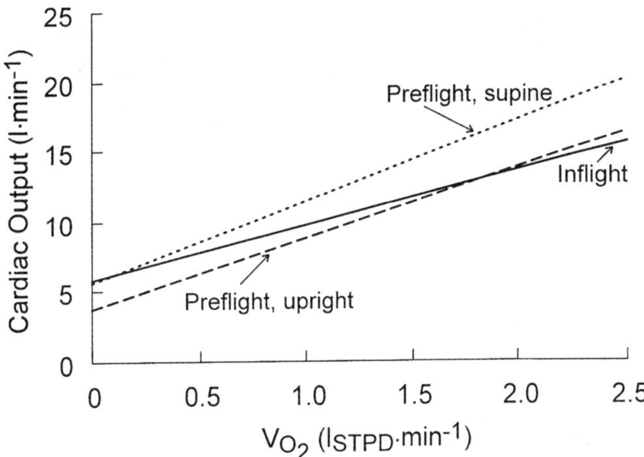

Figure 5   Cardiac output as a function of $\dot{V}_{O_2}$ in six astronauts before (upright, supine) and during sustained microgravity (inflight). Best-fit linear functions. Data obtained at rest and during exercise at 30 and 60% of the preflight maximum work load. The linear fit accounted for >95% of the total variance. Inflight, the slope of cardiac output on $\dot{V}_{O_2}$ was significantly reduced to 58% of preflight control ($p < 0.05$). (Data from Ref. 14.)

reviewer the most likely explanation. It might be argued that a similar pump problem would exist during the supine control experiments at 1 G, but here the alternating elevation of the legs during pedaling might have boosted venous return in a way it would not in the absence of gravity.

During the same flight and with the same subjects as Shykoff et al. (14), Levine et al. (18) determined the maximal exercise capacity and the peak O_2 uptake ($\dot{V}_{O_{2peak}}$) during an incremental work protocol. On the ground, but not in flight, they also determined the cardiac output by a CO_2 rebreathing method. Their main finding was that $\dot{V}_{O_{2peak}}$ was maintained during the space flight period despite typical decrements of plasma and blood volume (29) (Fig. 6). Upon return to Earth, however, $\dot{V}_{O_{2peak}}$ during upright leg exercise was decreased to 78% of preflight control. At the same time, peak $\dot{Q}$ was reduced to a similar degree. Even larger $\dot{Q}$ reductions were reported by Buderer et al. (30), studying astronauts performing submaximal exercise following Skylab missions. Within the next six to nine days, most of this decrement had been recovered (18,30). The postflight time course of $\dot{V}_{O_{2peak}}$ (18) was very similar to that observed after head-down tilt bed rest of similar duration (31,32), but the maintained $\dot{V}_{O_{2peak}}$ during space flight was an unexpected finding, since the plasma and blood volume decrements generally held to be the cause of the reduced aerobic capacity after bed rest and space flight already had been completed at the time when the inflight $\dot{V}_{O_{2peak}}$ determina-

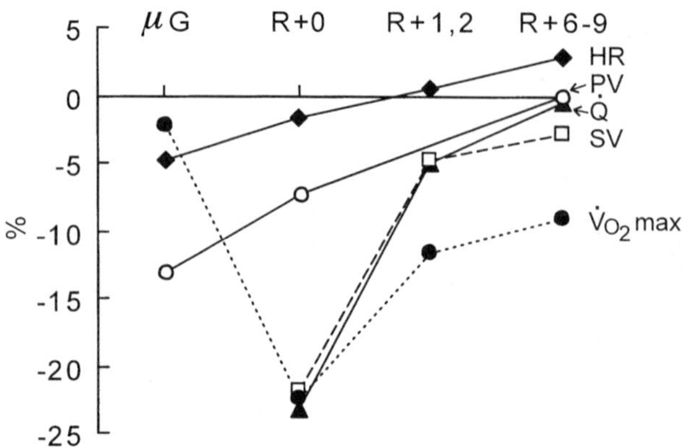

Figure 6 Peak O_2 uptake during an incremental work test ($\dot{V}_{O_2max}$), and simultaneous values for stroke volume (SV), cardiac output ($\dot{Q}$), and heart rate (HR). Same subjects as in Figure 5. µG, microgravity; R + 0, day of return from Space Shuttle flight; R + 1, 2, etc., days after return. (Hemodynamic data from Ref. 18.) Also shown are plasma volume (PV) data. (From Ref. 29.)

tions were made (29). Levine et al. (18) conclude that in the absence of gravity, skeletal muscle perfusion, and, likely, central blood volume and cardiac filling appear to be well maintained during maximal work in microgravity, possibly as a function of redistribution of the blood volume out of the capacitance vessels of the legs and splanchnic circulation. This conclusion, however, remains to be substantiated by actual measurements of cardiac output during classical maximum exercise tests including supramaximal loads to acertain level-off of $\dot{V}_{O_2}$ as a function of external power.

The operational constraints imposed on maximum exercise tests during space flight were circumvented by Stegemann et al. (15), who employed an innovative systems-analysis approach to assess aerobic capacity from submaximal exercise experiments. Their theoretical basis is the established relationship between the dynamics of $\dot{V}_{O_2}$ readjustments with sudden changes of power output on one hand and the relative work intensity on the other (33–35). Four subjects were studied during the D-2 shuttle flight in 1993. They performed sitting exercise with a pseudo-random binary sequence of submaximal workload changes before, during, and after 10 days of space flight. In the dynamic responses of breath-by-breath gas exchange, Stegemann et al. (15) found lag times between the muscles and lungs that were within the normal range. This suggested a maintained venous volume between the muscles and lungs, despite an overall decrease of plasma and blood volumes. Only upon return, a marked shortening of lag times suggested reduced volumes in the venous sections between the working muscles and the lungs. The time constants of $\dot{V}_{O_2}$ adjustment to sudden increases in external power were not changed, indicating a maintained aerobic capacity at the muscular level. In summary, the data of Levine et al. (18) and Stegemann et al. (15) point to a maintained capacity for peripheral O_2 utilization during and after sustained microgravity. The reduced plasma and blood volumes during and after sustained microgravity only impair O_2 transport after microgravity, but not during microgravity when blood volume and distribution appear optimized for that specific condition.

IV. Oxygen Requirements of Exercising Muscles

A. Hypergravity

The oxygen cost of a given external work intensity during sitting dynamic leg exercise increases with gravity in the head-to-foot direction (36). Rosenhamer (1) studied subjects at rest and while performing 50- and 100-W exercise and found an increase of $\dot{V}_{O_2}$ at 3 G compared with 1 G amounting to about 200 mL/min during exercise and about 50 mL/min at rest. Bjurstedt et al. (3) found a parallel increase of the O_2 cost of exercise at 50, 100, and 150 W of 240 to 280 mL/min, and 120 mL/min increase at rest (Fig. 7). Nunneley et al. (9) included zero-load pedaling in a similar study and found a parallel shift of the work load/

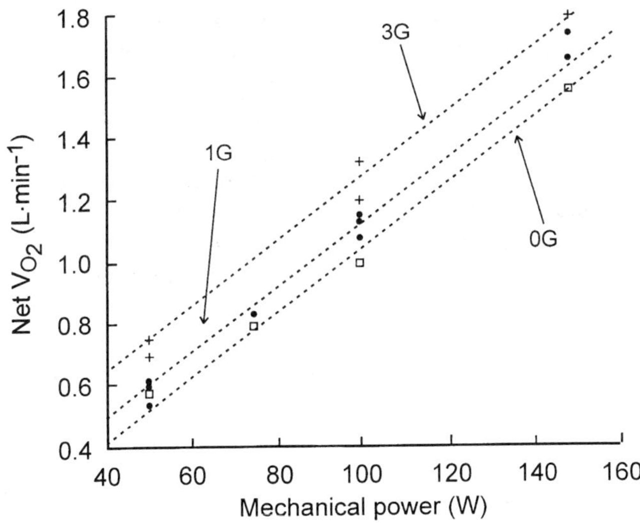

Figure 7 Net $\dot{V}_{O_2}$ (resting − exercise) as a function of external power output. (From Ref. 37.)

$\dot{V}_{O_2}$ relationship by $+300$ to 400 mL/min when comparing 3 G with 1 G data. A common feature in all three studies was that G-induced $\dot{V}_{O_2}$ differences were less at rest than during exercise. Among these investigators, however, there were differing conclusions as to the origin of the G-dependent increase of the O_2 cost of a given external intensity of dynamic leg exercise. Bjurstedt et al. (3) performed a model analysis of the energy cost of moving the legs, and concluded that the additional cost should be in the added postural and cardiorespiratory work. Nunneley et al. (9) on the other hand, considered the additional O_2 cost of exercise at 2 G and 3 G too large to be accounted for by only postural and cardiorespiratory work.

B. Microgravity

Exercise studies in microgravity would provide an ideal model for studies of the O_2 cost of moving the limbs in a gravitational field, providing that cardiorespiratory responses in microgravity are similar to those in normal gravity. As described above, the differences between cardiorespiratory responses to submaximal exercise between 1 G and microgravity are not such that they should significantly influence the overall O_2 cost of exercise. Michel et al. (13) reported that the O_2 cost of a given relatively heavy intensity is less at microgravity than during ground control. Data from Shykoff et al. (14) are not reported in such a way that a comparative analysis of the O_2 cost of exercise at 30 and 60% of $\dot{V}_{O_{2max}}$ could be performed. The data from Stegemann (15) from four subjects show great variability between flight days in one subject who was studied on three occasions, whereas one of the subjects was studied only once in flight, and consequently no statistical comparison was performed.

In microgravity, the body weight cannot be used to counterbalance the force of muscle action the pedals, so a comparison between normal gravity and microgravity must take into account to what extent an additional muscle effort has been required in microgravity to create such a counterforce. During the Skylab experiments (13), a head support was designed for that purpose, and in the experiments of Shykoff et al. (14) a shoulder support was utilized. In these two cases, therefore, postural muscles of the torso must have been engaged in microgravity in a way they probably would not have been on the ground. In the experiments of Stegemann (15), subjects were fastened on the saddle with a belt with the intention of creating a biomechanical similarity to sitting leg exercise in normal gravity.

Later, Girardis et al. (37) utilized an ergometer assembly in which the semirecumbent subject was fastened on a seat pad and against a back support with a belt. Two subjects were studied before and during a 6-month stay on the Russian Mir station. They performed dynamic leg exercise at 50, 75, and 100 W repeatedly; a total of six observations at each intensity was made at 1 G and total of 10 observations at microgravity. Oxygen uptake was consistently lower in microgravity, and Girardis et al. (37) used the data to propose a mathematical model in which the "internal" work of moving the legs against gravity was defined. The findings of Girardis et al. (37) together with the data of Michel et al. (13) and high-G results by Rosenhamer (1), Bjurstedt et al. (3), and Nunneley et al. (9) are compiled in Figure 7. Taken together, these data support the notion that the additional O_2 cost of exercise in a gravitational field is caused by the action of gravity on the limbs, where asymmetries in the ascending and descending pedal movements create gravity-dependent energy dissipation.

V.   Lessons from Microgravity: Exercise Gas Exchange

The most important lesson from microgravity is that our understanding of the physiological effects of gravity are not such that we readily can predict microgravity results by extrapolations based on experiments in normal gravity and hypergravity. This justifies a continued research effort in microgravity since it provides new insights that cannot be obtained otherwise.

A second lesson can be learned from a comparison between data obtained in hypergravity and microgravity with regard to completeness of measurements and data quality including number of subjects and experimental design. Such a comparison underscores that space research is vastly more complex than ground-based research and that consistency of procedures, high numbers of subjects, and complete sets of invasive measurements have been difficult, if not impossible, to achieve. For future research efforts in space physiology, scientific requirements must be given a higher priority over operational constraints than is presently the case.

The reviewed studies performed in hypergravity and microgravity show that alterations of the direction and magnitude of the force of gravity have multiple influences on the gas transport between the environment and the working muscles in humans. The need for O_2 flux through respiration and circulation for a given external power output has a minimum in weightlessness and increases with gravity in the head-to-foot direction. At the same time, systemic circulatory transport is curtailed in increased gravity by reduced venous return and increased hydrostatic pressure differences to cranial tissues. The efficiency of the lungs as a gas exchanger is markedly reduced at high gravity, and the shunt fraction increases with transition from rest to exercise and further with increasing exercise intensity. In summary, increased gravity acting in the head-to-foot direction has a profound negative impact on gas exchange in exercising humans. Not surprisingly, however, the overall function of the gas transport system in the exercising human appears to be optimized for normal gravity, so that no significant advantages are gained in microgravity.

References

1. Rosenhamer G. Influence of increased gravitational stress on the adaptation of cardiovascular and pulmonary function to exercise. Acta Physiol Scand 1967; 276(suppl):1–61.
2. Rosenhamer G. Antigravity effects of leg exercise. Acta Physiol Scand 1968; 72: 72–80.
3. Bjurstedt H, Rosenhamer G, Wigertz O. High-G environment and responses to graded exercise. J Appl Physiol 1968; 25:713–719.
4. Linnarsson D, Rosenhamer G. Exercise and arterial pressure during simulated increase of gravity. Acta Physiol Scand 1968; 74:50–57.
5. Bjurstedt H, Hesser CM, Liljestrand G, Matell G. Effects of posture on alveolar-arterial CO_2 and O_2 differences and on alveolar dead space in man. Acta Physiol Scand 1962; 54:65–82.
6. Wasserman K, Van Kessel AL, Burton GG. Interaction of physiological mechanisms during exercise. J Appl Physiol 1967; 22:71–85.
7. Riley RL, Lilienthal JL Jr, Proemmel DD, Franke RE. On the determination of the physiologically effective pressures of oxygen and carbon dioxide in alveolar air. Am J Physiol 1946; 147:191–198.
8. Fenn WO, Rahn H, Otis AB. A theoretical study of the composition of the alveolar air at altitude. Am J Physiol 1946;146:637–653.
9. Nunneley SA, Shindell DS. Cardiopulmonary effects of combined exercise and $+G_z$ acceleration. Aviat Space Environ Med 1975; 46:878–882.
10. Fowler KT, Read J. Cardiac oscillations in expired gas tensions, and regional pulmonary blood flow. J Appl Physiol 1961; 16:863–868.
11. Prisk GK, Guy HGB, Elliott AR, West JB. Inhomogeneity of pulmonary perfusion during sustained microgravity on SLS-1. J Appl Physiol 1994; 76:1730–1738.
12. Permutt S. Pulmonary circulation and the distribution of blood and gas in the lungs. In: Physiology in the Space Environment. Publication 1485B. Washington, DC: National Academy of Sciences and National Research Council, 1967.
13. Michel EL, Rummel JA, Sawin CG, Buderer MC, Lem JD. Results from Skylab medical experiment M 171—metabolic activity. In: Johnston RS, Dietlein LF, eds. Biomedical Results from Skylab. SP-377. Washington, DC: National Aeronautics and Space Administration, 1977:284–312.
14. Shykoff BE, Farhi LE, Olszowka AJ, Pendergast DR, Rokitka MA, Eisenhardt CG, Morin RA. Cardiovascular response to submaximal exercise in sustained microgravity. J Appl Physiol 1996; 81:26–32.
15. Stegemann J, Hoffmann U, Erdmann R, Essfeld D. Exercise capacity during and after spaceflight. Aviat Space Environ Med 1997; 68:812–817.
16. Prisk GK, Guy HJB, Elliott AR, Deutschman RA III, West JB. Pulmonary diffusing capacity, capillary blood volume, and cardiac output during sustained microgravity. J Appl Physiol 1993; 75:15–26.
17. Verbanck S, Larsson H, Linnarsson D, Prisk GK, West JB, Paiva M. Pulmonary tissue volume, cardiac output, and diffusing capacity in sustained microgravity. J Appl Physiol 1997; 83:810–816.

18. Levine BD, Lane LD, Watenpaugh DE, Gaffney FA, Buckey JC, Blomqvist CG. Maximal exercise performance after adaptation to microgravity. J Appl Physiol 1996; 81:686–694.

19. Torre-Bueno JR, Wagner PD, Saltzman HA, Gale GE, Moon RE. Diffusion limitation in normal humans during exercise at sea level and simulated altitude. J Appl Physiol 1985; 58:989–995.

20. Blomqvist CG, Stone HL. Cardiovascular adjustments to gravitational stress. In: Shepherd JT, Abboud FM, eds. Handbook of Physiology, Section 2: The Cardiovascular System. Vol. III, Part 2. Bethesda, MD: American Physiological Society 1983: 1025–1063.

21. Rowell LB, ed. Human Cardiovascular Control. New York: Oxford University Press, 1993.

22. Sjöstrand T. The regulation of the blood volume distribution in man. Acta Physiol Scand 1962; 26:312–327.

23. Gauer OH, Thron HL. Postural changes in the circulation. In: Hamilton WF, ed. Handbook of Physiology, Section 2: Circulation. Vol. III. Washington, DC: American Physiological Society, 1965:2409–2439.

24. Henry J, Gauer O, Kety S, Kramer K. Factors maintaining cerebral circulation during gravitational stress. J Clin Invest 1951; 30:292–301.

25. Stegall HF. Muscle pumping in the dependent leg. Circ Res 1966; 19:180–190.

26. Folkow B, Gaskell P, Waaler A. Blood flow through limb muscles during heavy rhythmic exercise. Acta Physiol Scand 1970; 80:61–72.

27. Folkow B, Haglund U, Jodal M, Lundgren O. Blood flow in the calf muscle of man during heavy rhythmic exercise. Acta Physiol Scand 1971; 81:157–163.

28. Linnarson D, Sundberg CJ, Tedner B, Haruna Y, Karemaker JM, Antonutto G, di Prampero PE. Blood pressure and heart rate responses to sudden changes of grvity during exercise. Am J Physiol (Heart Circ Physiol) 1996; 270:H2132–H2142.

29. Alfrey CP, Udden MM, Leach-Huntoon C, Driscoll T, Pickett MH. Control of red blood cell mass in spaceflight. J Appl Physiol 1996; 81:98–104.

30. Buderer MC, Rummel JA, Michel EL, Mauldin DG, Sawin CF. Exercise cardiac output following Skylab Missions: The second manned Skylab Mission. Aviat Space Environ Med 1976; 47:365–372.

31. Convertino VA, Bisson R, Bates R, Goldwater D, Sandler H. Effects of anti-orthostatic bed rest on the cardiorespiratory responses to exercise. Aviat Space Environ Med 1981; 52:251–255.

32. Saltin B, Blomqvist G, Mitchell JH, Johnson RL Jr, Wildenthal K, Chapman CB. Response to exercise after bed rest and after training. A longitudinal study of adaptive changes in oxygen transport and body composition. Circulation 1968; 7(suppl): 1–78.

33. Linnarsson D. Dynamics of pulmonary gas exchange and heart rate changes at start and end of exercise. Acta Physiol Scand 1974; 415(suppl):1–68.

34. Cerretelli P, Pendergast P, Paganelli WC, Rennie DW. Effects of specific muscle training on $\dot{V}_{O_2}$ on-response and early blood lactate. J Appl Physiol: Respirat Environ Exercise Physiol 1979; 47:61–69.

35. Hoffmann U, Essfeld D, Leyk D, Wunderlich HG, Stegemann J. Prediction of indi-

vidual oxygen uptake on-step transients from frequency responses. Eur J Appl Physiol 1994; 69:93–97.

36. Linnarsson D. Metabolic responses to gravitational changes. In: Cerretelli P, Whipp BJ, eds. Exercise Bioenergetics and Gas Exchange. Amsterdam: Elsevier/North-Holland Biomedical, 1980:297–302.

37. Girardis M, Linnarsson D, Moia C, Pendergast DR, Ferretti G. Oxygen cost of dynamic leg exercise on a cycle ergometer: Effects of gravity acceleration. Acta Physiol Scand 1999; 166:239–246.

10

Central Venous Pressure

JAY C. BUCKEY, Jr.

Dartmouth Medical School
Hanover, New Hampshire

I. Introduction

In 1987, the National Academy of Sciences report *A Strategy for Space Biology and Medical Science* recommended that central venous pressure be measured in space using an indwelling catheter (10). These measurements were deemed critical for understanding the early adaptation to weightlessness. At the time, the expectation was that central venous pressure would increase above standing and supine values due to the approximate 2-L shift of fluid from the legs to the chest upon entering weightlessness.

In 1991, the first direct measurements were made, but the expectations were not met. Central venous pressure declined (compared with preflight supine and seated values) rather than increased upon entering weightlessness (6). Subsequent measurements also did not measure increases above supine values (8,17), and most showed decreases below seated values. The reasons for these findings remain poorly understood today. The goal of this chapter is to review the changes in central venous pressure that have been seen in weightlessness and to provide possible explanations for these findings. Also, the space-flight results, measured in a setting where tissues have no weight, may provide insights into clinical conditions such as obesity, where tissue weight increases.

A.   Entry into Weightlessness: Physiological Effects

Leg volume decreases and the face becomes puffy shortly after achieving orbit (53). Leg volume decreases by approximately 1 L per leg (41)* and tissue thickness in the forehead increases by approximately 7% compared with a supine control (29). Pulmonary capillary blood volume increases by 25% (44). Intraocular pressure increases by 92% (12). Taken together, these changes reflect a significant shift of fluid from the lower body to the upper body.

Such a significant fluid shift would be expected to increase heart size and stroke volume. Early in flight (the first 24 to 48 hr), cardiac dimensions by echocardiography are increased (8,18,33). Data using rebreathing techniques also show a greater than 50% increase in stroke volume (44) and increased cardiac output. Qualitatively, these changes seem similar to what would occur when moving from the standing to supine position on Earth. Several findings, however, suggest that the early adaptation to weightlessness may differ from ground-based simulations using bed rest or water immersion.

First, the magnitude of the fluid shift may be quantitatively greater than what is seen with moving from upright to supine. A study by Thornton et al. (52) showed that simulations produced 50% or less of the fluid loss seen in the leg with weightlessness. Second, such a significant fluid shift might be expected to raise venous pressure, but measurements of peripheral venous pressure do not show this (28,30). Third, ground-based simulations all produce a diuresis, but a diuresis early in flight has not been seen in astronauts. Data from both the Apollo and Skylab programs showed decreased urine output during the first inflight day (35,36). Data from one Shuttle crewmember also showed decreased urine output and increased antidiuretic hormone (ADH) (34). In the Russian Soyuz program, two cosmonauts provided 24-hr urine samples on the day before flight and the first inflight day (20). They both showed reduced urine output. These combined reports provide urine output data for 15 people on their first day in space. Only one showed an increase above preflight levels.

B.   Rationale for Measuring Central Venous Pressure in Space

At the time of the National Academy of Sciences report in 1987, what was known was that a significant fluid shift occurred when entering weightlessness, but some

---

* For this chapter, the standing position will be used as the reference position. The rationale has been well-outlined by Gauer and Thron (19). Humans spend approximately two-thirds of their time upright, and adaptive mechanisms work to restore upright hemodynamics when people are exposed to continuous bed rest. Throughout the chapter, a change in weightlessness will be assumed to have changed with respect to a standing baseline, unless otherwise specified.

of the expected consequences of that shift (increased venous pressure, a diuresis) were not seen. This led to the recommendation to measure central venous pressure with an indwelling catheter, continuously from launch through orbital entry, to capture the pressure changes during the transition to weightlessness.

II. Measuring Central Venous Pressure in Space

Measuring central venous pressure (CVP) with an indwelling catheter on the shuttle presented several technical challenges. In contrast to a laboratory-based study, this research must take place in "the field," and the subjects cannot be limited in their activity. The volunteers for the study must be able to remain ambulatory and be able to respond to an emergency. The worst case would be if the crewmember had to perform an emergency egress during launch. They could not be constrained by cables and instrumentation. All the procedures and equipment had to be verified in ground-based simulations.

A. Ambulatory System with Minimal Drift Required

Since the venous system is very compliant, small pressure changes (1 to 5 mmHg) can represent large volume shifts. So any central venous pressure system for use in space should be able to measure small pressure changes. In addition, once a crewmember is instrumented, there is no guarantee that the launch will occur on time, or even on that day. The crewmember will need to be instrumented a few hours before entering the shuttle and may have to wait up to five hours on the launch pad. This means that drift is important and any system will have to have minimal drift. At the time preparations for the study began (early 1980s) the most reliable way to measure pressure long-term with minimal drift was with a saline-filled line connected to a pressure transducer. Also, using a saline-filled system allows a very flexible and thin polyurethane catheter to be used, which is important during movement and in emergencies.

B. Description of Space-Flight System

The system that was designed, built, and flown for the first central venous pressure measurement in space is described in the reference by Buckey et al. (7) and shown in Figure 1. The system was designed around a piezoresistive transducer with a measurement range of 0 to 100 mmHg. The drift was minimal (0.2 mmHg over 63 hr). The design incorporated several safety features to prevent electric shock. Batteries provided power for 28 hr of continuous measurement. A small pump provided a slow, continuous infusion of saline (1.5 mL/hr) through the measurement catheter to keep the tip open.

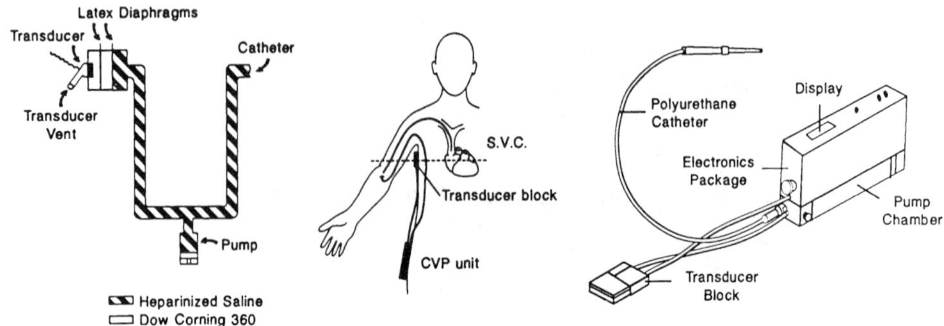

Figure 1   Schematic diagram of the central venous pressure used for measurements on the Spacelab Life Sciences 1 and 2 flights. A saline-filled catheter ran from the superior vena cava to the measurement unit. Latex diaphragms and silicon fluid provided two-failure tolerant electrical isolation. A pump provided a slow continuous flow of saline to keep the catheter tip open. (From Buckey Ref. 8.)

C.   Measurement Technique

To perform the measurements, a 4 French polyurethane catheter was advanced into the superior vena cava using fluoroscopic guidance. Once the catheter was in place, it was secured with an occlusive dressing and then the arm was moved throughout its range of motion under fluoroscopy to ensure that the catheter tip did not enter the right atrium or right ventricle. This was to prevent atrial or ventricular arrythmias. The position of the center of the right atrium was noted in both at anteroposterior and lateral dimensions and marked on the skin for proper positioning of the transducer. The pressure transducer was positioned in the axilla at the level of the right atrium and was maintained in a stable position with colostomy tape. The units themselves were sterilized and assembled using sterile technique. Each unit was individually tested in a hypobaric chamber to verify that it was insensitive to barometric pressure changes. Each unit was individually calibrated against a water column.

On launch morning, the catheter from the crewmember was routed to the outside of the launch suit by passing through a specially designed port in the suit. The catheter was connected to the measurement unit. A series of Valsalva and Mueller maneuvers were performed to verify the proper functioning of the unit. Once the crewmembers were instrumented, the measurement unit remained in a pocket on the launch suit, and the data were routed to a recorder underneath the launch seat. The data were routed through breakaway cables in case the crewmember had to egress in an emergency. Central venous pressure was measured continuously throughout launch and into orbit. Heart rate and blood pressure were

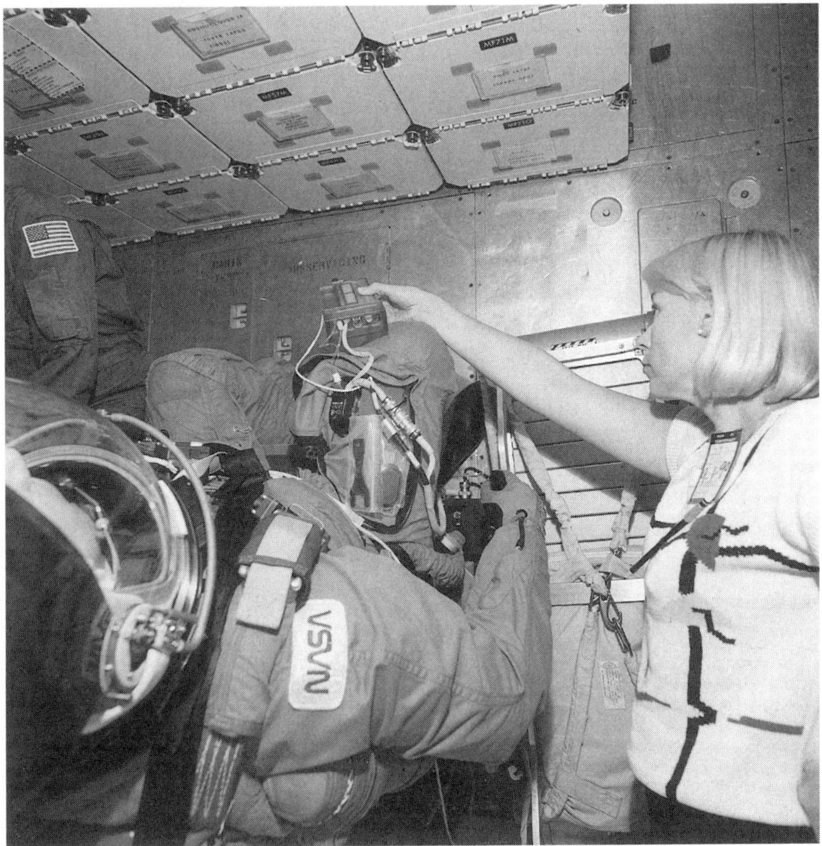

Figure 2 The central venous pressure measurement unit was placed in a pocket on the astronaut's thigh. In this picture, Lynda Lane R.N. a co-investigator on the project, holds the measurement unit to check central venous pressure during a training run. The astronaut launched in the supine, legs elevated position shown in this picture.

also measured. Figure 2 shows the arrangement of the measurement system in the mid-deck of the shuttle during a simulation.

D. First Measurement in Space

The first direct measurement of central venous pressure took place in October 1991 aboard the Spacelab Life Sciences-1 Spacelab mission. This marked the first time that an invasive cardiovascular measurement had been made on a human in space. On that flight, payload specialist F. Andrew Gaffney, M.D., a cardiologist and investigator on the project, rode the shuttle into space while instrumented with the central venous pressure line. Central venous pressure was low upon entering space and was confirmed by connecting the catheter to a backup measurement unit. Figure 3 shows mission specialist James P. Bagian, M.D., removing the catheter from Drew Gaffney during the SLS-1 mission in 1991. Central venous pressure was subsequently measured on two more individuals using this system. Two payload specialists on the Spacelab Life Sciences-2 flight (Martin Fettman, D.V.M., and Shannon Lucid, Ph.D.) also had central venous pressure lines inserted on their 1993 flight.

E. Fiber-Optic System Used Successfully

Not all central venous pressure measurements were made with the same system. Foldager et al. (17), had the opportunity to measure central venous pressure on the 1993 Spacelab D-2 mission. For their measurement system, they chose a fiber-optic central venous pressure catheter manufactured by Camino Labs (62). This system had excellent temperature stability (± 0.015 mmHg/C°) and drift

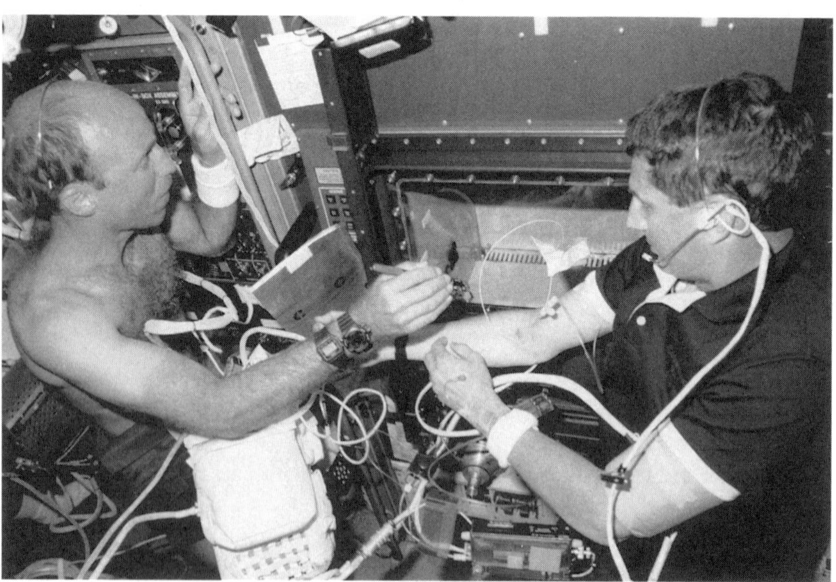

Figure 3 Dr. Drew Gaffney, the first individual to have central venous pressure measured in space has his catheter removed in space by Dr. James Bagian on the first day of the Spacelab Life Sciences 1 mission in 1993. Central venous pressure fell upon entering weightlessness. A backup unit onboard the mission confirmed the initial findings.

characteristics (± 0.1 mmHg/24 hr). One astronaut wore this system into space on the Spacelab Deutsche-2 Spacelab mission in April 1993.

III. Parabolic Flight and Space-Flight Results

A. Animal Studies

Central venous pressure was measured early in the space program during the Project Mercury ballistic flight program. Central venous pressure measured in a chimpanzee did not change appreciably upon entering weightlessness (25). Central venous pressure was again measured in one monkey during the Biosatellite 3 program. In this case, central venous pressure rose upon entering orbit and remained elevated throughout the flight (39,40). The monkey, however, became ill and died, and it is not clear if the changes on this flight represented normal physiology. Nevertheless, the results from these early flights did not refute the hypothesis that the fluid shift in space increased central venous pressure, just as in ground-based simulations such as bed rest, head-down bed rest, or head-out water immersion.

Latham et al. (31,32) measured central venous pressure in baboons during short periods of microgravity (20 to 30 sec) produced by parabolic flight. In this study, central venous pressure was reduced upon entry into microgravity in baboons who were volume depleted before the parabolas. Baboons who were fluid replete showed increases.

Rats have also been studied during parabolic flight. Somody et al. (51) measured heart rate, blood pressure, and central venous pressure continuously in four rats. The nature of any fluid shifts that might take place in a rat during microgravity exposure is not known, so it is difficult to postulate what changes in central venous pressure might take place in rat due to fluid shifts. Tissue compression and hydrostatic forces, however, may play a role. Since the heart in the rat is located below the lungs in the rat's normal posture, some hypotheses could be proposed about what might occur to cardiac pressures in the rat due to the removal of compressive and hydrostatic forces. Weightlessness might be expected to eliminate compressive forces on the heart and reduce central venous pressure. In the four animals studied by Somody et al., the short periods of microgravity produced rapid decreases in central venous pressure, and the hypergravity portions of the flight produced increases. Interestingly, blood pressure was increased and heart rate reduced in microgravity, suggesting an increase in stroke volume and cardiac output (although these variables were not measured).

B. Human Studies: Parabolic Flight

The first study to measure central venous pressure in humans during parabolic flight showed increases above both sitting and supine levels (42). In this study, subjects were in the upright sitting position when they entered the weightlessness portion of the parabola. Subsequent parabolic studies showed different results. Foldager et al. (17) measured central venous pressure in seven normal subjects using the fiber-optic system that was subsequently used in space on the Spacelab D-2 flight. In this study the subjects were supine when they entered weightlessness. Central venous pressure decreased from 6.5 ± 1.3 supine to 5.0 ± 1.4 mmHg during weightlessness.

Videbaek and Norsk (56) measured central venous pressure in seven more normal subjects using the fiber-optic system. In this study, left atrial diameter and esophageal pressures were also measured. Central venous pressure decreased from 5.8 ± 1.5 mmHg supine to 4.5 ± 1.1 mmHg in weightlessness. Left atrial diameter increased in weightlessness to 30.4 ± 0.7 mm from 26.8 ± 1.2 mm supine. The reason for the increased left atrial diameter at the decreased CVP was postulated to be due to a reduced intrathoracic pressure. Esophageal pressure, which was used to estimate intrathoracic pressure, decreased from 1.5 ± 1.6 mmHg supine in 1 G to −4.1 ± 1.7 mmHg in weightlessness. From these measurements, a transmural central venous pressure was calculated by subtracting the esophageal pressure from the CVP. When this was done, transmural pressure was noted to increase from 6.1 ± 3.2 mmHg supine to 10.4 ± 2.7 mmHg in weightlessness.

C. Human Studies: Space Flight

Kirsch et al. (30) measured peripheral venous pressure, to reflect central venous pressure, on the Spacelab 1 flight in 1983. Both subjects showed reductions below their supine values when the pressure was measured 22 hr in the mission. The authors hypothesized that venous pressure had increased upon entering weightlessness, but that by the time the measurements were made, the fluid shift was complete, plasma volume was reduced, and venous pressures were lower.

These investigators had a chance to test this hypothesis during the Spacelab D-1 mission in 1986 (28). On this mission, peripheral venous pressure data were collected on four astronauts within 20 to 40 min after launch. Again, venous pressure values in space were found to be below supine levels. One possible explanation was that central venous pressure had already leveled off while the

---

Figure 4 This summary figure presents all the central venous pressure data collected in space. During the launch, central venous pressure tracks the changes in the G load. Upon entering weightlessness, central venous pressure promptly fell and remained at a level below what it had been just minutes earlier on the launch pad. The top three tracings were made using a saline-filled catheter, the bottom tracing used a fiberoptic catheter. (Bottom tracing from Ref. 17.)

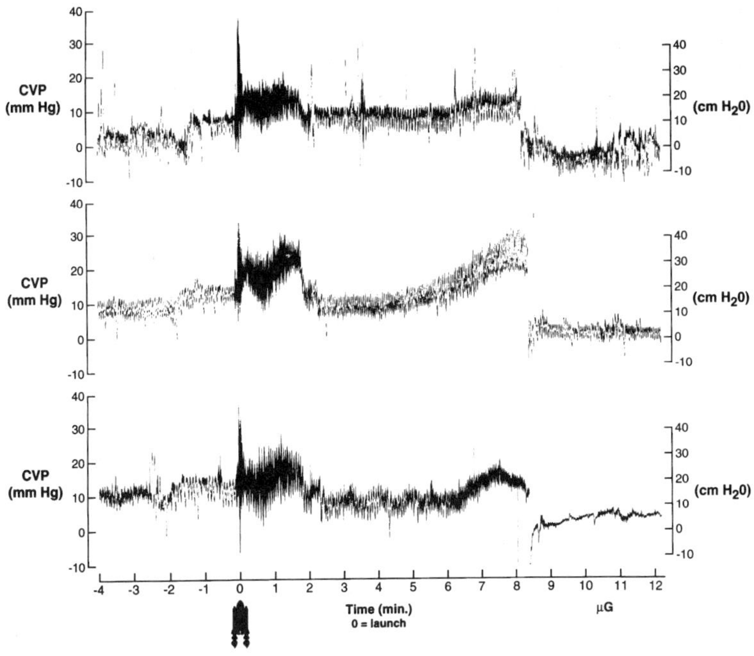

CVP
(mm Hg)
(cm H₂0)
Time (min.)
0 = launch
μG

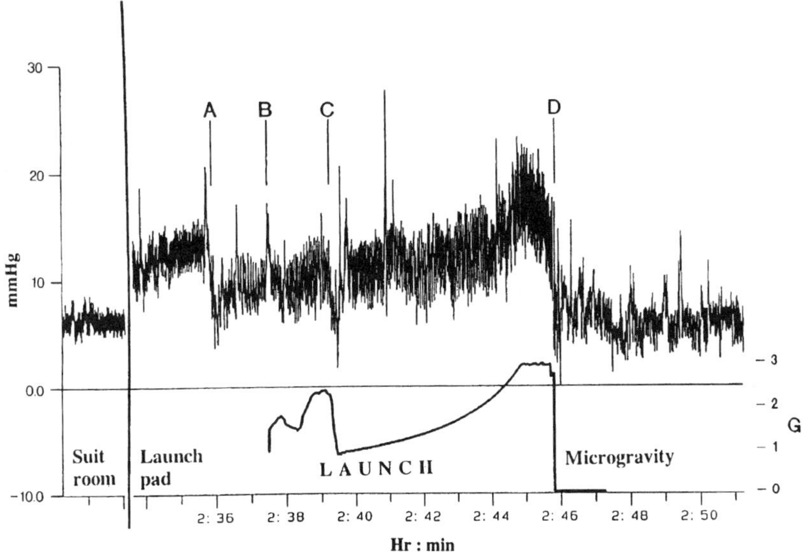

mmHg
A
B
C
D
G
Suit
room
Launch
pad
LAUNCH
Microgravity
Hr : min

Table 1 Central Venous Pressure Data (cmH_2O) on Three Subjects Before, During, and After Space Flight[a]

	Subject 1	Subject 2	Subject 3
Seated CVP	5.5	9.7	10.0
CVP 1 hr prelaunch	8.0	19.0	18.0
CVP after 10 min on orbit	2.0	3.5	2.0
CVP last measurement	1.0	3.5	8.0
Duration of CVP measurement (h)	9	10	44
CVP postflight supine	17.1		13.3

[a] In all cases, central venous pressure fell below the 1 G seated value while on orbit. The measurements made 1 h prelaunch were performed in the supine, legs-up position in the space shuttle.

astronauts were lying in the supine, legs-up position before launch on the pad. The only way to answer this question definitively was to measure central venous pressure from before launch, through the launch itself, and into orbit. This was the recommendation of the Committee on Space Biology and Medicine in 1987 (10).

Direct measurements of central venous pressure had been proposed by C. Gunnar Blomqvist and F. Andrew (Drew) Gaffney at the University of Texas

Table 2 Hemodynamic Data by Echocardiography on Three Subjects Supine Before Flight (Pre) and at the Time of Last Inflight Measurement (Inflight)[a]

| | Subject 1 | | Subject 2 | | Subject 3 | |
	Pre	Inflight	Pre	Inflight	Pre	Inflight
Heart rate (beats/min)	77	68	72	70	64	59
CVP (cmH_2O)	10.6	1.0	12.9	3.5	12.0	8.0
LVIDD (cm)	4.61	5.20	4.50	4.74	4.68	4.96
LVIDS (cm)	3.03	3.04	3.20	3.20	3.34	3.24
LAD (cm)	3.16	3.29	2.76	3.30	3.11	3.16
Diastolic volume (mL)	97.8	129.5	92.5	104.4	101.4	116.1
Systolic volume (mL)	35.9	36.2	41.0	41.0	45.4	42.2
Stroke volume (mL)	62.0	93.4	51.5	63.5	55.9	73.8
Cardiac output (L/min)	4.8	6.3	3.7	4.4	3.6	4.4
V_{cf} (circumferences/s)	1.11	1.30	1.00	1.02	0.87	1.12

[a] CVP = central venous pressure, LVIDD = left ventricular internal diameter diastole, LVIDS = left ventricular internal diameter systole, LAD = left atrial diameter, V_{cf} = velocity of circumferential fiber shortening.

Southwestern Medical Center (5,58). This proposal, first submitted in response to a NASA Announcement of Opportunity in 1978, would have to wait until 1991 to be realized on the Spacelab Life Sciences-1 mission. The results from this mission were surprising. Central venous pressure rose in the supine, legs-elevated position on the launch pad. During the launch, CVP rose further, as might be expected from previous studies on G_x acceleration (37). Upon entering orbit, however, CVP promptly dropped below the level established prelaunch on the launch pad and below the seated level. Once in orbit, a backup system was connected that confirmed the results. Hemodynamic measurements showed an increased stroke volume and cardiac output.

In 1993, another astronaut launched with a central venous pressure line in place on the Spacelab D-2 mission (17). In this case, a different measurement system was used (the Camino Labs system described above). Again, CVP fell promptly upon entering weightlessness to a level below what had been seen just before launch. Central venous pressure fell to a level at or below the supine level that had been seen in ground-based testing before flight. Hemodynamic data did not show an increase in heart rate. Later that same year, two more people wore CVP lines on the Spacelab Life Sciences-2 mission. Again, CVP decreased upon entering weightlessness, and hemodynamic data showed increases in end-diastolic volume, stoke volume, and cardiac output. Heart rate and blood pressure were unchanged. All the CVP data collected in space are shown together in Figure 4. A summary of CVP and hemodynamic findings on three people is shown in Tables 1 and 2.

IV. Components of Central Venous Pressure

Although central venous pressure is used clinically to determine volume status and to reflect right-sided filling pressure, CVP is affected by a variety of physiological events. Changes in intrathoracic pressure, cardiac function, and body position will alter CVP profoundly, so that any change in central venous pressure cannot be assumed to reflect solely a change in fluid status. These different components of central venous pressure are important for interpreting the unique results from space flight.

A. Hydrostatic Component

The presence of gravity introduces pressure gradients in all the fluid-filled compartments of the body. The gradients are due to the weight of the blood and tissues. Standing produces large venous pressure increases in the feet and pressure in the intracranial veins becomes negative. When lying supine, pressure gradients exist from the anterior to posterior chest. The pericardium and heart (4) also have pressure gradients. For work in physiology and clinical medicine, however, using a single pressure referenced to a standardized point, rather than a pressure gradient, has proved adequate. Central venous pressure is referred to the level of the tricuspid valve (23).

Although venous pressures can change dramatically as a consequence of gravity, the physiological effects are often not very profound. One part of the explanation for this is that gravitationally induced pressure gradients exist both in tissue and in blood (i.e., on both sides of vessels). For example, changes in cerebrospinal fluid and intraperitoneal pressures minimize the effects of gravitational forces on the cerebral and visceral circulations. As a result, gravitational effects can be minimized (3) since similar effects occur on both sides of the vessel wall. This underscores the importance of transmural pressure, not just the pressure within the vein or right atrium, for determining the true physiological effect of a pressure change.

B. Mean Circulatory Filling Pressure and Tissue Compression

One of the major determinants of central venous pressure is the static filling pressure within the cardiovascular system (mean circulatory filling pressure). Mean circulatory filling pressure is central venous pressure after cardiac arrest when the central arterial and venous pressures are equal (46).

Blood volume and the capacitance of the vascular system determine mean circulatory filling pressure. The capacitance properties of the vasculature are primarily determined by the characteristics of the veins and venules (46). Increases in blood volume will increase mean circulatory filling pressure. Changes in vascular capacitance will alter the pressure change seen for a given change in blood volume, with a stiffer system producing greater pressure changes for a given change in blood volume.

The role that gravity may play in determining vascular capacitance is not well understood. The weight of tissues generates compressive forces that can act on the walls of veins and venules (22). The magnitude of this effect is not known. If, however, all gravitational forces were removed from a supine person, this could remove compressive forces from the walls of veins and venules, increase vascular capacity, and increase unstressed volume (i.e., the volume of blood contained at 0 distending pressure). The net effect could be a reduction in mean circulatory filling pressure. This, in turn, could contribute to a reduction in central venous pressure.

This effect, removal of compressive forces, can be extended to larger vessels and to the heart itself. In the supine position, flow through the inferior vena cava can be impeded by compression of the vena cava from the weight of the abdominal contents (23). In pregnant women, this effect can be pronounced, leading to leg swelling and high leg venous pressures. If, however, the compression is removed, as would occur in space, this could remove this impediment to venous return, and lower venous pressures in the leg.

The heart itself, with a 4- to 5-cm internal diameter, has a hydrostatic gradient within it when supine of approximately 1 cmH_2O for every centimeter of vertical distance (61). This would be removed in space. Similarly, any compressive effects that result from the heart pressing on the lungs would disappear in space. For example, a balloon filled with water if placed on a table in Earth's gravity will deform due to its weight and assume an elliptical cross section. If the balloon becomes weightless, the balloon will assume a spherical shape and pressure in the balloon will drop. Studies in dogs during G_X (front to back) centrifugation show that the heart shape deforms with increased gravity, and that the sizes of the cardiac chambers decrease (49). Sandler (49) also constructed pressure–volume loops from the data collected during centrifugation and showed that the left ventricle became stiffer with transverse acceleration. The raises the possibility that if all gravitational forces were removed, the heart may become less stiff.

Other suggestive evidence for compression playing a role in cardiac filling comes from studies performed in the prone position. Supine, the heart is anterior in the chest and compressive forces could be generated by the chest wall and by the heart itself on the tissues below it. When prone, however, the heart is underneath the lungs. Hemodynamic measurements in the prone position show a decreased cardiac output and increased CVP (55,64). These results do not provide conclusive evidence for cardiac compression, however, since inferior vena caval compression and increased intrathoracic pressure due to chest wall and/or abdominal compression can also play a role. Also, when supine, the fluid surrounding the heart may be at a subatmospheric pressure, so that the heart is, in effect, suspended from the chest wall and not compressing the lungs. Nevertheless, these data do show how a simple movement within the gravity field can have major hemodynamic effects.

Overall, blood volume and venous capacitance affect central venous pressure. Tissue compression may have a role in determining venous and cardiac compliance. Loss of compressive forces could have a variety of effects throughout the cardiovascular system, including an increased vascular compliance, increased unstressed volume, reduced resistance in the inferior vena cava and increased distensibility of the heart. The effects of reduced compressive forces can only be hypothesized from ground-based studies, and space flight remains the only way to uncover these possible effects.

C. Dynamic Cardiac Component

Cardiac function affects central venous pressure profoundly. These relationships have been well described by Guyton et al. (21), and have been recently applied to the space-flight condition by White and Blomqvist (60). The most important of these relations for understanding central venous pressure changes in space is the mass conservation relation (or venous return curve), that shows the decrease in central venous pressure that occurs as cardiac output increases. Situations that increase cardiac output, such as might occur with increased cardiac contractility or an increase in heart rate, will also decrease central venous pressure. If space flight were to produce a primary increase in cardiac output, as might occur with increased sympathetic stimulation, this would decrease central venous pressure.

The other relationship, the cardiac function relationship, describes the change in cardiac output that occurs as central venous pressure varies. According to this relationship, a decrease in central venous pressure, such as might occur if mean circulatory pressure should fall, would decrease cardiac output. One of the most difficult problems with interpreting the central venous pressure results from space has been reconciling a decreased central venous pressure with increased cardiac output and stroke volume. The cardiac function relationship would predict a decrease in cardiac output (if nothing else changed) when CVP falls. The two relationships (cardiac function relationship and venous return curve) taken together define the equilibrium point for the circulation for a given condition.

Although these relationships have proved useful and robust over the years for describing cardiac performance, they may have limitations. The heart may have the ability to generate suction, and cardiac filling may not be totally dependent on passive filling due to central venous pressure (45,63). Elastic recoil during diastole may have a role in cardiac filling. It is interesting to speculate if this recoil may be enhanced by the absence of tissue compression in weightlessness. The other tissues within the thorax could impede recoil due to their weight and location near the heart. When the tissues are weightless, recoil might proceed unimpeded, leading to an increased diastolic volume for a given filling pressure.

D. Intrathoracic Pressure Component

Central venous pressure is extremely sensitive to pressure changes in the chest. The Valsalva maneuver, Mueller maneuver, coughing, and straining will all produce immediate and marked changes in CVP. Situations that reduce intrathoracic pressure, such as negative pressure breathing (26,27), also reduce central venous pressure. In the study by Kilburn and Sieker (27), an inspiratory pressure of -20 to -22 cmH$_2$O, decreased central venous pressure by 10.5 mmHg.

One effect of changing intrathoracic pressure would be to change the transmural filling pressure of the heart. When the thorax is opened, thereby increasing intrathoracic pressure, cardiac output is reduced (16). Conversely, decreases in intrathoracic pressure can increase cardiac transmural pressure and increase cardiac output. Data from studies using negative pressure breathing or negative pressure ventilation suggest this is the case. A continuous negative extrathoracic pressure of -10 cmH$_2$O increased cardiac output by 19% in the study by Torelli et al. (54), although a negative pressure of -20 cmH$_2$O in the study

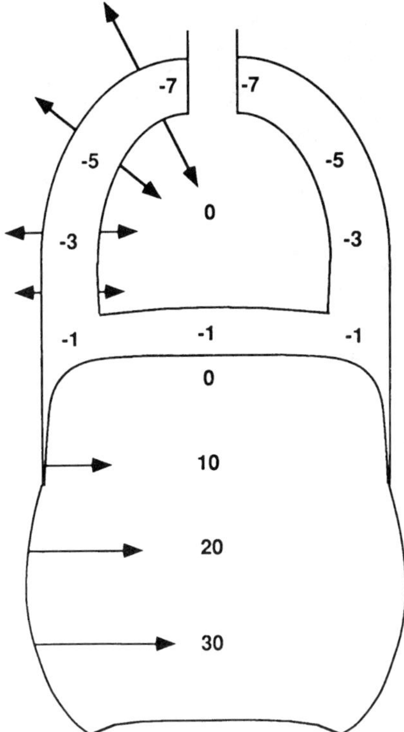

Figure 5 This diagram shows that gravity affects pressures in the abdomen, rib cage, and lung. The effect of weightlessness on the heart and lung must also take into account what will happen to pressures within the abdomen as well. (From Ref. 13.)

by Shiga et al. (50) increased left ventricular end-diastolic area, but not cardiac output. Negative pressure breathing at a level of -20 cmH$_2$O in the study by Kilburn and Sieker (27) increased cardiac index by 31%. Large decreases in intrathoracic pressure, however, decrease cardiac function. Mueller maneuvers at -60 cmH$_2$O decrease ejection fraction and V_{cf} (9). Negative intrathoracic pressures of -20 mmHg increased both end-systolic and end-diastolic volumes in dogs during steady-state right heart bypass (24).

From these studies, the relationship between intrathoracic pressure and cardiac output can be summarized as follows: Negative pressure breathing makes intrathoracic pressure more subatmospheric, which reduces lung volume, thoracic volume, and central venous pressure—but increases cardiac output. If the decrease in intrathoracic pressure is too large, then there is a decrease in cardiac

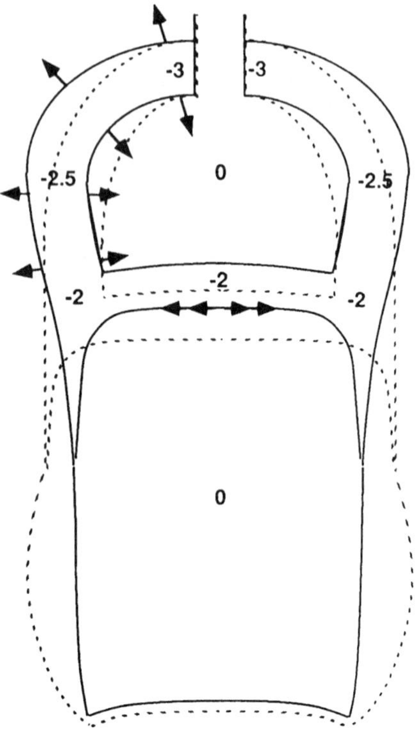

Figure 6 A proposed arrangement of abdominal and intrathoracic pressures in weight-lessness. Intrathoracic pressure is more uniform than when standing in 1 G and pressures at heart level are not changed dramatically. (From Ref. 17.)

function. One can also make intrathoracic pressure more subatmospheric by in-creasing lung (and thoracic) volume by decreasing extrathoracic pressure. This also decreases central venous pressure and increases cardiac output.

Normally, human intrathoracic pressure is negative compared with atmo-spheric pressure. The origin of this negative pressure is the elastic recoil of the chest wall. Hydrostatic effects superimpose a pressure gradient. When standing, intrathoracic pressure is most negative at the top of the lung and less so at the bottom. When supine, intrathoracic pressure is more negative anteriorly than pos-teriorly (48). In weightlessness, the chest wall is unloaded, and hydrostatic pres-sures are removed. Estenne et al. (15) noted that the rib cage at end-expiration is displaced in the cranial direction and adopts a more circular shape in microgravity produced during parabolic flight. A subsequent study suggested that findings in parabolic flight reflect what actually occurs in prolonged weightlessness (57). Such an expansion of the chest wall might lead to a more negative intrathoracic pressure. No data exist, however, on intrathoracic pressure in weightlessness. If intrathoracic pressure did become more negative, however, this would increase lung volume and decrease esophageal pressure. When weightless measurements of lung volume and esophageal pressure are compared with seated measurements in 1 G, they show decreased lung volumes and more positive (increased) esophageal pressure in space. This suggests intrathoracic pressure becomes more positive in weightlessness. If the comparison is made with supine values in 1 G, lung volume is increased in weightlessness, suggesting intrathoracic pressure becomes more negative. Controversy exists on how to interpret supine esophageal pressure measurements in 1 G to compare with weightless values.

With the loss of hydrostatic forces, intrathoracic pressure should become uniformly negative throughout the chest (13,61). Figures 5 and 6 show a proposed arrangement in intrathoracic pressure in weightlessness. A uniformly negative intrathoracic pressure, rather than a pressure gradient, will surround the heart. In comparison with the supine position, intrathoracic pressure anteriorly may become less negative and posteriorly it may become more negative. Similarly, when compared with the upright position, pressures at the apex may become less negative and at the bases more negative. The net effect of this change on cardiac transmural pressure is difficult to estimate.

E. Studies in Hypergravity

In the early 1960s considerable research was directed to the effects of G_X acceleration (acceleration from front to back on the chest). This was because of the G forces that astronauts would experience during reentry in a space capsule. These studies, however, provide data relevant for understanding central venous pressure during a shuttle launch. During a shuttle launch, the crewmembers also experience G_X loads, so that studies from centrifuge studies allow an understanding of the changes that occur during launch. Also, the data from hypergravity could be extrapolated to the 0 gravity state to formulate hypotheses.

Figure 7 shows data on right atrial pressure during G_X centrifugation taken from the study by Lindberg et al. (37). These data show right atrial pressure measured during the first minute of a 10-min G_X acceleration. With increasing G_X, right atrial pressure increased immediately. Whether the increase in right atrial pressure is primarily due to increased intrathoracic pressure, cardiac compression, or chest wall compression is not known. The overall change in pressure was approximately 5 cmH$_2$O per G of acceleration. Since the right atrial pressure change occurred immediately upon the initiation of acceleration, it is likely to reflect changes in hydrostatic and compressive forces rather than reflex hemodynamic changes in the circulation. If these data can be extrapolated to the weightless state, a reduction in central venous pressure of approximately 5 cmH$_2$O might be expected due to the loss of compressive forces upon entering weightlessness.

Animal data also showed similar results. Sandler (49) measured right atrial

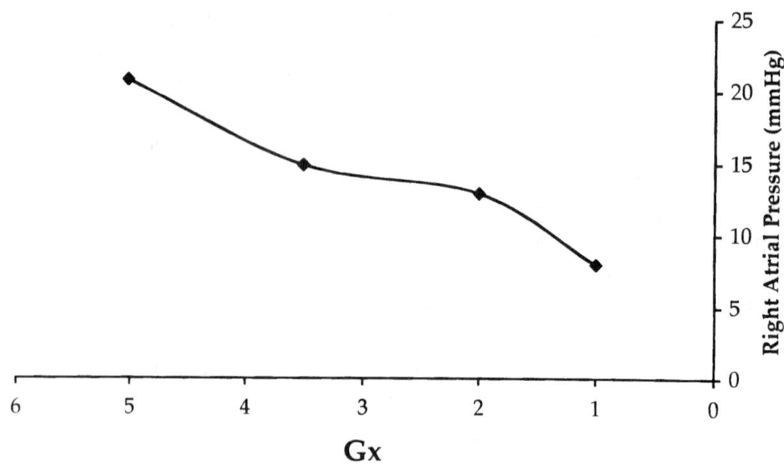

Figure 7   Right atrial pressure data from the article by Lindberg et al. are plotted against G level. The subjects rode in the centrifuge gondola in a supine, legs-up posture very similar to the one used during a Shuttle mission. Right atrial pressure rose promptly with the onset of G and fell as soon as the increased G force was removed. If these data were extrapolated to the weightlessness condition, an approximate 5 cmH$_2$O decrease in right atrial pressure would be expected.

pressure and cardiac volume via angiocardiography in dogs undergoing centrifugation at 5 G$_X$. Right atrial pressures were increased and stroke volume decreased during the runs. The angiograms showed deformation of the heart with acceleration. Pressure–volume curves plotted from the data showed that the heart had become stiffer during the centrifugation.

Transmural pressure during centrifugation has also been measured. Banchero et al. (4) measured both right atrial and pericardial pressures during G$_X$ acceleration in dogs. They found that the changes in pericardial pressure balanced the changes in right atrial pressure so that transmural right and left atrial pressures were unchanged by centrifugation. They concluded that ''since the heart is relatively flaccid and fluid filled, the maintenance of proper function of this pump would be much more certain if it were suspended in a hydrostatic system which automatically applied perfectly compensated pressures to all of its external surfaces whenever the gravitational or inertial forces acting on this organ were altered.'' On the basis of their findings, they concluded this was the case. They extrapolated their results to the weightless condition and predicted that the gradients in pericardial pressure would be eliminated in weightlessness.

The applicability of studies in hypergravity to weightlessness exposure is not known. If, however, the assumption is made that studies from high G exposures can be extrapolated to weightlessness, some interesting findings emerge. Based on hypergravity studies, weightlessness would be expected to produce (1) uniform intrathoracic pressures, (2) uniform pericardial pressures, (3) a reduced CVP, and (4) increased stroke volume.

F. Effect of Increased Body Weight

Another clinical situation that might be relevant to weightlessness is obesity. In obesity, tissue weight is increased, so that cardiovascular changes seen in obesity might allow for predictions about what would occur if tissue weight were 0. De Divitiss et al. (11) studied 10 normal, nonhypertensive, obese subjects with right- and left-heart catheterization. Right atrial pressure was positively correlated ($p <$ 0.05) with both weight and the degree of overweight. Both cardiac output and stroke volume were also positively correlated with weight, but cardiac index and stroke volume index were not. The authors also found that the ratio of stroke work index to left ventricular end-diastolic pressure (a measure left ventricular function) also declined significantly with both weight and the degree of over-weight. The authors concluded that right atrial pressure increased with body weight and that obesity produced an impairment in left ventricular function.

Agarwal et al. (1) found a similar situation when comparing a group of 10 obese (91.8 $\pm$ 18 kg) overweight surgical patients with 10 normal patients. The obese patients had significantly higher right atrial pressures (120% greater than the normal group). Cardiac output was increased in the obese patients, but not cardiac index. Plots of left ventricular function curves (left ventricular stroke work vs. pulmonary artery wedge pressure) showed decreased cardiac function in the obese group. The obese patients showed high filling pressures of both the right and left heart. Mean pulmonary artery pressure was higher in the obese patients but not mean arterial pressure.

Impaired cardiac performance was also demonstrated in the study by Alaud-din et al. (2) of 30 morbidly obese patients. This study demonstrated a significant positive correlation of pulmonary capillary wedge pressure with weight. In addition, this group gave a 1-L infusion of Ringer's lactate to the patients and measured the changes in central venous pressure and stroke volume. Although central venous pressure increased significantly, stroke volume did not. This group also studied 12 of their patients again after they lost an average of 55 kg. Filling pressures had decreased. In addition, echocardiographic measurements showed a reduction in left ventricular mass. The authors concluded that obesity produced elevated filling pressures and this might result from a noncompliant ventricle, perhaps due to hypertrophy.

Taken together, these studies all show a positive correlation between cardiac filling pressures and body weight. In addition, these higher pressures do not lead to increased stroke volume or cardiac output. With weight reduction, central

venous pressure declines and cardiac function improves. These results are difficult to apply directly to microgravity. Chronic obesity leads to multiple physiological changes (increased blood volume, increased blood pressure, left ventricular enlargement) that could affect central venous pressure independent of any
change in compression due to tissue weight. Nevertheless, the possibility exists
that some of the increase in central venous pressure seen with obesity is due to the
compression of thoracic contents. If this is true, the opposite might be expected to
occur with microgravity.

V. The Paradox: Increased Stroke Volume with Reduced Central Venous Pressure

The data collected on CVP in space clearly refuted the hypothesis that the fluid
shift that occurs in space would increase CVP. Also, heart rate and echocardiographic data showed no evidence for an increase in contractility or sympathetic
stimulation, which could decrease CVP while increasing cardiac output (see Tables 1 and 2). The results instead described a unique situation with an increased
stroke volume and reduced CVP. To date, these findings have defied an easy
explanation. While no definitive answer can be given, two major approaches to
explaining the results exist. In one, changes in intrathoracic pressure are para-
mount and in the other tissue compression plays the key role.

A. Reduced Intrathoracic Pressure Hypothesis

This hypothesis was recently well-outlined by White and Blomqvist (60). These
authors propose that ''during weightless spaceflight, the chest relaxes with a concomitant shape change that increases the volume of the closed chest cavity. This
leads to a decrease in intrapleural pressure, ultimately causing a shift of blood
into the vessels of the chest, increasing the transmural filling pressure of the heart,
and decreasing the central venous pressure.''

Evidence Supporting Reduced Intrathoracic Pressure Theory

In the study by White and Blomqvist (60), a three-compartment model based on
the principles used by Guyton was used to simulate the cardiovascular system.
In this model, a reduction in intrathoracic pressure was the one situation that
could best explain the findings from space flight. A reduction in intrathoracic
pressure will increase cardiac output and decrease CVP both in the model and
in human studies (see Section IV.D above). Data from space flight show that
chest wall configuration does change (57). Recent studies from parabolic flight
have shown a reduction in esophageal pressure in weightlessness, when compared
with supine esophageal pressure. When this esophageal pressure was subtracted

from central venous pressure, it could be demonstrated that cardiac transmural pressure was increased in weightlessness (56). Along with this calculated increase in transmural pressure, atrial dimension also increased, supporting the notion that atrial distension was due to an increased transmural pressure caused by a reduction in intrathoracic pressure induced by a change in chest wall configuration.

Also, this theory can fit with some of the lung volume data collected in space. Lung volume is intermediate between supine and seated values in weightlessness, with larger lung volumes in weightlessness than supine (14). So if a supine person is made weightless, the chest wall will expand and lung volume will increase. If alveolar pressure stays the same (which is the case), then the increase in lung volume would represent a decreased (more negative) intrathoracic pressure. In keeping with the description above, this should also produce a reduced CVP and increased cardiac transmural pressure.

Problems with Reduced Intrathoracic Pressure Theory

Currently, no data on pleural pressure exist from weightlessness, so information on pressure outside of the heart comes from measurements of esophageal pressure, such as in the study by Videbaek et al. How to interpret esophageal pressure measurements properly, however, has presented problems since the measurement was first performed. The evidence for increased transmural pressure in the study by Videbaek et al. comes from supine esophageal pressure measurements that were compared to microgravity values. Supine esophageal pressures, however, may be significantly more positive than pleural pressures in the supine position, as has been shown by Mead and Gaensler (38). Another study by Rutishauser et al. (48) showed that esophageal pressure exceeds intrapleural pressure at the same vertical height by 3 to 5 cmH$_2$O. Figure 8, from the study by Mead and Gaensle (38), shows dramatic changes in esophageal pressure when moving from supine to upright, compared with the modest change in intrapleural pressure. If these data are correct (and they have been questioned, since the situation at the catheter tip could not be precisely known), then the interpretation from the Videbaek study would be different.

In the study by Videbaek et al., it is possible that the pressures in weightlessness accurately reflect intrathoracic pressure, but that the ground-based measurements in the supine position are erroneously high as discussed above, perhaps due to compression by thoracic contents. As a result, transmural CVP would be underestimated in the supine position and transmural pressure would not actually increase in weightlessness. Another way to establish the change in intrathoracic pressure is to use seated esophageal pressures as the baseline. Measurements taken in the seated position in 1 G would not have the interpretation problems of supine measurements. When the seated esophageal pressure measurements from the Videbaek study are analyzed it turns out that esophageal pressure is

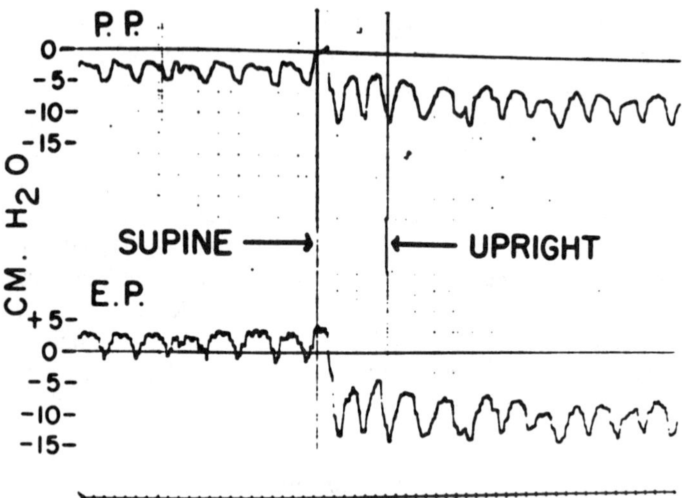

Figure 8   This figure shows simultaneous pressure tracings in the supine and seated position in a human subject where both esophageal and intrapleural pressure were measured. Supine esophageal pressure exceeds intrapleural pressure, and the change in esophageal pressure upon moving from the supine to the upright posture is much greater than the change in pleural pressure. Supine, esophageal pressure may be higher due to compression by the thoracic contents and may not reflect pleural pressure accurately. Seated esophageal pressures, however, show a good correlation. (From Ref. 38.)

significantly *less negative* upon entering weightlessness. This would not support the reduced intrathoracic pressure theory.

The data could be suspect for another reason. If the assumption is made that esophageal pressure supine does accurately reflect intrathoracic pressure, another problem emerges. In the supine position, due to the pressure gradients, intrapleural pressure will be higher posteriorly than anteriorly. The esophageal pressure transducer is below the heart when supine, and so the transducer is measuring a higher intrathoracic pressure that would be seen at the level of the CVP transducer. In effect, the transmural CVP measurement supine would be subtracting intrathoracic pressure below the heart from CVP at midheart level. In an adult, the distance between the esophagus and middle of the right atrium is in the range of 4 to 6 cm. Since the transmural pressure change measured in the Videbaek study was approximately 6 cm H_2O, compensation for the difference in location of the transducers might eliminate the reported change in transmural pressure.

Data on chest wall mechanics also do not necessarily support a major reduction in intrathoracic pressure, but the body position used as the baseline to interpret these data is extremely important. Most pulmonary physiology studies use a seated or standing baseline. In the study by Edyvean et al. (13), the rib cage moved outward (compared with the seated condition) in the apical zones in weightlessness. This outward movement could lead to a more negative intrathoracic pressure. In addition, however, they also noted *inward* displacement of the abdominal wall and abdominal rib cage and hypothesized that this would lead to slightly less negative pleural pressures at the lung bases.

Also, a change in intrathoracic pressure should be reflected as a change in lung volume. A reduced intrathoracic pressure would lead to increased lung volume and an increased intrathoracic pressure to decreased lung volume if alveolar pressure stays the same. Since lung volume decreases in weightlessness compared with the seated position in 1 G, this suggests that transpulmonary pressure is decreased. Transpulmonary pressure equals alveolar pressure minus intrathoracic pressure, and since alveolar pressure is unchanged, the reduction in lung volumes suggests that intrathoracic pressure has increased compared with the seated values. An increase in intrathoracic pressure is also supported by the esophageal pressure data from Videbaek and Norsk (56) and Prisk et al. (43). In both these studies, esophageal pressure increased in weightlessness compared with the seated 1 G value.

For the supine condition, however, the situation is slightly different. Lung volumes in weightlessness are increased compared with the supine values, which would indicate a decrease in intrathoracic pressure. As mentioned earlier, this supports the reduced intrathoracic pressure hypothesis. There is, however, still a question of whether the magnitude of this change could explain the data that have been collected to date. West and Prisk (59), estimated that the change in intrathoracic pressure that would explain the approximately 500-mL increase in lung volume seen in weightlessness would be <2 mmHg, which may be too low to explain the measured reductions in CVP, which have averaged approximately 6 to 7 mmHg.

Last, although modeling has proven to be extremely useful in understanding cardiovascular function, the standard models do not incorporate compressive forces, the removal of which may be important in weightlessness.

B. Tissue Compression Hypothesis

The unique feature of weightlessness exposure is that all tissues have no weight and that any forces or pressure gradients that exist due to the weight of tissues would disappear. The nature of these forces can be seen in hypergravity studies where the effects of tissue weight are exaggerated. The tissue compression hypothesis assumes that changes in cardiovascular variables that occur in hypergravity can be extrapolated into weightlessness. In this scenario, the heart is unloaded since it is weightless and no longer presses on surrounding tissues. In addition, the lungs and other tissues in the chest that interact with the heart are also weightless and would not interfere with cardiac function. The effect is similar to what would happen to a water-filled balloon resting on a table. In 1 G, the balloon is compressed by the table and the internal pressure is increased by this interaction. Once weightless, however, the pressure on the balloon is removed and the balloon accommodates the same volume at a lower pressure. If this effect were applied to the cardiovascular system, the net effect would be that the entire heart–lung system would become more compliant and accommodate a given volume at a lower pressure.

Evidence Supporting Tissue Compression Theory

The main support for this tissue compression theory comes from studies using front-to-back G_X acceleration. All the CVP measurements made to date have studied subjects who were lying in a supine, legs-up position in the shuttle. The subjects rested in 1 G_X then experienced up to 3 G_X, and then entered weightlessness. In this special case, a specific question can be asked: What is the effect of removing all gravitational forces from a person in the supine, legs-up posture? In all cases, CVP measured in weightlessness was decreased from the values seen in the supine, legs-up position just before launch. Also, in all cases, just as has been seen in centrifuge studies, CVP rose during ascent as G loads increased. As soon as the G forces were removed, CVP promptly fell.

Data from the study by Lindberg et al. (37) (see Fig. 7) when extrapolated to 0 G would predict a fall in CVP just as is seen in weightlessness. Also, the data on cardiac compliance from the study by Sandler (49), if extrapolated to 0 G, would predict an increase in cardiac compliance. The results of centrifugation studies would suggest that intrathoracic pressure becomes uniform, but would not show major decreases or increases at heart level (48). The net effect of weightlessness, extrapolated from G_X centrifugation, would be increased cardiac compliance, a relatively uniform intrathoracic pressure, and a decreased CVP.

Studies on obese patients also provide some support for the tissue compression hypothesis. As discussed earlier in the chapter, supine central venous pressure is increased in obese patients, and the increase in central venous pressure is proportional to the degree of obesity. When obese patients lose weight, the pressure normalizes.

Problems with Tissue Compression Theory

The main problem with the tissue compression theory is the same one that affects the intrathoracic pressure theory—lack of comprehensive data. In the absence of comprehensive pressure and volume measurements in the thorax, all theories explaining the CVP changes seen in weightlessness are speculative.

Also, it is not known if CVP measurements in weightlessness can be extrapolated from G_X centrifugation studies. If G_Z (head-to-foot) acceleration were used instead for comparison, a different conclusion would be reached. High G_Z levels decrease CVP, so if these data were extrapolated to weightlessness, the expectation would be that CVP would rise, just as was anticipated at the time of the 1987 National Academy of Sciences report (10). This highlights the problem of establishing the appropriate 1 G reference for weightless measurements.

VI. Conclusion

At the time that central venous pressure measurements were first made in space, the prediction was that CVP would rise, just as is the case with bed rest or water immersion. Based on this theory, upon entering weightlessness, a fluid shift would occur and this would be reflected in an increased CVP. The fact that CVP fell in weightlessness revealed that the effects of gravity on the cardiovascular system were not well understood, and that the standard thinking on the early adaptation to space flight was incorrect.

Two main hypotheses exist to explain the findings from space flight. In one theory, the changes in CVP and cardiac output can be explained by a reduction in intrathoracic pressure. In the other, the loss of gravitational compression of tissues is thought to change pressure–volume relationships acutely. Also, it is possible that components of both theories taken together can explain the findings. At present, insufficient data exist to determine which theory is correct. Without hard data on fluid shifts and on pressures surrounding the heart and lungs, any explanation remains speculative.

A. Remaining Questions

The lesson learned from studying the changes in central venous pressure in space is that some basic effects of gravity on the cardiovascular system are still not understood. Does intrathoracic pressure decrease in weightlessness compared with the standing position on Earth? Does it decrease compared with supine values? Is tissue compression an important factor, and is there a way to model this effect? Does mean circulatory filling pressure decrease? If tissue compression is an important factor, what are the effects on veins and venules throughout the cardiovascular system? All of these questions have a bearing on explaining the early adaptation to weightlessness.

B. Application to Clinical Medicine

The main application of the central venous pressure findings is in the situation where tissue compressive forces are increased—obesity. Reference values for monitoring patients do not necessarily take into account the effect that the weight of tissues may have on intravascular pressures. A pulmonary wedge pressure that would be interpreted as high for a normal-sized individual may, in fact, be within the normal range for an obese patient.

C. Summary

The truly surprising finding from the CVP measurements in space is how little is actually known about the effect of a changing force environment on cardiac and pulmonary function. To date, the CVP measurements in space have defied an easy explanation.

References

1. Agarwal N, Shibutani K, SanFilippo JA, Del Guercio LR. Hemodynamic and respiratory changes in surgery of the morbidly obese. Surgery 1982; 92:(2)226–234.
2. Alauddin A, Meterissian S, Lisbona R, MacLean LD, Forse RA. Assessment of cardiac function in patients who were morbidly obese. Surgery 1990; 108:(4)809–818; discussion 818–820.
3. Avasthey P, Wood EH. Intrathoracic and venous pressure relationships during responses to changes in body position. J Appl Physiol 1974; 37:(2)166–175.
4. Banchero N, Rutishauser WJ, Tsakiris AG, Wood EH. Pericardial pressure during transverse acceleration in dogs without thoracotomy. Circ Res 1967; 20:(1)65–77.
5. Blomqvist CG. Cardiovascular adaptation to weightlessness. Med Sci Sports Exer 1983; 15:428–431.
6. Buckey JC, Gaffney FA, Lane LD, Levine BD, Watenpaugh DE, Blomqvist CG. Central venous pressure in space. N Engl J Med 1993; 328:1853–1854.
7. Buckey JC, Goble RL, Blomqvist CG. A new device for continuous ambulatory central venous pressure measurement. Med Instrum 1987; 21:238–243.
8. Buckey JC Jr, Gaffney FA, Lane LD, Levine BD, Watenpaugh DE, Wright SJ, et al. Central venous pressure in space. J Appl Physiol 1996; 81:19–25.
9. Buda AJ, Pinsky MR, Ingels NB Jr, Daughters GT 2d, Stinson EB, Alderman EL. Effect of intrathoracic pressure on left ventricular performance. N Engl J Med 1979; 301:453–459.
10. Committee on Space Biology and Medicine. A Strategy for Space Biology and Medical Science for the 1980s and 1990s. Washington, DC: National Academy, 1987.
11. de Divitiis O, Fazio S, Petitto M, Maddalena G, Contaldo F, Mancini M. Obesity and cardiac function. Circulation 1981; 64:477–482.
12. Draeger J, Schwartz R, Groenhoff S, Stern C. Self-tonometry under microgravity conditions. Aviat Space Environ Med 1995; 66:568–570.
13. Edyvean J, Estenne M, Paiva M, Engel LA. Lung and chest wall mechanics in microgravity. J Appl Physiol 1991; 71:1956–1966.
14. Elliott AR, Prisk GK, Guy HJ, West JB. Lung volumes during sustained microgravity on Spacelab SLS-1. J Appl Physiol 1994; 77:2005–2014.
15. Estenne M, Gorini M, Van Muylem A, Ninane V, Paiva M. Rib cage shape and motion in microgravity. J Appl Physiol 1992; 73:946–954.

16. Fermoso JD, Richardson TQ, Guyton AC. Mechanism of decrease in cardiac output caused by opening the chest. Am J Physiol 1964; 207:1112–1116.

17. Foldager N, Andersen TA, Jessen FB, Ellegaard P, Stadeager C, Videbaek R, et al. Central venous pressure in humans during microgravity. J Appl Physiol 1996; 81: 408–412.

18. Frey MAB, Charles JB, Houston DE. Weightlessness and response to orthostatic stress. In: Smith JJ, ed. Circulatory Response to the Upright Posture. Boca Raton, FL: CRC, 1990:91.

19. Gauer OH, Thron HL. Postural changes in the circulation. In: Hamilton WF, ed. Handbook of Physiology. Circulation. Washington, DC: American Physiological Society, 1965:2409–2439.

20. Grigoriev AI, Kozyrevskaya GI, Natochin Yu V, Tigranyan RA, Balakhovskiy IS, Belyakova NI, et al. Metabolic endocrine processes. In: Gazenko OG, Kakurin LI, Kutznetsov AG, eds. Space Flights in the Soyuz Spacecraft. Biomedical Research. NASA Technical Translation TT F-17524 ed. Washington, DC: National Aeronautics and Space Administration, 1977:307–351.

21. Guyton AC, Coleman TG, Granger HJ. Circulation: Overall regulation. Annu Rev Physiol 1972; 34:13–46.

22. Guyton AC, Granger HJ, Taylor AE. Interstitial fluid pressure. Physiol Rev 1971; 51:527–563.

23. Guyton AC, Hall JE. Vascular distensibility, and functions of the arterial and venous systems. In: Guyton AC, Hall JE, eds. Textbook of Medical Physiology, 9th ed. Philadelphia: Saunders, 1996:171–181.

24. Hausknecht MJ, Brin KP, Weisfeldt ML, Permutt S, Yin FC. Effects of left ventricular loading by negative intrathoracic pressure in dogs. Circ Res 1988; 62:(3)620–631.

25. Henry JP, Mosely JD. Results of the Project Mercury Ballistic and Orbital Chimpanzee Flights. 1963; NASA SP-39 ed. Washington, DC: National Aeronautics and Space Administration, 1963.

26. Ikeda T, Iwase S, Saito M, Mano T. Effects of positive and negative pressure breathing on muscle sympathetic nerve activity in humans. Aviat Space Environ Med 1997; 68: 494–498.

27. Kilburn KH, Sieker HO. Hemodynamic effects of continuous positive and negative pressure breathing in normal man. Circ Res 1960; 8:660–669.

28. Kirsch K, Haenel F, Rocker L. Venous pressure in microgravity. Naturwissenschaften 1986; 73:447–449.

29. Kirsch KA, Baartz FJ, Gunga HC, Rocker L. Fluid shifts into and out of superficial tissues under microgravity and terrestrial conditions. Clin Invest 1993; 71:687–689.

30. Kirsch KA, Rocker L, Gauer OH, Krause R, Leach C, Wicke HJ, et al. Venous pressure in man during weightlessness. Science 1984; 225:218–219.

31. Latham RD, Fenton JW, Vernalis MN, Gaffney FA, Crisman RD. Central circulatory hemodynamics in nonhuman primates during microgravity induced by parabolic flight. Adv Space Res 1994; 14:349–358.

32. Latham RD, Fenton JW, Vernalis MN, Koenig SC. Circulatory filling pressures during transient microgravity induced by parabolic flight. Physiologist 1993; 36:S18–S19.

33. Lathers CM, Riddle JM, Mulvagh SL, Mukai C, Diamandis PH, Dussack LG, et al. Echocardiograms during six hours of bedrest at head-down and head-up tilt and during space flight. J Clin Pharmacol 1993; 33:535–543.

34. Leach CS. Fluid control mechanisms in weightlessness. Aviat Space Environ Med 1987; 58:A74–A79.

35. Leach CS, Alexander WC. Endocrine, electrolyte and fluid volume changes associated with Apollo missions. In: Johnston RS, DLFBCA, ed. Biomedical Results from Apollo. NASA SP-368 ed. Washington, DC: National Aeronautics and Space Administration, 1975:163–184.

36. Leach CS, Rambaut PC. Biochemical responses of the Skylab crewmen: An overview. In: Johnston RS, DLF, ed. Biomedical Results of Skylab. NASA SP-377 ed. Washington DC: National Aeronautics and Space Administration, 1977: 204–216.

37. Lindberg EF, Marshall HW, Sutterer WF, McGuire TF, Wood EH. Studies of cardiac output and circulatory pressures in human beings during forward acceleration. Aerospace Med 1962; 33:81–91.

38. Mead J, Gaensler EA. Esophageal and pleural pressures in man, upright and supine. J Appl Physiol 1959; 14:81–83.

39. Meehan JP, Rader RD. Cardiovascular observations of the Macaca nemestrina monkey in Biosatellite 3. Aerospace Med 1971; 42:322–336.

40. Meehan JP, Rader RD. Cardiovascular observations in the Macaca nemestrina monkey in Biosatellite 3. Physiologist 1973; 16:184–193.

41. Moore TP, Thornton WE. Space shuttle inflight and postflight fluid shifts measured by leg volume changes. Aviat Space Environ Med 1987; 58:A91–A96.

42. Norsk P, Foldager N, Bonde-Petersen F, Elmann-Larsen B, Johansen TS. Central venous pressure in humans during short periods of weightlessness. J Appl Physiol 1987; 63:2433–2437.

43. Prisk GK, Elliott AR, Guy HJB, West JB. Lung volumes and esophageal pressures during short periods of microgravity and hypergravity (abstr). Physiologist 1990; 33:A83.

44. Prisk GK, Guy HJ, Elliott AR, Deutschman RA 3d, West JB. Pulmonary diffusing capacity, capillary blood volume, and cardiac output during sustained microgravity. J Appl Physiol 1993; 75:15–26.

45. Robinson TF, Factor SM, Sonnenblick EH. The heart as a suction pump. Sci Am 1986; 254:84–91.

46. Rothe CF. Mean circulatory filling pressure: Its meaning and measurement. J Appl Physiol 1993; 74:499–509.

47. Rutishauser WJ, Banchero N, Tsakiris AG, Edmundowicz AC, Wood EH. Pleural pressures at dorsal and ventral sites in supine and prone body positions. J Appl Physiol 1966; 21:1500–1510.

48. Rutishauser WJ, Banchero N, Tsakiris AG, Wood EH. Effect of gravitational and inertial forces on pleural and esophageal pressures. J Appl Physiol 1967; 22:1041–1052.

49. Sandler H. Angiocardiographic and hemodynamic study of transverse (G_x) acceleration. Aerospace Med 1966; 37:901–910.

50. Shiga T, Takeda S, Nakanishi K, Takano T, Sakamoto A, Ogawa R. Transesophageal

echocardiographic evaluation during negative-pressure ventilation using the Hayek oscillator. J Cardiothorac Vasc Anesth 1998; 12:527–532.

51. Somody L, Fagette S, Frutoso J, Gharib C, Gauquelin G. Recording heart rate and blood pressure in rats during parabolic flight. Life Sci 1998; 63:851–857.

52. Thornton WE, Hedge V, Coleman E, Uri JJ, Moore TP. Changes in leg volume during microgravity simulation. Aviat Space Environ Med 1992; 63:789–794.

53. Thornton WE, Moore TP, Pool SL. Fluid shifts in weightlessness. Aviat Space Environ Med 1987; 58:A86–A90.

54. Torelli L, Zoccali G, Casarin M, Dalla Zuanna F, Lieta E, Conti G. Comparative evaluation of the haemodynamic effects of continuous negative external pressure (CNEP) and positive end-expiratory pressure (PEEP) in mechanically ventilated trauma patients. Intens Care Med 1995; 21:67–70.

55. Toyota S, Amaki Y. Hemodynamic evaluation of the prone position by transesophageal echocardiography. J Clin Anesth 1998; 10:32–35.

56. Videbaek R, Norsk P. Atrial distension in humans during microgravity induced by parabolic flights. J Appl Physiol 1997; 83:1862–1866.

57. Wantier M, Estenne M, Verbanck S, Prisk GK, Paiva M. Chest wall mechanics in sustained microgravity. J Appl Physiol 1998; 84:2060–2065.

58. West JB. Spacelab—the coming of age of space physiology research. J Appl Physiol: Respir Environ Exerc Physiol 1984; 57:1625–1631.

59. West JB, Prisk GK. Chest volume and shape and intrapleural pressure in microgravity [letter]. J Appl Physiol 1999; 87:1240–1241.

60. White RJ, Blomqvist CG. Central venous pressure and cardiac function during spaceflight. J Appl Physiol 1998; 85:738–746.

61. Wood EH, Nolan AC, Donald DE, Cronin L. Influence of acceleration on pulmonary physiology. Fed Proc 1963; 22:1024–1034.

62. Yablon JS, Lantner HJ, McCormack TM, Nair S, Barker E, Black P. Clinical experience with a fiberoptic intracranial pressure monitor. J Clin Monit 1993; 9:171–175.

63. Yellin EL. Concepts related to the study of diastolic function: A personal commentary. J Cardiol 1999; 33:223–236.

64. Yokoyama M, Ueda W, Hirakawa M, Yamamoto H. Hemodynamic effect of the prone position during anesthesia. Acta Anaesthesiol Scand 1991; 35:741–744.

11

Pulmonary Interstitial Fluid Balance

DANIELA NEGRINI

Università Degli Studi di Milano
Milan, Italy

GIUSEPPE MISEROCCHI

Università Degli Studi di Milano-Bicocca
Monza, Italy

I. Introduction

Changing G causes great modifications in the regional blood distribution and therefore is expected to affect fluid balance in the interstitial spaces. For example, it is well known that increasing G causes a marked accumulation of blood into the caudal part of the body, the reverse being true in 0 G where the general appearance of a subject is described as having ''spider legs and puffy face.'' In the latter condition, leg volume decreases by about 2 L (1) and tissue thickness of the forehead increases by approximately 7% with respect to normal (2). Although changes in interstitial fluid volume do occur in various organs depending on functional conditions, they certainly ought to be prevented in the lung tissue to assure optimal gas diffusion. This chapter deals with the pulmonary interstitial fluid balance in physiological conditions at 1 G and discusses the possible perturbations induced by 0 G exposure. Data obtained during space missions and parabolic flights indicate that 0 G exposure caused an increase in pulmonary blood volume and perfusion (3), conditions potentially leading to an increase in interstitial fluid volume; it will therefore be useful to discuss the series of events leading to interstitial fluid accumulation in the transition toward development of pulmonary edema. In parallel, we also present fluid dynamics in the pleural cavity, which is exquisitely sensitive to gravity and whose fate is closely associated with pulmonary fluid balance.

II. Structure of Pulmonary Interstitium

Pulmonary gas diffusion takes place across the alveolar–capillary membrane, which is well designed to fulfill this task; in fact, its specific anatomical features are a large ($\sim$70 m^2 in humans) and very thin ($\sim$0.1 to 0.5 μm) surface area for gas exchange. The alveolar and the capillary walls delimit the pulmonary interstitium (Fig. 1), a thin compartment made of a fiber system serving as a scaffold, and of other macromolecules forming the capillary and the alveolar basement membranes and filling the extravascular space. The extracellular matrix provides a strong but elastic framework for the delicate alveolar epithelial–capillary structure and consists mainly of collagen, elastic fibers, and proteoglycans (PGs) that are organized in the extracellular matrix through noncovalent interactions. Proteoglycans are important components of the macromolecular structure of the extracellular matrix; they are also involved in the regulation of cell adhesion, migration and proliferation, and in the modulation of the biological activities of matrix-bound growth factors and cytokines (4,5). Furthermore, PGs are responsible for two important aspects of microvascular and interstitial fluid dynamics, namely, the sieving properties of the capillary membrane and of the

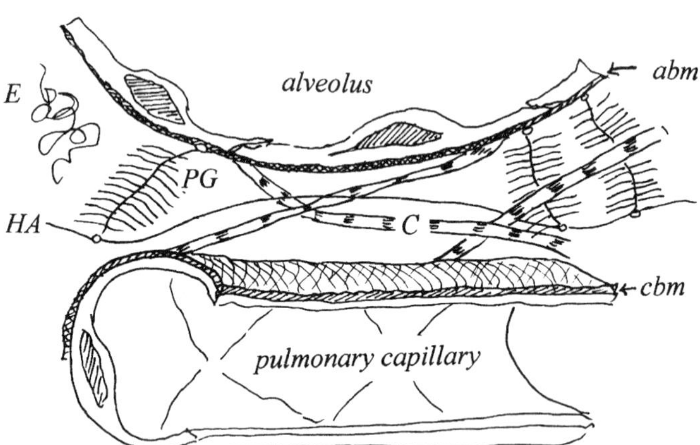

Figure 1 Schematic drawing of the pulmonary interstitium depicting some important macromolecules of the extracellular matrix. Abbreviations: abm, alveolar basal membrane; cbm, capillary basal membrane; E, elastic fibers; C, collagen bundle; HA, hyaluronic acid; and PG, proteoglycan.

extracellular matrix and the mechanical properties (compliance) of the interstitial tissue (6). Proteoglycans include families of multidomain core proteins covalently linked to one or more glycosaminoglycan (GAG) chains (4,5). Various PG families are present in the lung. Large chondroitin sulfate (CS)–containing PG (versican) is found in the interstitial matrix where it forms aggregates with the hyaluronic acid. The relatively high number of CS chains gives a high anion charge to the macromolecule, allowing it to display marked hydrophilic properties and to control the hydration of the interstitial tissues. Small dermatan sulfate (DS)–containing PG (decorin) is associated with collagen fibers (5). Heparan sulfate (HS)–containing PGs include perlecan and syndecans, found in the basal membrane and in the cell membrane, respectively (5). The copolymeric nature of HS chains permits specific interaction properties, which are involved in basement membrane organization, in receptor functions, and in cell-to-cell and cell-to-extracellular matrix interactions (5,7). Water is present as a free phase within the pores of the extracellular matrix and is subject to a continuous turnover due to transcapillary fluxes. As will be discussed later, the hydraulic pressure of the interstitial free water phase (P_i) is a key variable to discuss interstitial fluid dynamic.

III. The Pleural Cavity

The pleural cavity is a matrix-free interstitial space where the distribution of pleural liquid pressure (P_{liq}) values is characteristically affected by the gravitational field. In mammals, pleural liquid volume decreases with increasing body size, being about 2.5 mL/kg body weight in rats and ~0.4 mL/kg in sheep, the average pleural liquid thickness being ~10 to 20 µm (8). Pleural liquid ion content mirrors that of extracellular fluids, whereas its protein concentration decreased from 2.5 g/dL in rats to 1 g/dL in 40-kg mammals (8). A boundary type of lubrication between the sliding pleural mesothelia is provided by pleural fluid and by several layers of surface-active phospholipids adsorbed on the surface of the mesothelial cells (9,10).

IV. Measurements of Hydraulic Pressures in Intact Pleuropulmonary Compartment: The Pleural Window Technique

The hydraulic pressure of the liquid phase is a key variable in understanding the water balance at interstitial level. Indeed, interstitial pressure in the pleuropulmonary compartments relates to the tissue water content, resulting, in turn, from a complex interaction among several factors: (1) the transmembrane fluid fluxes, (2) the tissue forces related to the degree of lung expansion, (3) the forces arising from surface tension phenomena at the alveolar–air interface, and (4) the lymph fluid drainage. Any change in one set of forces would influence the other ones, so any of them can be regarded as either an independent or a dependent variable; the result of such a complex interaction might involve a perturbation of the steady-state extravascular water balance.

Most of the present knowledge on pleuropulmonary interstitial fluid balance relies on the data base of hydraulic interstitial (15) and microvascular (16) pressure values obtained through micropuncture technique with an experimental approach that preserved the integrity of lung–chest wall coupling and of the systemic and pulmonary circulation. This technique is based on opening a "window" in an intercostal space, after resecting the intercostal muscles down to the endothoracic fascia. Under stereomicroscopic view, the endothoracic fascia is gently stripped, exposing the underlying transparent parietal pleura through which a 2- to 3-μm tip glass micropipette can be advanced via a micromanipulator to record the hydraulic pressure from the extrapleural parietal interstitium, the pleural cavity, the pulmonary interstitium, and the pulmonary microvessels. Since the micropuncture technique is minimally invasive, it allowed us to monitor pressures in the physiological condition, as well as during the early transition phase toward development of interstitial hydraulic or lesional edema experimentally caused through discrete perturbations of pulmonary fluid balance.

V. Height Distribution of Hydraulic Pressure Values in Pleuropulmonary Compartments

Figure 2 shows the distribution of hydraulic pressures at the end-expiratory lung volume in the pleuropulmonary compartments as a function of lung height in supine rabbits (17); the right atrium level, identified as the indifferent pressure point of the vascular system, corresponds to a height of about 3 cm. Pleural liquid pressure (P_{liq}) is ~ -2 cmH$_2$O in the lowermost part of the cavity and becomes more subatmospheric by 0.7 cmH$_2$O/cm with increasing height within the pleural space. Pulmonary interstitial pressure (P_i) distribution displays the same slope, but is shifted toward more negative values as a function of height. The line labeled as P_{epl} refers to the distribution of fluid pressure values in the extrapleural parietal interstitium.

Hydraulic pressures in the vascular system are assumed to vary with a vertical gradient of 1 cmH$_2$O/cm height relative to pressure value at heart level of 23 and 10 cmH$_2$O in systemic capillaries of the parietal pleura (18) and in pulmonary capillaries (16), respectively. Figure 3 shows how a change in pleural surface pressure (that mirrors a change in lung volume) affects P_{epl}, P_{liq} and P_i measured at about 50% of lung height (19). The chest expansion causes pleural surface pressure to become more subatmospheric and the same trend

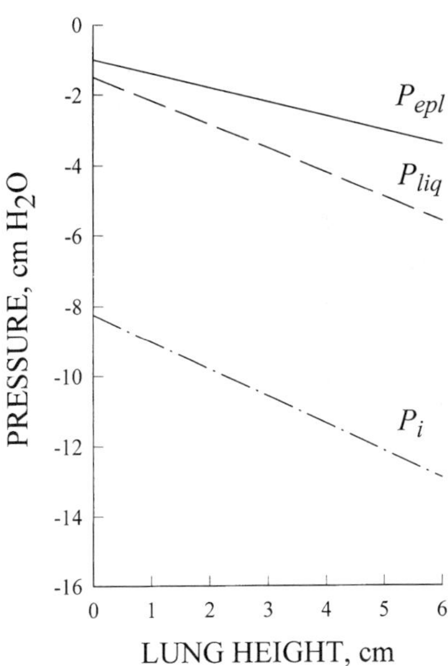

Figure 2 Distribution of parietal extrapleural interstitial pressure (P_{epl}), pleural liquid pressure (P_{liq}), and pulmonary interstitial pressure (P_i) as a function of lung height in supine rabbits. The slope of the linear regressions represents the vertical pressure gradient in each compartment and for all regressions is always less than hydrostatic (1 cmH_2O/cm height).

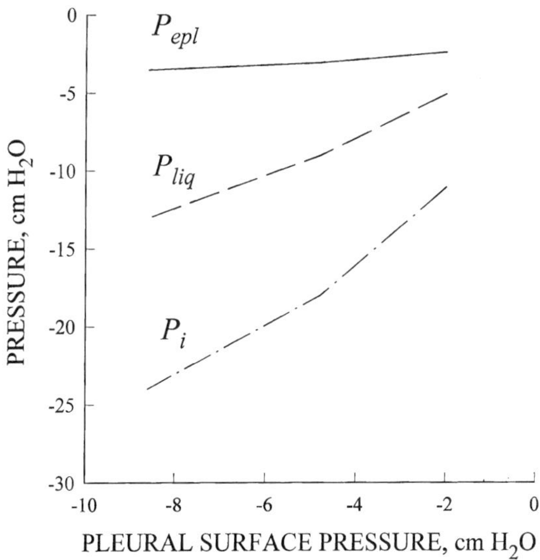

Figure 3 Dependence of P_{epl}, P_{liq}, and P_i measured at the right atrium level upon pleural surface pressure. P_{epl}, P_{liq}, and P_i become more subatmospheric when pleural surface pressure decreases as a consequence of an increase in lung volume.

is observed for pressures in the three compartments considered; note that the change in P_i is particularly large, reflecting the low compliance of pulmonary interstitial space.

VI. Transmembrane Fluid Fluxes in Pleuropulmonary Compartments

Transmembrane fluid fluxes (J_v) are described by the Starling equation: $J_v = L_p \cdot S [(P_a - P_b) - \sigma (\pi_a - \pi_b)]$, where L_p is the membrane hydraulic conductivity and S the membrane surface area; P and π are the hydraulic and colloidosmotic pressures in two adjacent compartments a and b separated by the membrane, and σ is the membrane reflection coefficient to total plasma proteins. Colloidosmotic pressures in the extrapleural parietal (20) and pulmonary (21) interstitium were measured directly on microsamples of interstitial fluid collected through the implanted "wick" technique. Figure 4 shows the net Starling balance of pressures calculated at heart level across the membranes delimiting the pleural and pulmonary compartments. According to the direction of net pressure gradients, pleural fluid is being filtered from the systemic capillaries of the parietal pleura into extrapleural interstitium and from the latter into the pleural space across the pari-

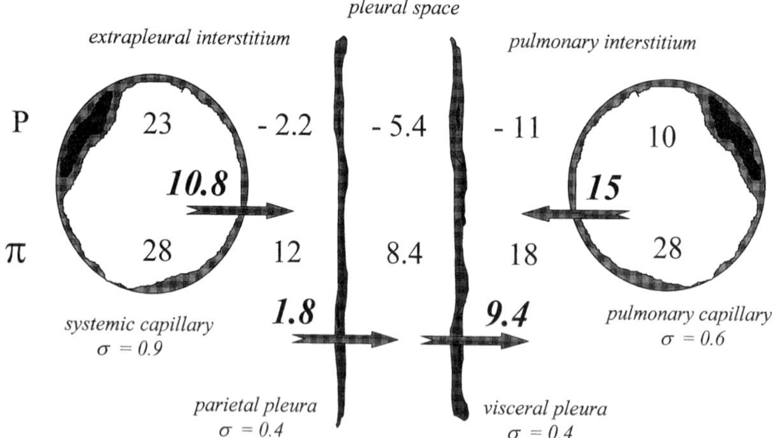

Figure 4 Net Starling pressure gradients calculated in the pleuropulmonary compartments at heart level in rabbits. Hydraulic (P) and colloidosmotic (π) pressures in each compartment and the net gradients across the membranes are expressed in cmH_2O. The arrows indicate the direction of the pressure gradients and therefore of the net flow. Protein reflection coefficient (σ) for systemic and pulmonary capillaries and parietal and visceral pleura are indicated.

etal mesothelium. At least 80% of pleural fluid drainage is accomplished via the parietal pleural lymphatics that directly connect the pleural space to an extended network of submesothelial lacunae (22,23). The capability of the pleural lymphatics to drain liquid from subatmospheric interstitial compartments was recently demonstrated by direct micropuncture measurements of intraluminal pressure in the initial pleural lymphatics, yielding an oscillatory pressure pattern with an average value of ~ -8 cmH$_2$O (24). Despite a relative large net Starling pressure gradient across the visceral pleura, only a minor share of pleural fluid is being drained into the pulmonary interstitium (14); this depends upon a fairly low hydraulic permeability of the visceral pleura as compared with the high conductance of the lymphatic draining route (25).

In physiological conditions, P_i is rather subatmospheric (see Figs. 2 and 4), averaging -10 cmH$_2$O at heart level (15), indicating low water content of the pulmonary interstitium; the balance of pressures determines a net fluid filtration from capillaries into the surrounding interstitium wherefrom interstitial fluid is eventually drained by pulmonary lymphatics (16). The fact that interstitial fluid pressure is subatmospheric in both the pleural and the pulmonary compartments indicates that the steady-state equilibrium between filtration and absorption is set at a "minimum" interstitial fluid volume. A thin layer of pleural fluid allows the greatest expansion of the lung during respiratory movements; a low hydration of the pulmonary interstitial compartment guarantees the thinness of the alveolar–capillary membrane allowing efficient gas exchange. The powerful action of the lymphatic draining system as opposed to the low permeability of the filtering membrane to liquid and proteins, actually sets the steady-state interstitial pressure at subatmospheric values both in the pleural cavity (24,26) and in the pulmonary interstitium.

Figure 5 shows the calculated Starling pressure gradients in the pleuropulmonary compartments as a function of lung height. By assuming that the permeability coefficients of the endothelial and mesothelial layers do not change with lung height, net fluid filtration from systemic and pulmonary capillaries increases from top to bottom. This ought to be expected on the basis of the differences in vertical pressure gradients in the vessels (1 cmH$_2$O/cm) compared with extrapleural and pulmonary interstitial spaces (<1 cmH$_2$O/cm). Conversely, pleural fluid filtration through the parietal pleura is higher in the nondependent regions of the cavity, as expected from the difference in slopes of P_{liq} and P_{epl}, respectively, as a function of lung height (see Fig. 2). Most of the pleural fluid is drained by lymphatics at the bottom of the cavity, particularly in the diaphragmatic and mediastinal regions (14). The preferential filtration of pleural fluid at top of the cavity and its drainage at the bottom results in a top-to-bottom intrapleural flow of liquid (27) in line with a vertical gradient of pleural liquid pressure that is different from hydrostatic (11–13,17). As shown in Figure 3, both pulmonary interstitial and pleural liquid pressures become more subatmospheric on increas-

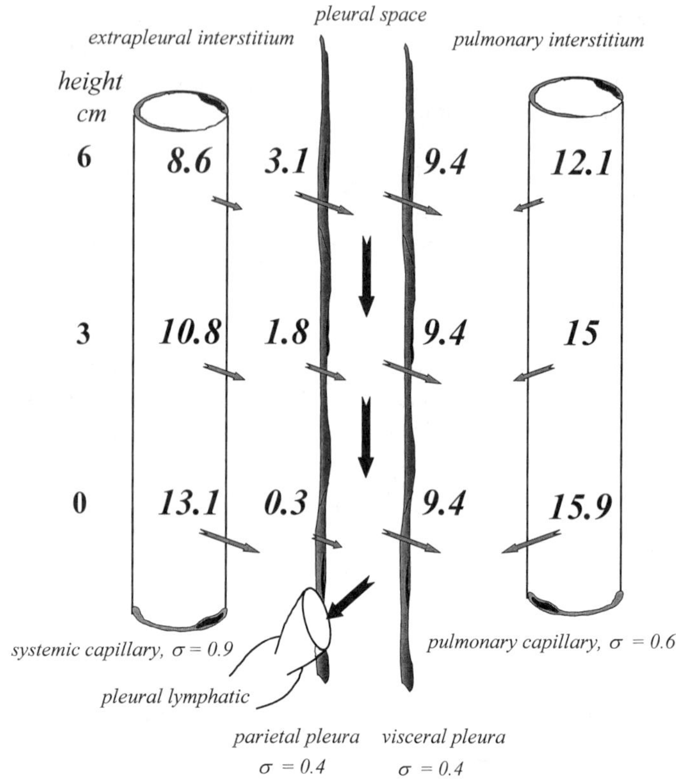

Figure 5 Net Starling pressure gradients calculated at the top (6 cm lung height), at the heart level (3 cm) and at the bottom (0 cm) in the pleuropulmonary compartments. All gradients are expressed in cmH_2O. Black arrows in the pleural space indicate the direction of intrapleural top-to-bottom flow and drainage into the pleural parietal lymphatics.

ing lung volume (19); accordingly, this would favor an increase in filtration rate from pulmonary capillaries into the lung interstitium and from extrapleural interstitium into the pleural space.

On the average, the turnover of pleural fluid in rabbits amounts to 0.02 mL/(kg · h) (14) and the turnover of pulmonary interstitial fluid amounts to 0.05 mL/(kg · h) (28). Extrapolation to humans is difficult; furthermore in humans, unlike other mammals including rabbits, the blood supply to the visceral pleura derives from the bronchial circulation, which is of systemic origin, and therefore capillary hydraulic pressure is higher than in the pulmonary capillaries. Accordingly, in humans, pleural fluid filtration might occur even through the visceral pleura.

VII. Transition from Physiological Condition to Pulmonary Interstitial Edema

The lung is subject to wide changes in capillary perfusion, as for example when cardiac output increases during physical exercise or hypoxia exposure. Microgravity is another condition leading to increased pulmonary blood flow and capillary recruitment (3). All conditions of increased microvascular filtration potentially induce an increase in interstitial volume and pressure; the increase in interstitial pressure is, in fact, the mechanical cause leading to an increase in lymphatic flow. As long as lymphatic flow matches the increased filtration rate, a new steady state is achieved with an interstitial volume that is larger than the physiological one.

To detect early events leading to interstitial fluid accumulation we have developed, in the experimental animal, two models of interstitial lung edema that cause a slow and modest increase in extravascular lung water: (1) a "hydraulic" type of edema obtained through an intravenous infusion of saline solution [0.5 mL/(min kg)], causing only $\approx 15\%$ increase in plasma volume in 60 min (29), and (2) a "lesional" type of edema induced by injecting a single bolus of pancreatic elastase (200 µg, 7 IU) (30). Elastase is an omnivorous proteolytic enzyme with broad affinity for a variety of soluble and insoluble protein substrates, including the components of the extracellular matrix (31). The left panel of Figure 6 shows the time course of P_i as microvascular filtration is increased in "hydraulic" or "lesional" type of edema (29,30). In this plot, the time scale is expressed relative to the attainment of peak interstitial pressure value. One can appreciate that, although the time course of P_i is not exactly the same in the two edema models, it nevertheless displays two common features. At the onset of edema development (mild edema), P_i increases from the control value (-10 cmH$_2$O) to about $+5$ cmH$_2$O; thereafter P_i decreases toward 0 as edema formation progresses with time. The right panel of Figure 6 shows that, in both edema models, the initial marked increase in P_i is accompanied by only a negligible increase in tissue hydration, indexed by the wet weight to dry weight ratio of the lung (W/D). Interstitial compliance of the pulmonary tissue during the rising phase of P_i amounts to 0.5 mL/(mmHg · 100 g wet weight), a value about 20-fold lower compared with other tissues (29). Clearly, a low compliance provided by the structure of the extracellular matrix represents an important "tissue safety factor" to counteract further progression of pulmonary edema; in fact, the attainment of a positive P_i value buffers and even nullifies the Starling pressure gradient that

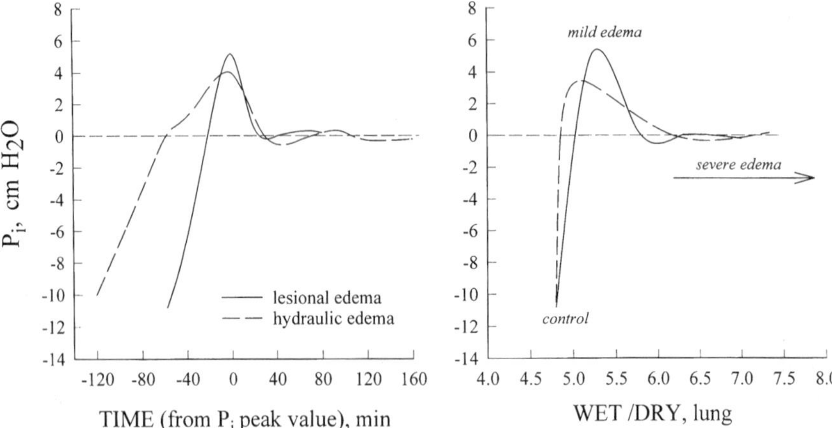

Figure 6 (Left) Time course of P_i during development of hydraulic or lesional lung edema. Time is expressed relative to the attainment of the peak positive P_i value. (Right) P_i data are plotted as a function of the corresponding wet weight to dry weight ratio of the lung.

causes microvascular fluid filtration. The right panel of Figure 6 also shows that, as the severity of edema progresses, P_i drops back to 0 and subsequently remains unchanged despite a marked increase in W/D ratio. On mechanical grounds, this can be attributed to a loss of integrity of the interstitial scaffold resulting in an increased tissue compliance. Note that in the late phase of edema development, fluid filtration occurs down a smaller transendothelial Starling pressure gradient compared with control condition. In fact, pulmonary tissue pressure at heart level is -10 cmH$_2$O in control and ~ 0 cmH$_2$O as W/D increases beyond 6. Since the filtering transcapillary pressure gradient is reduced, an increase in fluid filtration rate can be achieved only through an elevated microvascular permeability (29). Hence, data of Figure 6 indicate that at least two mechanisms interact to determine the development of pulmonary edema: the loss of the ''tissue safety factor'' reflecting an increase in tissue compliance and the increase in microvascular permeability.

Pleural liquid pressure remained unchanged during the development of lung edema (29). This finding confirms that the conductance of the visceral pleura in conditions close to the physiological one is very low. In fact, because of the increase in pulmonary interstitial pressure, the Starling pressure gradients across the visceral pleura is reversed, yet this does not result in an appreciable increase in pleural liquid pressure and volume.

VIII. Alteration in Extracellular Matrix Integrity During Development of Pulmonary Edema

To relate pulmonary interstitial pressure to the functional state of the extracellular matrix, we studied the structure of pulmonary interstitial PGs in control conditions and during the development of hydraulic and lesional edema (29,31–33). We focused on the involvement of two important families of PGs found in the extracellular matrix, namely, HS-containing PGs of the basal membrane (perlecan) and CS-containing PGs of the interstitial matrix (versican). The first important finding was that in both hydraulic or lesional mild edema, the extractability of PGs from the lung tissue was increased, suggesting a weakening of the noncovalent bonds of PGs with other components of the extracellular matrix (29,31–33).

The modifications induced by edema development on the tissue proteoglycans content were detected by gel-filtration chromatography of PGs extracted from tissue samples. A mild hydraulic edema (W/D ratio up to ~6), caused a marked fragmentation of large-matrix proteoglycans of the versican family (31,33). In elastase-treated animals (lesional edema model, W/D ~5.6), the degradation of basal membrane proteoglycans (perlecan) prevailed over that of the matrix proteoglycans (versican). The biochemical investigations also allowed us to evaluate that the binding properties of tissue proteoglycans to other macromolecules of the extracellular matrix such as collagen type I, collagen type IV, and hyaluronic acid, were markedly reduced both in hydraulic and lesional edema.

One possible hypothesis for the fragmentation of PGs relates to the tissue stresses developing within the extracellular matrix with increasing extravascular lung water; such a mechanism may also progressively weaken the noncovalent bonds of PGs with other matrix molecules. A further mechanism is based on the activation of metalloproteinases (gelatinase A, also indicated as MMP-2, and gelatinase B, also indicated as MMP-9) whose activity was found to increase in lung edema (32,33). It was indeed found that purified MMP-2 and MMP-9 extracted from edematous lung tissue could cleave versican, the main proteoglycan that is being fragmented in the initial stage of hydraulic edema (32). Finally, we could identify in lungs with hydraulic edema a form of MMP-9 (135 and >200 kDa) that is also expressed by neutrophils (34–36). Therefore, even in mild hydraulic edema, a proteolytic lesional component might be triggered by neutrophil activation, suggesting that an upregulation of this system occurs for a very moderate perturbation of microvascular and interstitial fluid dynamics.

In summary, interstitial edema occurs as a result of a combined action of increased microvascular permeability and loss of tissue safety factor due to early fragmentation of proteoglycans of the basal membrane and of the interstitial matrix.

IX. Simulated Microgravity in Experimental Animals

We have measured the time course of P_i and P_{liq} at heart level in supine rabbits maintained at 20° head-down tilt position up to 3 hours (Miserocchi, personal communication, 1999). Head-down tilt has been proposed as a good model of microgravity as it causes a decrease in lung volume and a blood shift into the lung, similar to what happens with 0 G exposure. The tilting maneuver caused a change in esophageal pressure (that mirrors pleural surface pressure) from -2 cmH$_2$O to about -0.5 cmH$_2$O, indicating a decrease in lung volume due to a cranial movement of the diaphragm. Figure 7 shows that P_i shifts already after 1 h toward 0 and attains about 2 cmH$_2$O at 3 h. The decrease in lung volume would be expected to cause a shift in P_i from -10 cmH$_2$O to -7.5 cmH$_2$O; therefore, the observed change in P_i following head-down tilt suggests fluid accumulation into the interstitial matrix. Pleural liquid pressure averaged -4 cmH$_2$O in control and increased to -2 cmH$_2$O when the animal was tilted, remaining thereafter steady. Since pleural liquid volume was unchanged at three hours compared with control, we interpreted the increase in P_{liq} as mainly due to the decrease in lung volume.

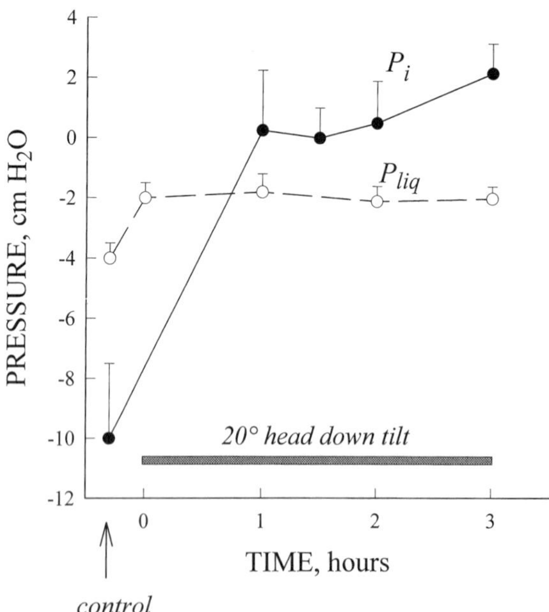

Figure 7   Time course of P_i and P_{liq} during simulated microgravity attained in experimental animals by 20° head-down tilt maintained up to 3 hours.

X. Pleural and Pulmonary Compartments in Microgravity

In humans, changing G_z (cranial to caudal direction) is expected to induce a change in the functional residual capacity of the respiratory system, as reflected by the modification of the pressure volume relationships of the lung and chest wall. In particular, the volume pressure relationship of the chest wall is influenced by the weight of the abdomen and therefore is expected to be greatly modified when changing G_z. In general, relative to 1 G_z, microgravity determines a craniad displacement of the diaphragm and a decrease in lung volume of the order of ~ 300 ml. Opposite changes are observed at 1.8 G_z during parabolic flight (37,38). A prediction on how pleuropulmonary fluid dynamics is affected by microgravity is at present difficult. Factors to be considered include the dependence of P_i and P_{liq} from pleural surface pressure (and therefore lung volume) and the different time constant of equilibration of hydraulic pressures in the various compartments. Under 0 G_z, the decrease in lung volume causes pleural surface pressure and therefore also P_i and P_{liq} to become less subatmospheric (see Fig. 3); accordingly, one would expect fluid filtration into the pulmonary interstitium and the pleural cavity to decrease with respect to 1 G_z.

After exposure to 0 G_z, a rapid equilibration is expected in the microcirculation and in the pleural cavity due to the relatively low fluid dynamic resistance; pressures are expected to equilibrate at the value existing at the right atrium level (about 50% lung height). Conversely, a longer equilibration time ought to be required for the interstitial compartments, which display a higher viscous flow resistance due to the solid fiber matrix component. After pressure equilibration is attained, regional differences in fluid filtration and absorption should disappear: compared with control at 1 G_z, fluid filtration ought to be reduced at top of the pleural cavity, but increased at the bottom.

Our data indicate that at 1 G_z pleural fluid is preferentially drained by the lymphatics in a gravity-dependent fashion, so that one might hypothesize that the lymphatic network would be better developed in the more dependent regions of the respiratory system. A similar anatomical design likely exists in other body compartments and it would explain why no tissue swelling occurs in 1 G_z in the lowermost body regions, although liquid continuously drains downward. Conversely, swelling occurs in the uppermost body segments when, as in microgravity, the regional-interstitial-fluid-filtration-to-absorption ratio is shifted in favor of filtration. On the other hand, if the lymphatic stomatas are homogeneously distributed over the pleural surface, no hindrance of pleural fluid drainage will occur in microgravity.

As far as the pulmonary compartment is concerned, experimental data from simulated microgravity in animals demonstrated an increased fluid filtration leading to pulmonary interstitial edema. Hence, the protective effect of a decreased lung volume is more than offset by conditions leading to increased filtration, namely capillary recruitment. One may recall that in patients lying in bed for a long time, edema formation is observed in the more dependent regions of the lung where two conditions coexist, namely, blood engorgement and a decreased regional lung volume. A possible parallel between a condition of sustained sub-edematous condition of the lung, such as that presumably occurring in microgravity, could be done with the transplanted lung, which is basically devoid of lymphatics, the main route for interstitial fluid drainage. In fact, lung transplant implies resection of pulmonary lymphatics and, although some regeneration may occur, it appears difficult to envisage a complete restoring of the parenchymal lymphatic network. Accordingly, in the transplanted lung, interstitial fluid volume control mainly depends upon the "tissue safety factor" provided by the extracellular matrix structure. Relatively recent reports stem for the development of focal interstitial fibrosis in transplanted lungs, a condition that may indeed represent the tissue reaction to sustained interstitial edema. Hence, one may speculate that a state of subclinical interstitial pulmonary edema might develop in humans exposed to long-term microgravity, leading to progressive development of pulmonary fibrosis. In this respect, we may comment that our studies in experimental animals reveal that a negligible increase in the amount of extravascular lung water triggers a cascade of events activating interstitial matrix remodeling whose entity is instead not negligible. A challenging question remains on the functional behavior of the pulmonary tissue matrix under a sustained condition of pulmonary interstitial edema. Despite such concern, no clinically detectable complications relative to lung fluid balance have been reported so far in humans exposed to prolonged microgravity. One may comment that our studies describe the transition phase from physiological conditions to the onset of lung edema, characterized by a modest increase in extravascular lung water, certainly not detectable in humans with standard methods. The extent to which lung tissue resists a prolonged condition of increased hydration reveals that the organ is well designed to fulfill the main task of accommodating wide changes in capillary recruitment.

References

1. Moore TP, Thornton, WE. Space shuttle inflight and postflight fluid shift measured by leg volume changes. Aviat Space Environ Med 1987; 58:A91–A96.
2. Kirsch KA, Baartz FJ, Gunga HC, Rocker L. Fluid shift into and out of superficial tissue under microgravity and terrestrial conditions. Clin Invest 1993; 71:687–689.
3. Prisk GK, Guy HJ, Elliott AR, Deutschman RA, West JB. Pulmonary diffusing capacity, capillary blood volume and cardiac output during sustained microgravity. J Appl Physiol 1993; 75:15–26.
4. Hardingham T, Fosang J. Proteoglycans: Many forms and many functions. FASEB J 1992; 6:861–870.
5. Roberts CR, Wight TN, Hascall VC. Proteoglycans. In: Crystal RG, West JB,

Barnes PJ, Weibel ER, eds. The Lung: Scientific Foundations. 2d ed. Philadelphia: Lippincott-Raven, 1997:757–767.

6. Juul SE, Wight TN, Hascall VC. Proteoglycans. In: Crystal RG, West JB, Barnes PJ, Cherniack NS, Weibel ER, eds. The Lung: Scientific Foundations. New York: Raven, 1991:413–420.

7. Crouch EC, Martin GR, Brody JS, Laurie GW. Basement membranes. In: Crystal RG, West JB, Barnes PJ, Weibel ER, eds. The Lung: Scientific Foundations. 2d ed. Philadelphia: Lippincott-Raven, 1997:769–791.

8. Miserocchi G, Negrini D, Mortola J. Comparative features of Starling-lymphatic interaction at the pleural level in mammals. J Appl Physiol 1984; 56:1151–1156.

9. Hills, BA. Graphite-like lubrication of mesothelium by oligolamellar pleural surfactant. J Appl Physiol 1992; 73:1034–1039.

10. Hills BA, Butler BD, Barrow RR. Boundary lubrication imparted by pleural surfactants and their identification. J Appl Physiol 1982; 53:463–469.

11. Miserocchi G, Mariani E, Negrini D. Role of the diaphragm in setting liquid pressure in serous cavities. Respir Physiol 1982; 50:381–392.

12. Miserocchi G, Nakamura T, Mariani E, Negrini D. Pleural liquid pressure over the interlobar, mediastinal and diaphragmatic surfaces of the lung. Respir Physiol 1981; 46:61–69.

13. Miserocchi G, Pistolesi M, Miniati M, Bellina CR, Negrini D, Giuntini C. Pleural liquid pressure gradients and intrapleural distribution of an injected bolus. J Appl Physiol 1984; 56:526–532.

14. Negrini D, Pistolesi M, Miniati M, Bellina CR, Giuntini C, Miserocchi G. Regional protein absorption rates from the pleural cavity in dogs. J Appl Physiol 1985; 58: 2062–2067.

15. Miserocchi G, Negrini D, Gonano C. Direct measurement of interstitial pulmonary pressure in in situ lung with intact pleural space. J Appl Physiol 1990; 69:2168–2174.

16. Negrini D, Gonano C, Miserocchi G. Microvascular pressure profile in intact in situ lung. J Appl Physiol 1992; 72:332–339.

17. Miserocchi G, Negrini D. Pleural space: Pressure ad fluid dynamics. In: Crystal RG, West JB, Barnes PJ, Weibel ER, eds. The Lung: Scientific Foundations. 2d ed. Philadelphia: Lippincott-Raven, 1997:1217–1225.

18. Guyton AC, Taylor AE, Granger HJ. Circulatory Physiology II. Dynamics and Control of Body Fluids. Philadelphia: Saunders, 1975.

19. Miserocchi G, Negrini D, Gonano C. Parenchymal stress affects interstitial and pleural pressures in in-situ lung. J Appl Physiol 1991; 71:1967–1972.

20. Negrini D, Del Fabbro M, Venturoli D. Fluid exchanges across the parietal peritoneal and pleural mesothelia. J Appl Physiol 1993; 74:1779–1784.

21. Negrini D, Passi A, Bertin K, Wiig H. Isolation of pulmonary interstitial fluid by a modified wick technique. Am J Physiol 2001; 280(5): in press.

22. Negrini D, Mukenge S, Del Fabbro M, Gonano C, Miserocchi G. Distribution of diaphragmatic lymphatic stomata. J Appl Physiol 1991; 70:1544–1549.

23. Negrini D, Del Fabbro M, Gonano C, Mukenge S, Miserocchi G. Distribution of diaphragmatic lymphatic lacunae. J Appl Physiol 1992; 72:1166–1172.

24. Negrini D, Del Fabbro M. Subatmospheric pressure in the rabbit pleural lymphatic network. J Physiol (Lond) 1999; 520:761–769.
25. Miserocchi G, Venturoli D, Negrini D, Del Fabbro M. Model of pleural fluid turnover. J Appl Physiol 1993; 75:1798–1806.
26. Negrini D, Ballard ST, Benoit JN. Contribution of lymphatic myogenic activity and respiratory movements to pleural lymph flow. J Appl Physiol 1994; 76:2267–2274.
27. Miserocchi G, Venturoli D, Negrini D, Gilardi MC, Bellina CR. Intrapleural fluid movements described by a porous flow model. J Appl Physiol 1992; 73:2511–2516.
28. Miserocchi G, Negrini D. Pleural lymphatics as regulators of pleural fluid dynamics. News Physiol Sci 1991; 6:153–158.
29. Miserocchi G, Negrini D, Del Fabbro D, Venturoli D. Pulmonary interstitial pressure in the intact in-situ lung: Transition to interstitial edema. J Appl Physiol 1993; 74: 1171–1177.
30. Negrini D, Passi A, De Luca G, Miserocchi G. Proteoglycan involvement during development of lesional pulmonary edema. Am J Physiol 1998; 274 (Lung Cell Mol Physiol 18):L203–L211.
31. Negrini D, Passi A, De Luca G, Miserocchi G. Pulmonary interstitial pressure and proteoglycans during development of pulmonary edema. Am J Physiol 1996; 270 (Heart Circ Physiol 39):H2000–H2007.
32. Passi A, Negrini D, Albertini R, Miserocchi G, De Luca G. The sensitivity of versican from rabbit lung to gelatinase A (MMP-2) and B (MMP-9) and its involvement in the development of hydraulic lung edema. FEBS Lett 1999; 456:93–96.
33. Passi A, Negrini D, Albertini R, De Luca G, Miserocchi G. Involvement of lung interstitial proteoglycans in development of hydraulic- and elastase-induced edema. Am J Physiol 1998; 275(Lung Cell Mol Physiol 19):L631–L635.
34. Finlay GA, O'Driscoll LR, Russel LR, D'Arcy EM, Masterson JB, FitzGerald MX, O'Connor CM. Matrix metalloproteinase expression and production by macrophages in emphysema. Am J Respir Crit Care Med 1997; 156:240–247.
35. Hibbs MS, Hasty KA, Seyer JM, Kang AH, Mainardi CL. Biochemical and immunological characterization of the secreted forms of human neutrophil gelatinase. J Biol Chem 1985; 260:2493–2500.
36. Kjeldsen L, Johnson AH, Sengelov H, Borregaard N. Isolation and primary structure of NGAL, a novel protein associated with human neutrophil gelatinase. J Biol Chem 1993; 268:10425–10432.
37. Elliot AR, Prisk GK, Guy HJB, West JB. Lung volumes during sustained microgravity on Spacelab SLS-1. J Appl Physiol 1994; 77:2005–2014.
38. Paiva M, Estenne M, Engel LA. Lung volumes, chest wall configuration, and pattern of breathing in microgravity. J Appl Physiol 1989; 67:1542–1550.

12

Control of Ventilation

G. KIM PRISK

University of California, San Diego
La Jolla, California

I. Introduction

Until recently, the factors that control ventilation have been essentially unstudied in the microgravity (μG) environment. There are likely a number of reasons for this, principal among them being the perception that there was likely to be little change in the control of ventilation in μG. In addition, in the early periods of space flight, the cabin atmospheres of the spacecraft differed significantly from that of air breathed at sea level. This change in atmosphere presented such a confounding factor that measures of changes in ventilatory control would be unable to be interpreted in terms of the effect of μG per se.

A. Gaseous Environments in Spacecraft

Early Spacecraft

Primarily for engineering reasons, the cabin atmospheres of the earliest U.S. manned spacecraft were pure O_2 at a reduced pressure. For example, the Project Mercury and Project Gemini spacecraft were 100% O_2 at a pressure of ~260 mmHg, giving a $P_{I_{O_2}}$ of ~260 mmHg (21). Having a pure O_2 atmosphere meant that the absolute cabin pressure could be kept low with substantial savings in the weight of the capsule as a lighter pressure vessel was required. In addition, it was then only necessary to carry liquid O_2 to top-off the cabin atmosphere as it leaked out, resulting in a simpler (and therefore lighter and more reliable) environmental control system. In contrast, the early Russian Vostok and Voskhod capsules had a 21% O_2 atmosphere at a normal barometric pressure ($PI_{O_2} \sim 160$ mmHg) (21).

Unfortunately, the launch pad fire in the Apollo 1 capsule in 1967 that killed three astronauts showed that a pure O_2 atmosphere presented a serious fire risk. The Apollo command module was designed such that a pressure differential between the cabin and exterior of the spacecraft was maintained. This resulted in conditions on the launch pad in which the cabin pressure exceeded 1 atmosphere, and under such conditions, many substances thought to be nonflammable will burn readily. As a consequence, the environmental control system was redesigned so that while the capsule was still on the ground, a gas mixture of 60% N_2 and 40% O_2 was used and this was replaced in flight by pure O_2. Since the absolute pressure outside the cabin was essentially 0 during flight, the pressure differential resulted in a cabin pressure of ~ 260 mmHg, 100% O_2. Because of the lower absolute pressure, the atmosphere of 100% O_2 did not present an unacceptably high fire risk during flight. During the long-duration Skylab missions, an atmosphere of 70% O_2 and 30% N_2 was used at an absolute pressure of ~ 260 mmHg ($PI_{O_2} \sim 180$ mmHg), lowering the fire risk, and reducing the chance of absorption atalectasis (21).

Current Spacecraft (Shuttle/ISS/Space Suits)

Currently, the only manned spacecraft in operation are the U.S. Space Shuttle, the Russian Soyuz, and the International Space Station (ISS). All these vehicles operate at a normal atmosphere of 760 mmHg, 21% O_2, i.e., a sea-level normoxic environment, as did the now-abandoned Mir space station.

All the spacecraft, however, have a significant difference from the terrestrial environment in that the CO_2 levels are elevated. Because the cabin is closed, it is necessary to remove the CO_2 produced by metabolic activity of the crew. In the past, this was done by using CO_2 absorbers, usually lithium hydroxide. More recently, regenerative CO_2 removal systems have been used (12). However, whatever the system in use, it is impractical to scrub the CO_2 level to 0. Shuttle standards (and future ISS standards) set an upper limit on CO_2 levels of 7.6 mmHg (1% at 1 atm) in the cabin. This is the same limit set by the U.S. Navy in their submarine fleet, and appears to produce only small physiological changes (24,33).

The other spacecraft in routine use are the space suits used during extravehicular activity (EVA) or spacewalks. Both the Russian Orlan suits and the U.S. suits operate a low absolute pressure (~ 290 and ~ 220 mmHg, respectively) with a 100% O_2 atmosphere. The reason for the striking difference between the suit environment and the cabin environment is that suits running at a higher pressure would be so stiff as to be almost impossible to move. While there has been considerable research into higher-pressure suits, the engineering challenges are formidable, especially in the design of a suitable glove. The marked difference in absolute pressure from the cabin to the suits brings a risk of nitrogen bubble formation in the blood. This presents a risk of possible decompression sickness if the transition from the cabin atmosphere to the suit is too rapid (see Chapter 13).

B. Possibilities for Changes in Ventilatory Control

There are a number of observations that suggest that there may be some changes in ventilatory control in humans exposed to µG. There are small but significant changes in the resting end-tidal CO_2 levels in humans in space flight, with a slight increase of ~2 mmHg in the crews of SLS-1 and SLS-2 (25) although this is somewhat variable. There are also anecdotal reports of breathlessness (dyspnea) in response to exercise, particularly during EVA. However, it is difficult to assess how much of this (if any) is due to a real physiological change, and how much is a result of having to work in a difficult and uncomfortable space suit and to the psychological factors associated with being outside the spacecraft cabin. In addition, sleep has been reported to be of poor quality in µG (10,11,22,31), and it was thought possible that changes in ventilatory control might contribute to this.

To date, the only studies of ventilatory control in µG have been those performed on LMS in 1996, in which the response to CO_2 was measured in flight and the response to low O_2 preflight and postflight only, and on Neurolab in 1998, in which both were measured inflight.

II. Response to Lowered Oxygen

A. Chemoreceptors

The hypoxic response in humans is primarily a result of the carotid chemoreceptors (17). These receptors are located in the carotid bodies at the carotid bifurcation in the neck, and sense a lowering of the P_{O_2} in the arterial blood. In addition to the carotid bodies, there are the aortic receptors in the aortic arch. While the aortic receptors play a large role in the hypoxic response of many animals, such as the cat (18), it appears that in humans, their role is less important, with most of the response resulting from the carotid receptors.

B. Alterations in Microgravity

The only measurements performed to date of the hypoxic ventilatory response (HVR) in µG were performed using the rebreathing technique originally described by Rebuck and Campbell (30). The subject rebreathes from a bag initially containing a gas mixture of (in the case of Neurolab) 17% O_2, 7% CO_2, and the balance N_2. As the rebreathing continues, metabolic consumption of O_2 reduces the O_2 in the bag, and the resulting hypoxia stimulates ventilation. In order to avoid ventilatory stimulation due to hypercapnia, a fan circulates gas in the bag through a canister filled with soda lime, and a control algorithm maintains a constant end-tidal P_{CO_2} by varying the fan speed. The rebreathing continues until the desired level of hypoxia is reached (6% inspired O_2 in the case of Neurolab). An example of the data collected and the analysis is provided in Figure 1.

Microgravity causes a large reduction in the ventilatory response to hypoxia (26). In studies performed during the 16-day Neurolab flight, the hypoxic ventilatory response, measured as the slope of the increase in ventilation with lowered arterial O_2 saturation was approximately halved (Fig. 2). There was a similar change in the intercept of the fitted line, and the combined effect of both these changes was to greatly reduce the increase in ventilation resulting from a given hypoxic stimulus in µG. For example, the ventilation at an arterial oxygen saturation of 75% in µG was only 65% of that in the same group of subjects standing in 1 G during the preflight period.

The reduction in the HVR in µG was almost identical to that seen when the subjects were acutely placed in the supine position (see Fig. 2). The supine position has previously been noted to reduce the HVR (35,40), but, somewhat surprisingly, this does not appear to be widely known.

The similarity of the changes in the HVR between upright and supine, and between upright and µG, suggested that the cause of the reduction was likely an increase in the blood pressure at the carotid bodies of the subjects. Because the carotid bodies are located at a level above the heart, there is a hydrostatic blood pressure difference between them, with carotid blood pressure being ~15 to 20 mmHg lower than heart-level blood pressure in an upright adult. When the subjects are placed in the supine position, or when gravity is removed, the hydrostatic difference is abolished, and the heart-level and carotid-level blood pressures become similar. Since systemic blood pressure is largely unchanged by either the supine position (an increase of ~3 mmHg) or in µG (a decrease of ~4 mmHg) (9), in either case, there is a substantial increase in blood pressure at the carotid level.

There is ample evidence for changes in blood pressure modifying the output of the chemoreceptors, and hence the ventilation. Heymans (15), in his 1945 Nobel lecture noted that ''variations in arterial blood pressure exert an effect on the respiratory center . . . by a reflex mechanism involving the aortic and carotid sinus receptors.'' In cats, decreases in systemic blood pressure markedly increase the output of the aortic chemoreceptors and, to a lesser extent, the carotid chemoreceptors (16,18). The effect on the carotid chemoreceptors was, however, much more pronounced under hypoxic conditions than under normoxic conditions. Un-

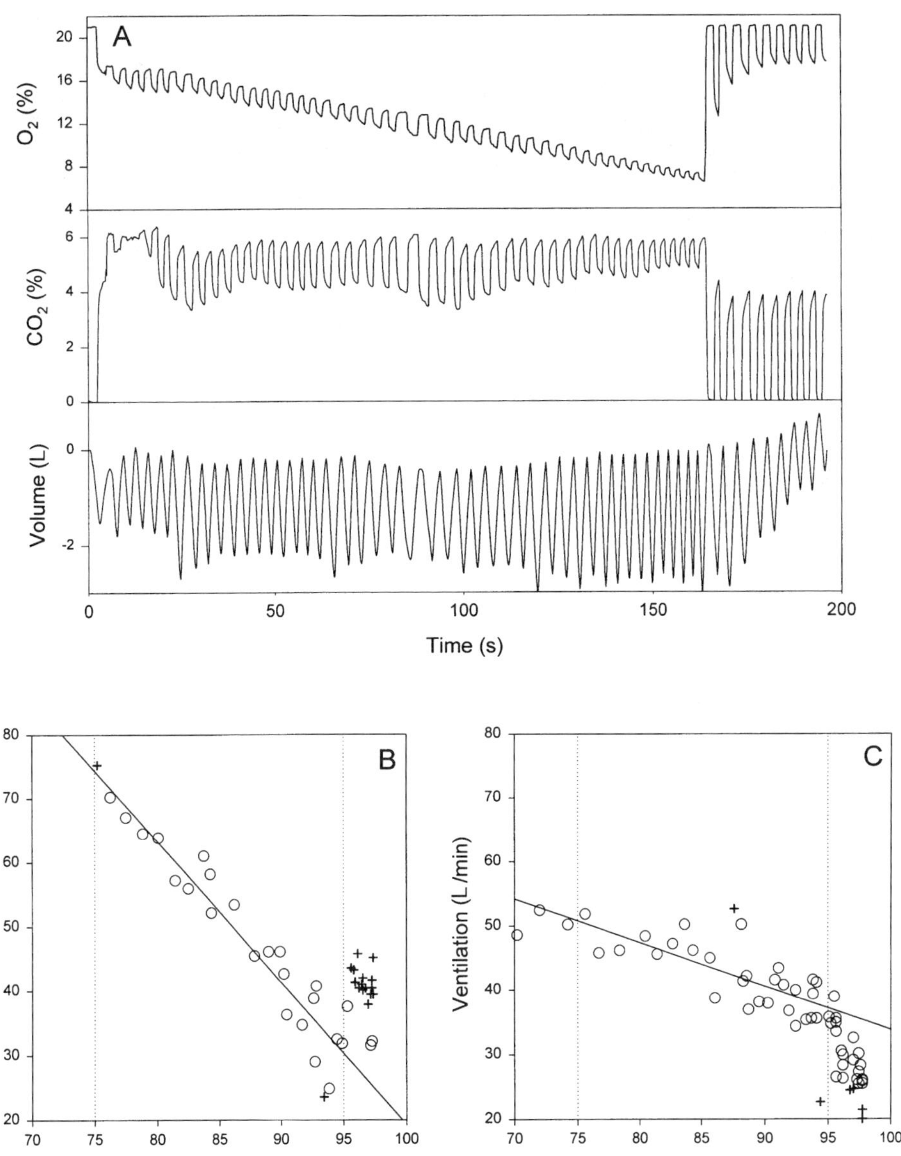

Figure 1   (A) The upper plot shows raw data from the measurement of the hypoxic ventilatory response (HVR). Subjects rebreathed from a bag for up to 4 min during which time the O_2 in the bag fell and CO_2 was held constant by a computer-controlled CO_2 removal circuit. The HVR analysis for one subject for data collected standing in 1 G (B) and in μG (C) is shown below. Breath-by-breath ventilation is plotted as a function of Sa_{O_2}. The line is the least-squares best fit to the points lying within 2 SD of the best fit to all data points (see Ref. 26 for details). Points marked with a "+" were excluded from the fit, as were points with an Sa_{O_2} below 75% or above 95%. (From Ref. 26.)

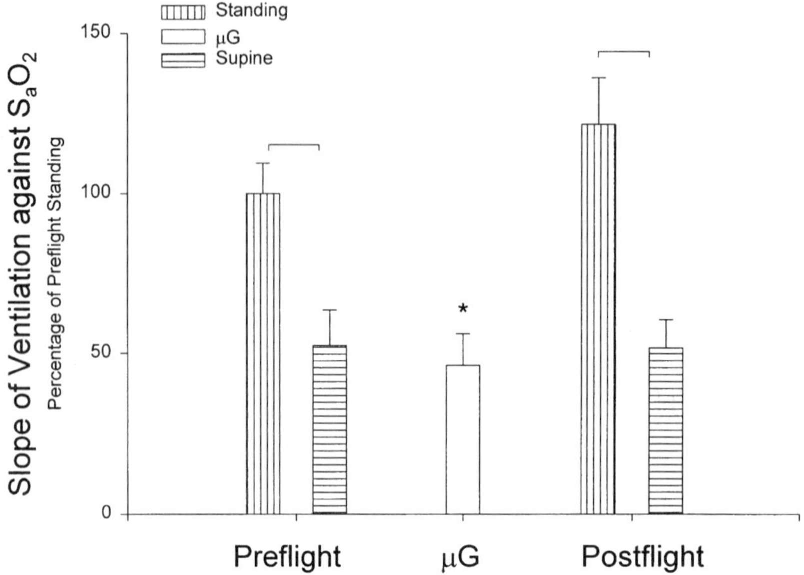

Figure 2   The slope of the ventilatory response to hypoxia calculated as the rise in ventilation resulting from a decrease in arterial oxygen saturation. Data are normalized to each subject's preflight standing control. Vertical shading, standing; horizontal shading, supine; open bars, μG. * Indicates $p < 0.05$ compared with preflight study. A marker between adjacent bars shows $p < 0.05$. Error bars are SE. (From Ref. 26.)

der hypoxic conditions, the carotid chemoreceptors were virtually unaffected by a reduction in blood pressure, but when the animal was hypoxic (Pa$_{O_2}$ of 62 mmHg), their output rose markedly as blood pressure fell.

Similarly, in dogs, systemic hypotension increases carotid chemoreceptor output. This occurs by a central pathway. Studies by Heistad et al. (13,14) showed that when carotid distending pressure was decreased on one side of an animal in an isolated preparation, carotid body firing rate was altered on the contralateral side. This indicates that some central integration is involved in the process. As was the case in cats, the effects are more pronounced under hypoxic than under normoxic conditions.

In humans, the effects are less well documented. Sommers et al. (38) showed that an increase in systemic blood pressure of ~10 mmHg induced using phenylepherine resulted in a decrease of ~33% in the degree of ventilatory stimulation resulting from a given hypoxic challenge (breathing 10% O$_2$). There seem to be few other studies that directly address these two effects, but the results of Xie et al. (40) and Serebrovskaya et al. (35) generally fit the observations of the other studies. In both these cases, subjects were studied in both the upright and supine postures and there were substantial decreases in the HVR measured in the supine position (43% and 30 to 43%, respectively).

In addition to the direct measures of ventilation, Prisk et al. (26) also measured inspiratory occlusion pressures both during normoxic breathing and during the HVR tests. The inspiratory occlusion pressure is that generated when the subject makes an inspiratory effort against an airway that has been unexpectedly closed. The pressure 100 ms after the airway closure (a time period before conscious realization has occurred) provides a measure of respiratory muscle activation (39). This is referred to as the P_{100}. When the P_{100} was measured on the Neurolab crew there were no differences among standing, supine, or μG when the subject was breathing air (Fig. 3). Hypoxia (in this case, an inspired oxygen level of between 75 and 85 mmHg) resulted in a substantial increase in the P_{100}

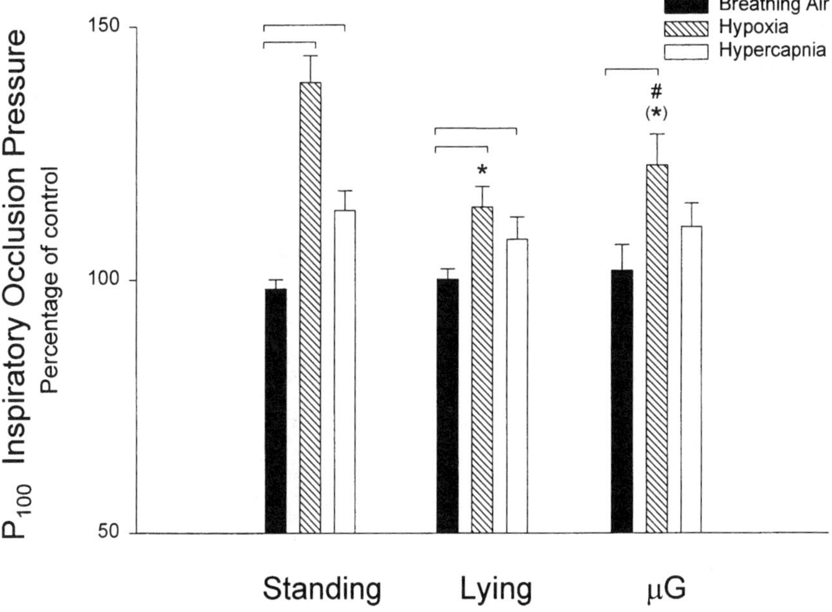

Figure 3 The inspiratory occlusion pressure measured 100 msec after the closure of a valve at the beginning of inspiration. The black bars show the P_{100} values breathing air. The shaded bars show the P_{100} measured during the HVR test when the P_{O_2} was between 75 and 85 mmHg. The open bars show the P_{100} measured during the HCVR test when the P_{CO_2} was between 43 and 50 mmHg. A marker between adjacent bars indicates $p < 0.05$. * Indicates $p < 0.05$ compared with preflight standing. # Indicates $p < 0.05$ compared with preflight supine. Parentheses indicate $0.10 < p < 0.05$. (From Ref. 26.)

under all conditions. However, the increase was much less in both the supine posture and in μG than it was with the subjects upright. This indicates that the increase in the ventilatory drive in response to a given hypoxic stimulus was less in the supine posture and in μG than upright.

The P_{100} measurements also serve to refute the notion that the decrease in the HVR seen in μG (and supine) was a result of the respiratory muscles being placed in a less advantageous position to respond to a given stimulus. Such a mechanism has previously been suggested as a possible cause of the lower HVR seen in the supine posture (2,6). The P_{100} was also measured in the Neurolab crew during hypercapnia (see below, Sec. III.B). Hypercapnia (a P_{CO_2} between 43 and 50 mmHg) resulted in an increase in the P_{100} that, although increased above that measured under normocapnic conditions, was not different among the upright and supine postures and μG (Fig. 3). The lack of a difference in P_{100} in different postures during normoxia and hypercapnia suggests that a mechanical disadvantage is not responsible for the differences seen in the P_{100} during hypoxia.

The changes in the HVR appear to be sustained in μG, at least over the course of 16 days. Figure 4 shows the HVR measured at different time periods preflight, during flight, and postflight. The results indicate that there is no alteration in the HVR as a result of time spent in μG, with the lower response persisting throughout the course of the flight. Data from Fritsch-Yelle et al. (9) indicate that the blood pressure remains essentially unchanged over the course of flights of similar duration. Thus, it seems that the persisting reduction in the HVR is evidence that the changes do not result from rapidly adapting receptors. This is in contrast to the situation in cats reported by Biscoe et al. (4) who showed that the response to abrupt changes in carotid sinus pressure was relatively short-lived (lasting much less than 1 min). The basis of the differences between these studies is presently unknown.

Postflight, there was a small increase in the HVR in the upright position (see Fig. 4). It seems likely that is a result of a small decrease in the systemic blood pressure during the period postflight. During this period, circulating blood volume is decreased (1) and cardiac output is lower when the subjects are upright (27). Certainly in the Neurolab crew, cardiac output remained somewhat depressed in the week following flight (Prisk, unpublished observations), and this is likely to be associated with a lower blood pressure, and hence an increase in the HVR.

There is a degree of difficulty associated with the technique of Rebuck and Campbell (30) that was used to measure the HVR. The rebreathing technique maintains the end-tidal CO_2 level constant, but slightly elevated above that seen during resting breathing. However, there are substantial CO_2 stores in the body, and it has long been known that it takes ~10 min for the body stores to become equilibrated following a step change in CO_2 (28). As a consequence, the CO_2 receptors are exposed to a changing stimulus over the course of the short hypoxic challenge. More recent advancements in the technique call for a long period (~6

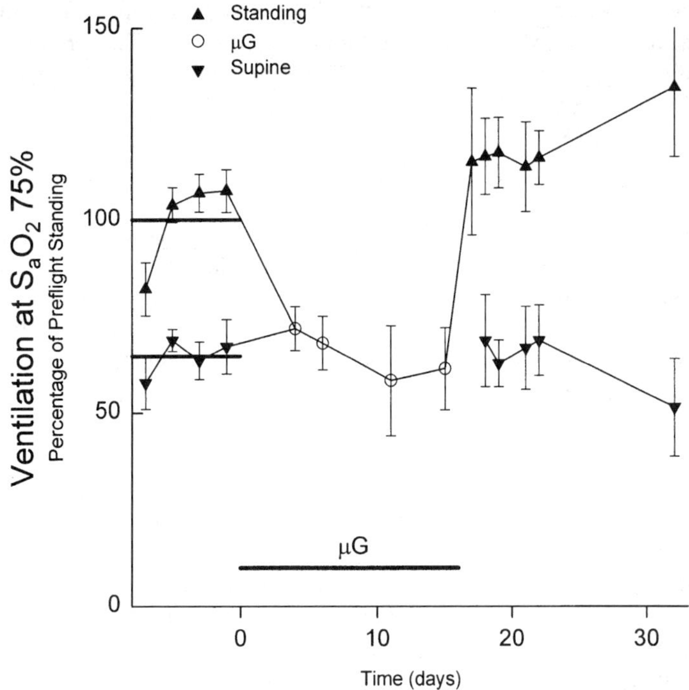

Figure 4 The ventilation calculated at an arterial oxygen saturation of 75% from the HVR response. The lines on the left show the average preflight control values. Note that the preflight data have been given arbitrary times and were collected in the 3 months preceding space flight. Errors bars are SE. ▲, Standing; ▼, supine; ○, μG. (From Ref. 26.)

min) of normoxic breathing at the desired end-tidal CO_2 level, immediately followed by the hypoxic challenge (20). However, the constraints of space flight (both crew time and compressed gas storage) precluded this technique being implemented. However, there is little doubt about the HVR results from Neurolab. First, the change in HVR was very large. Second, there was no substantial change in the ventilatory response to CO_2 caused by either the supine position or by μG (see below, Sec. III.B), and so any effect relating to the equilibration of the CO_2 stores is constant between the conditions studied.

III. Response to Raised Carbon Dioxide

A. Carbon Dioxide Sensors

Unlike the sensors for hypoxia, which are entirely in the peripheral circulation, the majority of the ventilatory response to hypercapnia is due to the central chemoreceptors, located in the brain stem. There is some component of the CO_2 response due to both the carotid and aortic receptors (17). In normal subjects the, ventilatory response to increased CO_2 is much more vigorous than that to hypoxia.

B. Alterations in Microgravity

The hypercapnic ventilatory response (HCVR) was also measured during Neurolab. The test developed by Read (29) is similar to the hypoxic rebreathing test. The subject rebreathes from a bag initially filled with 7% CO_2, 60% O_2, and the balance N_2. As rebreathing continues, the CO_2 in the bag and in the subject's lungs rises, stimulating ventilation. While the O_2 also falls, the initial concentration is sufficiently high that even by the end of the 4-min test, the gas mixture is still hyperoxic, ensuring no contribution from the hypoxic chemoreceptors. Similar to the HVR (see Fig. 1), the breath-by-breath ventilation is plotted as a function of end-tidal P_{CO_2}. Following a period of little response, a threshold level is reached and ventilation begins to increase rapidly.

In sharp contrast to the ventilatory response to hypoxia, which is approximately halved by μG (see preceding section), the ventilatory response to CO_2 is unaltered by exposure to μG (26). Figure 5 shows the results of a rebreathing CO_2 response test in the Neurolab crew, the same crew that participated in the hypoxic challenges shown in Figures 2 and 4. There was a nonsignificant increase in the slope of the response, and a corresponding increase in the P_{CO_2} at 0 ventilation, which suggests a slight steepening of the response with a concomitant shift to the right. When combined, these changes serve to produce no alteration in the ventilation at a P_{CO_2} of 60 mmHg either in the supine posture or in μG.

The HCVR was also measured during an earlier flight. Again, these results showed no change in the HCVR in response to μG, although there were no measurements made in the supine position in that case. In both missions, there was no change with time in the HCVR as the time spent in μG increased.

The measurements of inspiratory occlusion pressure (P_{100}) also match those of the HCVR (see Fig. 3). When the P_{100} was measured during mild hypercapnia (a P_{CO_2} between 43 and 50 mmHg) it was significantly elevated compared with that measured while breathing air. However, the elevation was the same in each condition studied, indicating that the respiratory muscle drive was similar upright and supine in 1 G and in μG.

There are small but significant changes in the resting end-tidal CO_2 levels in humans in space flight, with a slight increase of ~2 mmHg in the crews of SLS-1 and SLS-2 (25). However, this increase was not found to be consistent, with there being no change in end-tidal CO_2 in the SLS-1 crew, and a marked rise (~4.5 mmHg) in the SLS-2 crew. While there were some differences in the environmental CO_2 levels between these two flights (SLS-2 was 1 to 3 mmHg

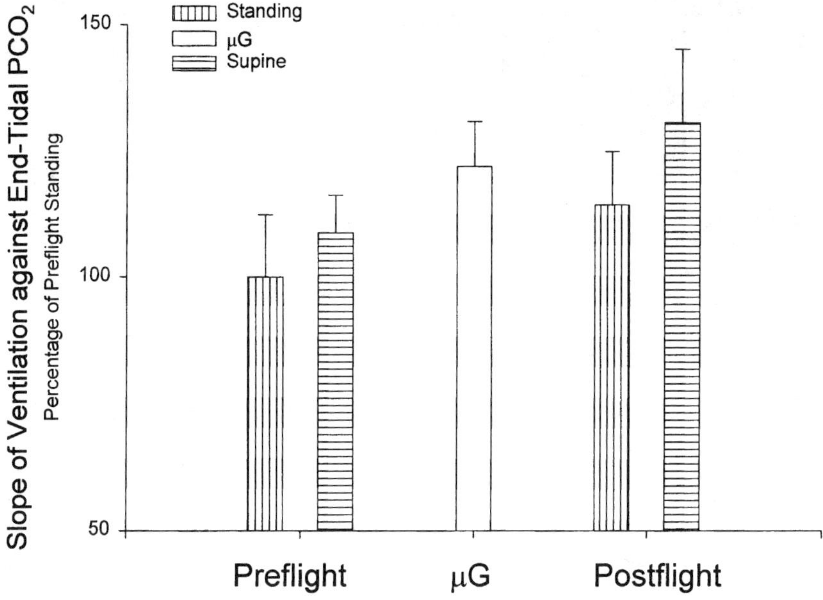

Figure 5 The slope of the ventilatory response to carbon dioxide (HCVR). Data are normalized to each subject's preflight standing control. Format same as Figure 2. (From Ref. 26.)

higher than ground controls and higher than SLS-1), the difference seems too large to be attributable to environmental factors alone (see below).

C. Long-Term Exposure

The major difference between modern space flight and ground control studies in terms of atmosphere is the slightly elevated CO_2 levels within the closed environment of the spacecraft. Even in the Spacelab missions in which extra efforts were made to maintain a low environmental CO_2 level, the P_{CO_2} was often around 4 mmHg (~0.5%).

Extensive studies of the effects of raised environmental CO_2 were made relating to the closed environment of nuclear submarines (24,33,34). More recently, a joint American–German study examined the effects of low levels of environmental CO_2 on a number of physiological factors. The results are reported in a series of articles published together in 1998 (3). The study initially examined the effects of 23 days of 0.7% CO_2 (~5 mmHg) on six subjects. The initial study design called for a subsequent 21-day study of 0.3% CO_2, but preliminary analysis

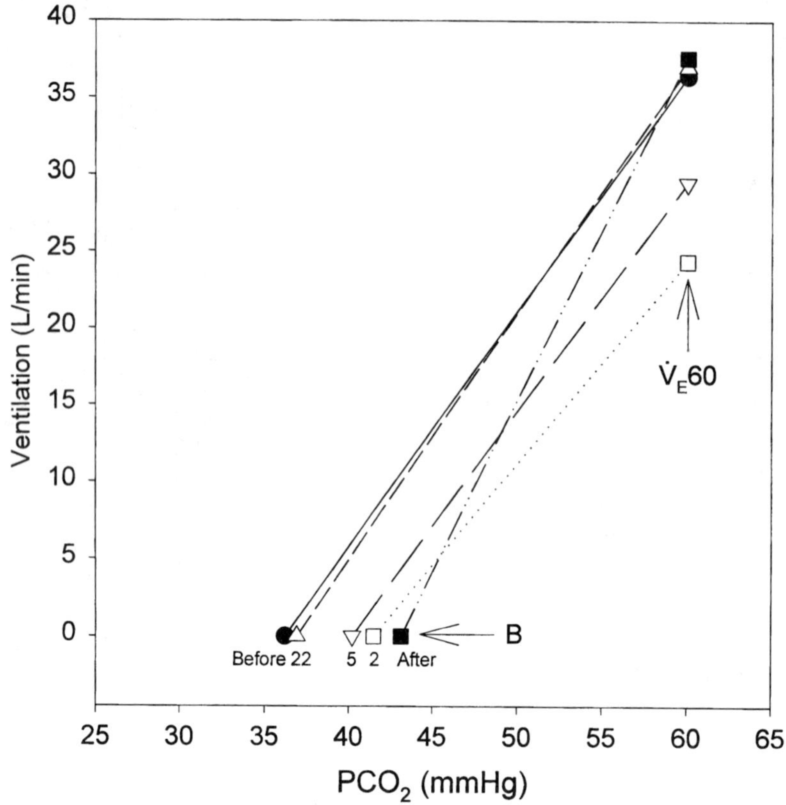

Figure 6 Synopsis of the changes in HCVR before, during, and after 23 days of exposure to 1.2% CO_2. Each line represents the average HCVR line plotted between the intercept of the response (B) and the ventilation at a PCO$_2$ of 60 mmHg (VE$_{60}$). Before exposure (●) represents a normal HCVR curve. Following 2 days (□) and 5 days of exposure (▽) there is a significant shift of the curve to the right which lowers VE$_{60}$; however, this change abated after 22 days of exposure (△). After return to air breathing (■) there is a shift in the intercept, but this is offset by a steeper response, leaving VE$_{60}$ the same as control values. (From Ref. 8.)

showed only very modest changes at 0.7%, and so 1.2% was used for the second 21-day period, which studied the same subjects.

There were only modest changes in ventilation as a result of these exposures (8). Expired ventilation was elevated by ~22% after 5 days of exposure to the higher CO_2 level (1.2%) and decreased thereafter. There was also a significant and persisting increase in end-tidal CO_2 of ~5 to 6 mmHg with no change in the end-tidal O_2. The hypercapnic ventilatory response showed a modest (but not significant) increase in slope and a significant shift to the right of the response line. These changes were maximal when measured on the fifth day of the exposure and decreased thereafter. The combined effect of these changes was to increase the breakpoint of the response (the P_{CO_2} above which ventilation increases with increasing CO_2), and thus lower the ventilation at a fixed P_{CO_2} of 60 mmHg (V_{E60}). The V_{E60} was decreased at day 2 by 28% and by 18% at day 5, but was not significantly different thereafter. These results are summarized in Figure 6. The results are consistent with previous studies of chronic CO_2 exposure (5).

At the lower CO_2 level (0.7%), there were no significant changes in ventilation, although end-tidal CO_2 was altered by a similar degree to that seen at the higher (1.2%) exposure level. The fact that both exposure levels resulted in similar changes in end-tidal P_{CO_2} is interesting in terms of the measurements of end-tidal CO_2 from SLS-1 and SLS-2. On SLS-1, in which CO_2 was very well controlled (not significantly elevated compared with ground controls) there was no change in the end-tidal P_{CO_2}. However on SLS-2, environmental P_{CO_2} averaged 1 to 3 mmHg and end-tidal CO_2 was on average elevated by 4.5 mmHg (25). At the time, Prisk et al. (25) were unable to explain the differences in end-tidal CO_2 between the missions. The results from the CO_2 exposure study raise the possibility that even low levels of environmental CO_2 result in a change in the setpoint for end-tidal PCO_2 that seems to be independent of CO_2 level. It may be that the environmental CO_2 levels on SLS-2 were sufficiently high to cause this change, although the mechanism of such a change, if it occurs, remains unclear.

In the chamber study, 0.7% CO_2 resulted in highly variable changes in the CO_2 response curve between subjects. Overall, these changes failed to reach the level of statistical significance, and it appears that chronic exposure to CO_2 at these low levels has little physiological effect in terms of the control of ventilation. Given these results, it seems unlikely that the levels of CO_2 present in the Neurolab study (which were typically less than 0.5%) played any major role in the lack of changes seen in the HCVR results.

IV. Physiological Implications

A. Exercise

The changes in exercise capacity in space appear to be mostly a result of changes in cardiac performance, fluid distribution in the body, and muscle performance (19,36). There have, to date, been no studies that examine if the possible changes in the CO_2 setpoint have an effect on exercise ventilation in space. On the ground, exercise while breathing an elevated CO_2 results in a lower exercise tolerance (32).

Perhaps the biggest source of change might be expected to found in the exercise encountered during extravehicular activity. In this situation, the person is enclosed in a space suit of very small volume, and so the potential for CO_2 buildup and subsequent CO_2 reinhalation is high. This is especially true during bouts of high activity when it might be expected that CO_2 could rise rapidly within the suit before it can be adequately removed, and could be locally high in the helmet.

Since there seems to be no change in the HVCR, it is unlikely that exercise in space is affected differently by this response. In general, unless exercise is performed using hypoxic gas mixtures, the HVR does not contribute to the ventilatory response. Thus, it seems unlikely that the changes observed in the HVR in μG significantly affect exercise.

B. Sleep

During sleep, there are considerable changes in ventilation (23). For example, during non-REM sleep end-tidal CO_2 decreases by ~4 mmHg (37), indicating a considerable change in ventilatory control. Despite previous reports of poor sleep in μG (10,11,22,31), when it was quantitatively measured during Neurolab, sleep disruption in space seemed to be largely similar to that on the ground, based on the number of arousals detected in the EEG record (7). However, there was a reduction the number of arousals resulting from upper airway obstruction consistent with the removal of gravitational effects on the upper airway. The cause of the other arousals from sleep is unknown.

The HVR, while reduced compared with upright controls in 1 G, is the same as that measured supine (26). From this, it would seem that there is no basis to suggest any change in sleep resulting from the HVR per se. Since there is no change in the HCVR in μG (26), it again falls to environmental factors to determine whether or not sleep is altered. However, it is not a simple matter to determine whether or not these play a role. Measurements of the average cabin P_{CO_2} probably fail to adequately reflect the exposure of the space traveler, especially in situations in which sleeping compartments are used. A recent flight to the fledgling International Space Station in 1999 highlighted this. Crewmembers became temporarily ill when working in a particular portion of the cabin. The symptoms (including nausea and headache) were alleviated by relocating to better-ventilated cabin areas. While no precise cause could be identified, it was generally considered to be due to either off-gassing of materials or CO_2 buildup in a location with poor cabin ventilation. Clearly, CO_2 buildup within corners of a sleep compartment (for example) could result in locally high areas of CO_2, which might cause a triggering of the CO_2 response. Such triggering may be contributory to the reported sleep disruption in space.

References

1. Alfrey CP, Udden MM, Leach-Huntoon C, Driscoll T, Pickett MH. Control of red blood cell mass in spaceflight. J Appl Physiol 1996; 81:98–104.
2. Attinger EO, Monroe RG, Segal MS. The mechanisms of breathing in different body postures. J Clin Invest 1954; 35:904–911.
3. Bassett Frey MA, Sulzman FM, Oser H, Ruyters G. The effects of moderately elevated ambient carbon dioxide levels on human physiology and performance: A joint NASA-ESA-DARA study overview. Aviat Space Environ Med 1998; 69:282–284.
4. Biscoe TJ, Bradley GW, Purves MJ. The relation between carotid body chemoreceptor discharge, carotid sinus pressure and carotid body venous flow. J Physiol 1970; 208:99–120.
5. Dempsey JA, Forster HV. Mediation of ventilatory adaptations. Physiol Rev 1982; 62:262–346.
6. Duggan CJ, Watson A, Pridel AB. Increases in nasal and pulmonary resistance in the supine posture in asthma and normal subjects (abstr). Am Rev Respir Dis 1990; 141.
7. Elliott AR, Shea SA, Dijk D-J, Wyatt JK, Riel E, Neri DF, Czeisler CA, West JB, Prisk GK. Microgravity reduces sleep-disordered breathing in humans. Am J Respir Crit Care Med. In press.
8. Elliott AR, Prisk GK, Schöllman C, Hoffman U. Hypercapnic ventilatory response in humans before, during and after 23 days of low level CO_2 exposure. Aviat Space Environ Med 1998; 69:391–396.
9. Fritsch-Yelle JM, Charles JB, Jones MM, Wood ML. Microgravity decreases heart rate and arterial pressure in humans. J Appl Physiol 1996; 80:910–914.
10. Frost JD, Shumante WH, Salamy JG, Booher CR. Sleep monitoring: The second manned Skylab mission. Aviat Space Environ Med 1976; 47:372–382.
11. Gundel A, Polyakov VV, Zulley J. The alteration of human sleep and circadian rhythms during space flight. J Sleep Res 1997; 6:1–8.
12. Guzenberg AS. Air regeneration in spacecraft cabins. In: Rummel JD, Kotelnikov VA, eds. Life Support and Habitability. Washington, DC: American Institute of Aeronautics and Astronautics, 1994:175–207.
13. Heistad D, Abboud F, Mark AL, Schmid PG. Interaction of baroreceptor and chemoreceptor reflexes. J Clin Invest 1974; 53:1226–1236.
14. Heistad D, Abboud FM, Mark AL, Schmid PG. Effect of baroreceptor activity on ventilatory response to chemoreceptor stimulation. J Appl Physiol 1975; 39:411–416.
15. Heymans C. The part played by vascular presso- and chemo-receptors in respiratory control. In: Nobel Lectures—Physiology or Medicine (1922–1941). Amsterdam: Elsevier, 1965:460–481.
16. Lahiri S. Role of arterial O_2 flow in peripheral chemoreceptor excitation. Fed Proc 1980; 39:2648–2652.

17. Lahiri S, Gelfand R. Mechanisms of acute ventilatory responses. In: Thomas, ed. Regulation of Breathing. Part II. New York: Marcel Dekker, 1981:773–843.
18. Lahiri S, Nishino T, Mokashi A, Mulligan E. Relative responses of aortic body and carotid body chemoreceptors to hypotension. J Appl Physiol 1980; 48:781–788.
19. Levine BD, Zuckerman JH, Pawelczyk JA. Cardiac atrophy after bed-rest deconditioning: A nonneural mechanism for orthostatic intolerance. Circulation 1997; 96: 517–525.
20. Mahutte CK, Rebuck AS. Influence of rate of induction of hypoxia on the ventilatory response. J Physiol 1978; 284:219–227.
21. Malkin VB. Barometric pressure and gas composition of spacecraft cabin air. In: Rummel JD, Kotelnikov VA, eds. Life Support and Habitability. Washington, DC: American Institute of Aeronautics and Astronautics, 1994:1–36.
22. Monk TH, Buysse DJ, Billy BD, Kennedy KS, Willrich LM. Sleep and circadian rhythms in four orbiting astronauts. J Biol Rhythms 1998; 13:188–201.
23. Pack AI. Changes in respiratory motor activity during rapid eye movement sleep. In: Dempsey JA, Pack AI, eds. Regulation of Breathing. New York: Marcel Dekker, 1995:983–1010.
24. Pingree BJW. Acid-base and respiratory changes after prolonged exposure to 1% carbon dioxide. Clin Sci Mol Med 1977; 52:67–74.
25. Prisk GK, Elliott AR, Guy HJB, Kosonen JM, West JB. Pulmonary gas exchange and its determinants during sustained microgravity on Spacelabs SLS-1 and SLS-2. J Appl Physiol 1995; 79:1290–1298.
26. Prisk GK, Elliott AR, West JB. Sustained microgravity reduces the human ventilatory response to hypoxia but not hypercapnia. J Appl Physiol 2000; 88:1421–1430.
27. Prisk GK, Guy HJB, Elliott AR, Deutschman RAI, West JB. Pulmonary diffusing capacity, capillary blood volume and cardiac output during sustained microgravity. J Appl Physiol 1993; 75:15–26.
28. Read D, Nickolls P, Hensley M. Instability of the carbon dioxide stimulus under the "mixed venous isocapnic" conditions advocated for testing the ventilatory response to hypoxia. Am Rev Respir Dis 1977; 116:336–339.
29. Read DJC. A clinical method for assessing the ventilatory response to carbon dioxide. Aust Ann Med 1967; 16:20–32.
30. Rebuck AS, Campbell EJM. A clinical method for assessing the ventilatory response to hypoxia. Am Rev Respir Dis 1974; 109:345–350.
31. Santy PA. Analysis of sleep on shuttle missions. Aviat Space Environ Med 1988; 59:1094–1097.
32. Schaefer KE. A concept of triple tolerance limits based on chronic carbon dioxide toxicity studies. Aerospace Med 2000; 32:197–204.
33. Schaefer KE, Hastings BJ, Carey CR, Nichols GJ. Respiratory acclimatization to carbon dioxide. J Appl Physiol 1963; 18:1071–1078.
34. Schaefer KE, Nichols GJ, Carey CR. Acid-base balance and blood and urine electrolytes of man during acclimatization to CO_2. J Appl Physiol 1964; 19:48–58.
35. Serebrovskaya T, Karaban I, Mankovskaya I, Bernardi L, Passino C, Appenzeller O. Hypoxic ventilatory responses and gas exchange in patients with Parkinson's disease. Respiration 1998; 65:28–33.
36. Shykoff BE, Farhi LE, Olszowka AJ, Pendergast DR, Rokitka MA, Eisenhardt CG,

Morin RA. Cardiovascular response to submaximal exercise in sustained microgravity. J Appl Physiol 1996; 81:26–32.

37. Skatrud JB, Dempsey JA. Interaction of sleep state and chemical stimuli in sustaining rhythmic ventilation. J Appl Physiol 1983; 55:813–822.

38. Somers VK, Mark AL, Abboud FM. Interaction of baroreceptor and chemoreceptor reflex control of sympathetic nerve activity in normal humans. J Clin Invest 1991; 87:1953–1957.

39. Whitelaw WA, Derenne J-P, Milic-Emili J. Occlusion pressure as a measure of respiratory center output in conscious man. Respir Physiol 1975; 23:181–199.

40. Xie A, Takasaki Y, Popkin J, Orr D, Bradley TD. Influence of body position on pressure and airflow generation during hypoxia and hypercapnia in man. J Physiol 1993; 465:477–488.

13

Decompression Sickness in Extravehicular Activities

WILLIAM T. NORFLEET

National Aeronautics and Space
  Administration
Houston, Texas

BRUCE D. BUTLER

The University of Texas–Houston Medical
  School
Houston, Texas

I. Introduction

Extravehicular activities (EVAs), better known as ''spacewalks,'' involve exposures of individuals to the environmental conditions of space with a minimum of protective equipment. This environment includes many factors that threaten crew safety and health, such as thermal extremes, ionizing radiation, objects traveling at extraordinary relative velocities, and, of direct relevance to this book, extremely low ambient atmospheric pressures. This chapter focuses on the physiological consequences and limits of exposures to low ambient pressure, and the methods used to prevent violations of these physiological limits while still maintaining the capability for performing useful work. Additionally, because Earthbound training in water immersion facilities is an essential part of successful EVA programs, this aspect of space flight preparations will also be considered. Taken together, these environments represent ambient pressure extremes from near-vacuum to several atmospheres.

During the era in which the International Space Station (ISS) is constructed and maintained, two types of ''space suits'' will be utilized, the extravehicular mobility unit (EMU) developed by the United States, and the Russian Orlan suit. To maximize the flexibility of these suits while maintaining an adequate alveolar oxygen partial pressure, the suits are supplied with nearly pure oxygen at sub-atmospheric pressures. Specifically, the absolute pressure within the EMU is 29.6 kPa, while that in the Orlan suit is 38.6 kPa. Under most circumstances, the cabin atmosphere within all spacecraft (the Space Shuttle, the Soyuz spacecraft [the "lifeboat" spacecraft], and ISS itself) approximates sea-level conditions with a composition of 21% oxygen, balance nitrogen and trace gases, at an absolute pressure of 101 kPa. These conditions may not be optimal from the point of view of spacecraft systems engineering, but they do facilitate performance of Earth-based control experiments. Consequently, when crew transition from a spacecraft cabin to the exterior of the spacecraft, even while wearing a space suit, they experience a substantial decompression equivalent to an altitude of approximately 9100 m (approximating the summit of Mount Everest). If this decompression were accomplished without specific preparation, serious decompression sickness (DCS) would be virtually inevitable, including neurological and/or cardiopulmonary manifestations of this disease process. Similarly, EVA training programs in water immersion facilities involve long dives to depth equivalents of 15 m; without proper technique, serious DCS would be a common feature of these activities as well. This chapter discusses the nature of DCS, the critical role the lungs play in protecting systemic vascular beds from decompression-induced intravascular bubbles, and the means used to prevent and treat DCS. Since some of these preventive measures involve risks of their own to the pulmonary system, primarily in the form of pulmonary oxygen toxicity, this disease process is discussed.

II. Decompression Sickness

Although procedures for hypobaric exposures with a low risk of DCS have been developed and effective treatments for hypobaric DCS exist, not a great deal is known for certain about the pathophysiology of this disorder. The principal etiology of DCS involves the formation of nitrogen bubbles in the interstitial spaces of various tissues and in venous blood. This section discusses the process of bubble formation and resolution, the probable sites of bubble formation within the body, the interactions of bubbles with nearby tissue, the uptake and elimination of the inert gas that drives bubble formation, and the pathophysiology of DCS in specific organ systems, with emphasis on the lungs.

A. Nomenclature

The spectrum of disorders arising from ambient pressure changes has been classified by a variety of schemes. Many authors utilize the term "decompression *illness*" (DCI) to refer to *all* of these disorders, including such heterogeneous entities as arterial gas embolism and trauma to enclosed gas spaces like the thorax, sinuses, and middle ear. The subset of diseases caused by evolved gas bubbles arising from inert gas dissolved within tissues is commonly referred to as "decompression *sickness*" (DCS). Decompression sickness that occurs during diving is termed hyperbaric DCS, and disease that arises from aerospace operations is called hypobaric or altitude DCS.

Traditionally, DCS is further subdivided into Type I and Type II forms. The exact demarcation line between these two types is drawn differently by various organizations such as the U.S. Navy, U.S. Air Force, and civilian diving organizations. As a result, comparisons between databases can be very difficult, and crisp communication among practitioners regarding diagnosis, prognosis, and return-to-duty considerations is hampered. Originally, Type II DCS meant, simply, serious disease, and Type I DCS was considered not so serious (1). The U.S. Navy (2) has defined these two types as follows:

1. Type I decompression sickness includes joint pain (musculoskeletal or pain-only symptoms) and symptoms involving the skin (cutaneous symptoms), or swelling and pain in lymph nodes.
2. Type II, or serious symptoms, are divided into neurological and cardiorespiratory symptoms. Type I symptoms may or may not be present at the same time.

Within the U.S. Air Force, this scheme has been modified to include the "Type I peripheral nervous system case" (3). The motivation for this change arises more from administrative concerns than biomedical knowledge: because inclusion of a diagnosis of Type II DCS in the medical history of a U.S. Air Force aviator might have serious career implications, medical practice within the Air Force has evolved such that paresthesias of limited anatomical distribution that resolve without sequelae are included with "Type I peripheral nervous system DCS" rather than Type II DCS. The *U.S. Navy Diving Manual* has also created a category of "patchy peripheral paresthesias" that is treated separately in determining when return to duty is permissible (4).

Recently, several new schemes have been proposed and discussed to replace the Type I/Type II designations (5–7). However, no clear consensus has yet been reached concerning the best nomenclature for DCS. For this chapter, the Type I/Type II classification scheme quoted above from the current version of the *U.S. Navy Diving Manual* is utilized.

B. Bubble Formation

Bubbles can form when the sum of partial pressures of gases dissolved in a liquid, plus the vapor pressure of the liquid itself, exceeds the hydrostatic and hydrodynamic pressures in the liquid. However, when this criterion is met, bubble formation is not instantaneous, and substantial supersaturation can occur without gas bubble formation. In fact, in pure water at 1 atmosphere absolute (ata) without preexisting bubbles, dissolved gas, previously equilibrated with a gas phase at over 1000 ata, can remain in solution (8,9). In contrast, bubbles have been observed in humans after very modest dives (7.8 m) (10) and flights (3700 m) (11). These data indicate that the process of bubble formation in humans differs from that which occurs in a beaker of still, pure water.

Several factors, including preexisting gas micronuclei, modify bubble development in vivo and obviate the need for de novo gas formation; gas molecules may simply diffuse into and enlarge existing gas nuclei. Experimental evidence to support this notion can be found in studies with rats in which dives began with a very short, deep pressure spike designed to "crush" preexisting nuclei; this pressure excursion protected against DCS in a subsequent decompression (12). Similar findings have been obtained in shrimp (13). In contrast, McDonough and Hemmingsen (14) have reported that a hydrostatic pressure spike had no effect on bubble formation in adult crabs. Currently, the role of preformed nuclei in the generation of DCS is not well defined, at least not in higher animals (15,16).

Another factor that may influence the development of a gas phase in humans following decompression and explain the genesis of gas micronuclei is the generation of negative pressures in tissue as the result of movement and locomotion. When two closely opposed tissue surfaces are forced to separate, fluid must flow into the widening gap to fill it, but the viscous properties of the fluid tend to oppose this flow. The resulting negative pressure in this gap can be tremendous, large enough to cavitate the fluid and generate bubbles. A similar phenomenon occurs when two tissue surfaces slide against each other, a process called tribonucleation (17). McDonough and Hemmingsen (18) demonstrated the importance of locomotion in the generation of bubbles in vivo with elegantly simple experiments in which the development of bubbles within crabs was compared in animals left free to scurry around following decompression vs. those whose legs were immobilized: more bubbles were seen within the free-roaming crabs. These investigators did similarly imaginative studies throughout the development of fish from hatching to adulthood that also demonstrated the importance of movement in the generation of bubbles (19). For an excellent review of in vivo and in vitro bubble formation, see Hemmingsen (16).

C. Sites of Bubble Formation In Vivo

Where do bubbles form in the body? The precise location of generation of bubbles that cause disease is largely unknown. When considering where bubbles might form, tissue compartments can be categorized as follows:

    Intravascular
        Arterial
        Capillary
        Venous

Extravascular
   Intracellular
   Interstitial
   Lymphatic

Regarding intravascular bubbles, the arterial circulation seems an unlikely site for bubble *formation*. Inert gas tensions in arterial blood are, in most circumstances, in equilibrium with alveolar gas, so arterial blood is not supersaturated with inert gas. Also, arterial blood is pressurized hydraulically by the heart, further impeding any gas-phase formation. Despite this, circulating bubbles have been detected in arterial blood, and these microemboli may be responsible for certain more serious forms of DCS.

Venous blood has been shown to contain significant numbers of circulating microbubbles following even modest, asymptomatic decompressions (10). However, formation of bubbles within venous blood itself seems unlikely, at least when blood-filled vessels are isolated from the circulation: when blood-filled venae cavae of several species were excised, placed in saline, and decompressed to altitudes well in excess of those that produce DCS, no bubble formation occurred (20). The ultimate fountainhead for venous bubbles is unclear; they may form in capillary beds and be swept into the central venous circulation or they may arise in extravascular tissues and migrate into the circulation. Regardless of their source, circulating venous microbubbles, in most circumstances, do not seem to be the proximate cause of Type I DCS, since the magnitude of their numbers correlates only very loosely with the likelihood of the development of symptoms (21–24).

Within the extravascular compartment, bubble formation within cells seems uncommon. Cells and unicellular organisms are very resistant to the formation of bubbles (25). The intracellular environment is not conducive to bubble formation as demonstrated by the fact that microscopic particles that serve as a nidus for bubble formation in water fail to do so when ingested by *Tetrahymena* (26).

This leaves the interstitial space as a possible site for the formation of bubbles that cause symptoms. In some tissues, experimental evidence exists to support this notion. As an example, in situ interstitial gas-phase formation—so-called autochthonous bubbles—has been observed in spinal cord tissue following decompression (27). In contrast, for the brain and kidney, experimental evidence exists that excludes bubble formation within these tissues under operationally realistic pressure profiles (28); disease in these organs is probably caused by the embolic effects of arterial microbubbles that originate in other areas. Additionally, exposures of rabbits to hypobaric conditions did not generate extravascular bubbles (29,30). It appears that bubbles formed in situ may cause some forms of DCS, whereas emboli originating in remote sites may cause other manifestations. For many forms of DCS, such as common, pain-only limb "bends," the site of formation of the provocative bubbles has not been clearly demonstrated.

D. Interactions of Bubbles with Blood and Tissue

What is it about the presence of a bubble that causes disease? Broadly speaking, these effects can be thought of as mechanical and nonmechanical in nature. Mechanical effects include embolization of capillary beds by circulating microbubbles with consequent hypoxemia and ischemia. Bubble accumulation in venules and veins may inhibit venous drainage with consequent tissue edema, ischemia and, perhaps, interstitial bubble formation as dissolved inert gas that otherwise would have been swept out of the tissue in solution remains sequestered in the tissue and comes out of solution. Ferris and Engel (31) argued, however, that embolic ischemia from intravascular gas does not cause Type I DCS pain based on the observation that, if a second altitude exposure takes place within 6 hours of a bends-producing flight, pain usually recurs in the same site—if the original pain had been caused by ischemia from a bubble embolus, then the interflight compression would have squeezed the bubble, permitting the bubble to be swept out of the area and, thereby, eliminating a source of pain in a subsequent altitude exposure.

Expanding interstitial bubbles may compress surrounding tissue causing ultrastructural damage, stimulate localized nerve endings, and raise local tissue pressure to the point that circulation is compromised. Inman and Saunders (32) provided evidence that pain similar to that of Type I DCS could be created by injection of small volumes of normal saline into the muscles and ligaments in joint areas. They further reported that the pressure of the injectate, not the volume, produced the pain.

In many clinical situations, the nonmechanical effects of bubbles may be more important than their mere mechanical presence. The presence of bubbles in venous blood and tissues is often without clinical consequence (so-called "silent bubbles") (33–35). Conversely, the onset of symptoms sometimes occurs so long after a pressure excursion that the continued presence of actual gas bubbles seems unlikely at the time when symptoms finally begin (36). Apparently, like boiling chips in a distillation apparatus, the presence of bubbles in body tissues serves as a nidus for the initiation of mechanical, biochemical, and cellular processes that, once begun, can lead to morbidity and mortality.

Some of the nonmechanical effects of bubbles may be initiated at the interface of bubbles with blood or interstitial fluid, a boundary that demarcates regions of vastly different physicochemical properties. Proteins that interact with this discontinuity may undergo conformational changes (37). The presence of bubbles in blood affects many cascading systems, such as complement factors, coagulation factors, kinins, and fibrinolytic systems (38–41). The results can include pain, edema, coagulation, reduction in local tissue perfusion, chemotaxis, and activation of platelets and leukocytes.

Leukocytes and platelets have been demonstrated to adhere to bubbles (42–44) as well as to sequester in the lungs after experimental air embolism (45). Staub et al. (46) demonstrated that venous gas bubbles in sheep increase the neutrophil count in lung parenchyma by twofold and in pulmonary vasculature by sevenfold. Circulating neutrophils likely play a role in DCS-mediated lung injury as well. Neutrophil activation was reported with hyperbaric decompression by Benestad (47). Once activated, neutrophils become larger and less deformable (48,49), which reduces their flow through the lungs in acute injury (50). Attraction of neutrophils to the gas bubbles with subsequent sequestration in the pulmonary capillaries is triggered by a number of chemotactic factors (51). Wang et al. (52) reported that venous gas emboli (VGE) in experimental conditions did not increase the neutrophil count in the lung perfusate but did elevate the number of activated cells. Activated neutrophils are also reported to stimulate production of oxygen radicals (44,53) that can damage endothelial cells, thereby altering the normal fluid balance of the lungs, leading to edema formation. These substances, including superoxide anions, hydrogen peroxide, and hydroxyl radicals, have been implicated in the acute injury associated with air embolism (54). Attenuation of the increased microvascular membrane permeability that caused edema was shown by Flick et al. (53) with removal of neutrophils from the circulation prior to embolization. Such effects were not observed with either fibrinogen or platelet depletion before gas embolism (55,56).

Experimental observations that bridge the mechanical–nonmechanical paradigm include denuding or other direct damage to capillary endothelium by bubbles (57–59) with subsequent inflammation (42,60), and peroxidation of myelin initiated by free iron (61) released when hemorrhage occurs around autochthonous bubbles in spinal tissue (62).

The combined effects of bubbles may explain the many-faceted clinical presentation of DCS and the observation that symptoms may persist or even begin in a time frame beyond the expected longevity of a bubble within a tissue. Blood–bubble interactions are summarized in Table 1.

E. Inert Gas Elimination

As discussed above, the precise microenvironments within tissues that produce bubbles that lead to DCS are largely unknown. Consequently, measurements of inert gas tensions within these regions have not been reported. Analysis of whole-body inert gas elimination has been studied by many investigators (63) but whole-body gas elimination does not necessarily reflect gas exchange in the tissues of interest. Gas exchange has also been investigated in experimentally induced subcutaneous gas pockets (64), but these studies require a similar leap of faith

Table 1   Direct and Indirect Blood–Bubble Interactions

Adsorption/denaturation of plasma proteins
Adsorption of phospholipids
Adsorption of fibrinogen
Activation of Hagemen factor
Complement activation
Clumping of red cells
Leukocyte activation/adherence
Lipid peroxidation
Microthrombi production
Platelet activation/adherence
Thrombin activation
Phospholipase activation
Endothelial cell damage

Source: Ref. 201.

in order to draw conclusions regarding gas exchange in clinically relevant tissues. Despite this dearth of fundamental knowledge, humans have already been exposed to a wide variety of pressure–time profiles—with varying degrees of success. Use has been made of this database to develop mathematical models of inert gas exchange. These models are useful in extrapolating between known regions of the pressure–time continuum and, less successfully, in venturing into new extremes. A wide variety of mathematical models have been advanced, and a description of these inventions is well beyond the scope of this chapter (65). However, brief consideration of one method is relevant to the current discussion, not because it necessarily represents the state of the art, but because it presently governs Space Shuttle operations (66).

This method is applied primarily to modeling inert gas elimination during oxygen prebreathing and step-reductions in ambient pressure that occur prior to an EVA from the Space Shuttle or ISS. This approach is not new; it has its origins in the work of Haldane (67). Briefly stated, the method assumes (based upon empirical evidence) that inert gas elimination can be modeled by dividing the body into a number of conceptual compartments, each with its own rate constant for elimination of inert gas:

$$P_t = P_0 + [(P_a - P_0)(1 - e^{-kt})] \tag{1}$$

where

P_t = inert gas partial pressure in tissue after t minutes
P_0 = initial inert gas partial pressure
P_a = inert gas partial pressure in inspired gas (although use of alveolar or arterial gas pressure would be more correct in physiological terms)

  t = exposure time in minutes
  k = compartment rate constant (k is related to the inert gas half-time $t_{1/2}$ by: $k = 0.693/t_{1/2}$)

In the typical operational scenario, only the longest of these compartments (empirically determined to have a half-time of 360 min) governs decompression.

Inert gas partial pressure in tissue can be compared to ambient atmospheric pressure as an indication of the decompression "stress" as follows:

$$TR = P_t/P_{amb}$$

where

  TR = tissue ratio
  P_{amb} = ambient atmospheric pressure

This modeling method is useful in determining conditions under which bubbles will form. As previously discussed, bubbles can form when the sum of partial pressures of gas dissolved in a liquid exceeds the hydrostatic and hydrodynamic pressure in the liquid. However, the process of bubble formation is subject to many other factors that make substantial supersaturation possible before bubble formation takes place, and the presence of bubbles in tissue does not invariably lead to DCS. The net result of these deliberations is the realization that, for a significant number of symptom-producing bubbles to form at all, TR > 1.0. Similarly, when TR ≫ 1.0, symptoms are likely. In other words, the risk of DCS is a function of TR. This rather simplistic approach to quantitating the risk of DCS has some utility, although it does not include many factors that are known to modify this risk. In the early years of the Shuttle era, NASA senior managers determined that an effective compromise between mission objectives and DCS risk could be achieved by limiting the decompression stress in nominal operations to TR < 1.7.

F.   Pathophysiology of Decompression Sickness

The manifestations of DCS are often diffuse, multifocal, and protean. Multiple organ systems can be simultaneously involved. Although bubbles serve as the initiating agent, disease can persist and probably even progress after the bubbles themselves have been resorbed. Clinicians who resolutely seek the "one lesion" are likely to be frustrated in their attempts to characterize the clinical manifestations of a case of DCS (68,69). Therefore, the following discussion of the impact of DCS on organ systems should be read with the understanding that simultaneous, interacting disease in many systems can occur.

Hyperbaric vs. Hypobaric Decompression Sickness

Much of our current understanding of the pathophysiology of hypobaric DCS has actually been derived by a process of extrapolation from experience with *hyperbaric* DCS. While the two situations are somewhat analogous, it is important to understand that fundamental differences do exist:

1.   Pressure profile: The usual dive begins with an individual at sea level whose tissues contain a dissolved mass of inert gas that is in equilibrium with inert gas in the surrounding atmosphere—the diver is said to be "saturated" with air at 1 ata that contains approximately 80% inert gas (nitrogen, in this case). During the course of a typical dive, the individual descends to some depth, then returns to the surface in short order. The diver has accomplished a downward excursion from saturation conditions (in this discussion, "down" refers to an increase in ambient pressure as is encountered when a diver descends into the ocean). In terms of inert gas uptake and elimination, divers absorb inert gas from their breathing air during the course of the dives, and subsequently eliminate this nitrogen during a process that begins upon initiation of ascent, but continues for some period of time following return to their usual sea-level environment.

The astronaut's EVA pressure profile is fundamentally different. These individuals also begin their day saturated in specified conditions, but the pressure profile involves a *reduction* in ambient pressure—an *upward* excursion from saturation conditions of the Space Shuttle or ISS atmospheric pressure of 101 kPa (14.7 psia, 1 ata, or sea-level equivalent). Inert gas elimination takes place *during* the course of the pressure excursion, not afterward. Most importantly, for the spacewalker, every exposure ends with a potentially therapeutic compression, not a decompression. In this respect, hypobaric and hyperbaric operations are quite different.

2.   Scenario differences at onset of symptoms: This consideration is closely related to the above topic: divers are typically at risk of developing DCS only upon return to the surface when the job has been completed and they are "home." In contrast, the onset of symptoms in spacewalkers is most likely to occur while they are still engaged in EVA. Consequently, if DCS occurs, completion of the job of a spacewalker is more likely to be disrupted by DCS, and symptom onset is more likely to occur in the midst of already hazardous operations.

3.   Importance of gases other than inert gases: In addition to inert gases such as nitrogen, other gases are also dissolved in body tissues. For example, carbon dioxide is generated by cellular processes and is maintained at a tension of approximately 5.9 kPa by the circulatory and ventilatory systems. The tension of water vapor is a function of temperature, so it is found at a constant pressure of 6.3 kPa in tissues. In the diver, these tensions are trivial compared to that of

inert gas (as high as 6600 kPa in very deep dives) (70). Consequently, when considering the physics of bubble formation in divers, gas species other than inert gas can largely be ignored. In contrast, the ambient atmospheric pressure of the spacewalker is typically only about 30 kPa, so consideration of the contribution of water vapor and carbon dioxide (and, to a lesser extent, oxygen) to bubble formation and resolution is very important. As an example, under equilibrium conditions breathing air at a depth of 10 m, 8% of the volume of a bubble consists of gas species other than nitrogen, whereas at an altitude of 9100 m, 56% of the bubble volume encompasses these gas species. This consideration may be responsible for the fact that some mathematical models of inert gas uptake and elimination that work well for divers fail when applied to hypobaric conditions.

4.   Time course of bubble formation: The fundamental physical principles underlying the formation of a bubble from dissolved gas are quite complex. For the purpose of this discussion, it is noteworthy that a bubble of a given volume will form more slowly under hypobaric conditions than in a hyperbaric environment. For example, Piccard (71) observed that the same volume of gas was liberated from water saturated with air at 5 ata when it was decompressed to 1 ata as when water at 1 ata was decompressed to 0.2 ata, but that: ''In the first case the bulk of dissolved air had escaped after 5 seconds, while in case 2, several minutes elapsed before the gas production even approximately ceased.'' Bubble formation may be sufficiently slowed under hypobaric conditions to permit physiological processes to eliminate some inert gas from tissues before significant gas has time to evolve.

5.   Qualitative differences in DCS natural history: The typical case of hypobaric DCS is less severe than hyperbaric DCS, is less likely to cause neurological or other sequelae, and is more amenable to treatment. The explanations for these empirical observations are not clear, but they may involve the differences in the pressure profiles and the time course of bubble formation discussed above, as well as the fact that a bubble in hypobaric conditions is more compliant than a bubble in hyperbaric conditions. Meaningful comparisons between databases from the hyperbaric and hypobaric communities require careful definition of what, exactly, constitutes a decompression ''hit.'' For example, operational managers in the hypobaric arena may be willing to accept an overall probability of DCS that far exceeds that which would be considered responsible by their colleagues in the diving world because the incidence of the subset of cases with serious, lasting injury is less in hypobaric operations.

Despite these distinctions between hypobaric and hyperbaric decompression physiology, much can be learned about the pathophysiology of hypobaric DCS by considering experience from diving operations, especially decompression from saturation diving. In fact, the microgravity environment may blend elements of both flying and diving, involving, for example, fluid shifts analogous to those of the immersed diver and pressure excursions similar to those encountered by the aviator. Consequently, the following discussion draws heavily from diving and hyperbaric physiology.

Signs and Symptoms

Information regarding the manifestations of hypobaric DCS can be derived from the many thousands of human exposures conducted each year by the world's armed forces in hypobaric chambers as part of the training of flight crew members. Table 2 depicts the manifestations and incidence of altitude DCS and related disorders encountered by the U.S. Navy from 1981 to 1988 (72). In addition to these phenomena, fatalities have been reported (73–75). In the years 1985 to 1987, the U.S. Air Force experienced 282 cases of DCS in the course of 239,343 hypobaric chamber exposures; 89% of these were classified as Type I, and 11% were Type II (76). During 1984 to 1989, the U.S. Army suffered 42 cases of DCS in 21,498 exposures, but a breakdown of clinical presentation was not reported (77). Interestingly, military organizations outside of North America report extremely few cases of altitude DCS (78), perhaps due to differing reporting methods. Few diseases are as diverse in their manifestations as DCS.

The following discussion considers the expressions of DCS in the cardiopulmonary system, beginning with an assessment of the effects of bubbles in blood.

Blood

The nature of the interaction of bubbles with blood has been discussed above. Blood also plays many roles in the pathophysiology of DCS by serving as a conduit for gaseous microemboli to the pulmonary and systemic vasculature, transporting leukocytes and platelets to bubble-damaged vascular endothelium and tissues, and participating in local inflammatory reactions (44,45,79,80). Hemoconcentration and a decline in intravascular volume are frequently observed in patients with DCS (31,42,81,82); consequently, fluid therapy is a cornerstone of treatment for the disease. Additional therapeutic maneuvers to counteract observed changes in blood, such as administration of heparin and low-molecular-weight dextran, have not yet proved useful.

Cardiopulmonary System

Bubbles carried in the venous circulation will eventually reach the pulmonary capillary vasculature and embolize the lung (Fig. 1). In severe cases, these gaseous microemboli produce a syndrome know as "the chokes," which is characterized by dyspnea, cough, retrosternal pain, and cardiovascular collapse (83). The lungs are highly effective in removing bubbles from the circulation and pre-

Table 2 Manifestations of DCS from 140 Patients in 136,696 Exposures in U.S. Navy Altitude Chambers, 1981–1988

Clinical manifestations[a]	Number of patients	Percent of patients
Joint and limb pain	99	70.7
Extremity paresthesia	46	32.9
Numbness	33	23.5
Muscular weakness	24	17.1
Dizziness	22	15.7
Headache	12	8.6
Nausea and vomiting	11	7.8
Visual disturbances	10	7.9
Fatigue and malaise	8	5.7
Apprehension	7	5.0
Mental confusion	7	5.0
Disorientation	7	5.0
Hyperventilation	5	3.6
Paralysis	4	2.9
Pruritis	3	2.1
Muscle spasm	3	2.1
Skin mottling	2	1.4
Ataxia	2	1.4
Chokes	1	0.7
Unconsciousness	1	0.7
Slurred speech	1	0.7
Vertical nystagmus	1	0.7
Abdominal pain	1	0.7
Hot and cold flashes	1	0.7
Difficulty forming words	1	0.7

[a] Note: More than one manifestation could occur in a given individual patient.
Source: Ref. 36.

venting embolization of important systemic capillary beds in tissues, such as the brain and spinal cord. However, large bubble loads (in excess of 0.3 mL/kg/min in anesthetized dogs) can overwhelm this filtering effect (84–89). Numerous clinical reports of venous gas embolism, including altitude decompression, also describe arterialization of the bubbles, even in the absence of interconnecting vessels or anatomical cardiac defects, such as a patent foramen ovale (90–95). Various pharmacological agents may also adversely affect the filtering capacity of the lung, thereby increasing the risk for systemic embolization (85,87,96–98).

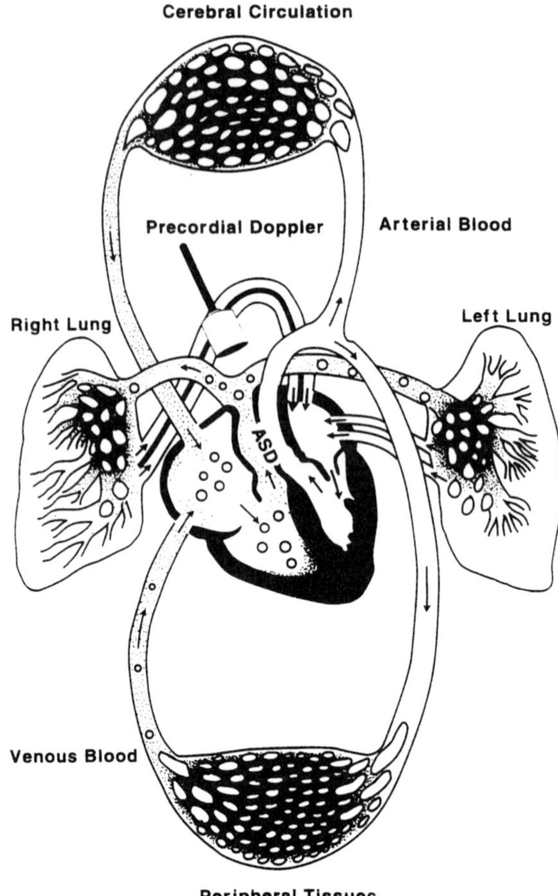

Figure 1 Schematic of venous bubbles circulating through the right heart chambers and into the left and right lungs where they are usually filtered out. The presence of an atrial septal defect (ASD) may allow venous bubbles to cross over to the arterial circulation, bypassing the lung filter. The common site for venous bubble detection via Doppler ultrasound in decompression studies is over the precordial position. (Adapted from Ref. 203.)

Bubbles that lodge in the lungs remain in place for many minutes, at least during air breathing (99,100), providing an adequate period of time for the development of significant interactions with surrounding pulmonary tissues. Resolution rates are dependent upon such factors as bubble size, total gas volume, and the gas composition of the bubble and inspired gas (100–106). Microvascular permeability has been observed to increase after gas embolization, resulting in reversible pulmonary edema (107). Also, leukocyte cell count and lysophosphatidylcholine levels are increased in bronchoalveolar lavage fluid following decompression in rats (108). These interactions of gaseous microemboli with the pulmonary vasculature can impair oxygen transport by the lung (109). Pulmonary emboli are also thought to impair oxygen transport by altering the matching of regional ventilation and perfusion, and opening intrapulmonary arteriovenous shunts (106,110–112).

As venous bubbles are carried into the pulmonary circulation, their principal effects are mechanical obstruction and resultant hemodynamic and biochemical responses. Both obstruction and vasoconstriction cause elevations in pulmonary artery, right ventricular, and central venous pressures, and increases in pulmonary vascular resistance. These elevated resistances have been attributed to both direct vascular reflexes as well as local mediator release (Table 3). Active constriction of pulmonary vessels likely occurs in the small muscular arteries ($<$500 μm) where obstruction exits. Bubble-induced vasoconstrictive mediators are released from activated platelets, macrophages, and leukocytes. Activation of granulocytes results in the formation of vasoactive and microvascular permeability–altering mediators, especially from the lungs, the principal target organ where venous bubbles accumulate. These agents include serotonin, histamine, and kinins, as well as free arachidonic acid and its metabolic byproducts resulting from both the cyclooxygenase and 5-lipoxygenase pathways (52,113,114). The cyclooxygenase byproducts include prostacyclin and other prostaglandins, as well as thromboxanes, which cause vasoconstriction, bronchoconstriction, and further aggregation of platelets. Lipoxygenase pathway products include the leukotrienes that cause contraction of bronchial smooth muscle and increased microvascular permeability leading to edema formation (115). Elevated thromboxane and leukotriene levels have been reported in plasma and alveolar lavage following experimental air embolism (116–118) and with both hyperbaric and hypobaric decom-

Table 3 Mediator Release
from Gas Bubble Embolization

Adenosine diphosphate
Serum catecholamines
Hydrogen peroxide
Hydroxy radicals
Leukotrienes
Prostaglandins
Serotonin
Superoxide anions
Thromboxanes

Source: Ref. 201.

pression where pulmonary edema was also evident (119,120). Recent studies have also shown alterations in plasma nitric oxide levels with DCS and venous air embolism (121). As with many inflammatory responses, these mediator levels were shown to display a circadian rhythmicity following experimental decompression sickness (122).

When sufficient venous bubbling occurs, the net hemodynamic response results in increased afterload to the right heart causing a decrease in cardiac output and systemic blood pressure (86,103,123–125). In severe cases, right ventricular failure can be attributed to myocardial ischemia resulting from reduced coronary artery blood flow following the drop in aortic diastolic pressure and the increased right ventricular pressure (126,127).

Generally, the lungs provide excellent filtering of venous bubbles (85,86). In fact, it is this filtration function of the pulmonary microcirculation that enables mammals to experience bubble-provoking decompressions usually without serious systemic embolization. However, as mentioned previously, bubbles have been detected in the systemic arterial circulation following decompression stress. The source of these arterial bubbles is likely to be transpulmonary or "right-to-left" shunting of bubbles from the venous circulation (84–86,88,89,109, 128,129). Such "shunting" of venous bubbles can occur principally via three routes: normal vessels within the lung tissue, larger venous-to-arterial shunts, and anatomical defects within the heart (84–86,88,90,94,124,130–132). These routes of arterialization are depicted in Figure 2.

Bubbles circulating in blood that bypasses the lungs through anatomical shunts, such as a patent foramen ovale (PFO), will not be captured by the lungs. Delivery of these bubbles into the systemic arterial vasculature might result in gas embolism of peripheral vessels, including the coronary and cerebral circulations. Consequently, attention has recently been focused on determining if a PFO or other interatrial or intrathoracic "right-to-left" shunts are a risk factor for neurological and cardiovascular forms of altitude DCS. This complex topic draws upon information concerning the origins of cryptogenic thromboembolic stroke in which PFO seems to play an important role (133–135), and the etiology of neurological DCS in divers in which PFO also is a factor in some cases (136,137). Data concerning the contribution of anatomical shunts to altitude DCS are limited (138,139). Clark and Hayes (138) reported no increased prevalence of PFO in a group of military personnel who developed neurological altitude DCS, but the pressure profiles to which these individuals were exposed are not likely to have generated venous gas emboli. In contrast, ground-based simulations of EVA-like pressure profiles suggest that these exposures generate venous gaseous emboli in many persons (140). Whether these bubbles can gain access to the systemic arterial circulation through anatomical shunts, thereby generating unusual forms of DCS such as cerebral dysfunction and, perhaps, cutaneous marbling, seems possible, but the likelihood of such events remains to be resolved.

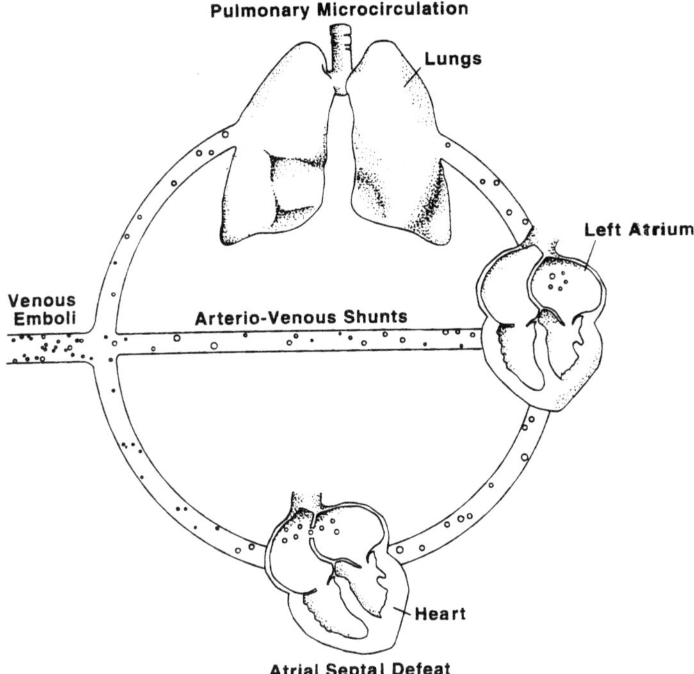

Figure 2 Schematic representation of routes of transpulmonary or "right-to-left" shunting of bubbles from the venous circulation into the systemic arterial vasculature.

Clinical Course of Decompression Sickness

The clinical course of hypobaric DCS is influenced by a number of factors, including

Duration of oxygen prebreathe
Duration of altitude exposure
Altitude reached
Magnitude and type of exercise (if any) during both prebreathe and altitude exposure
Time until initiation of treatment
Type of treatment
Rapidity of depressurization (ascent) and repressurization (descent)

Just as all dives are not equal, all hypobaric exposures are not the same. The vast majority of acute human exposures to hypobaric conditions that might cause DCS occur during physiological training courses for aviators, primarily armed forces personnel. Additional DCS risk occurs with loss of cabin pressure or high-altitude flying of nonpressurized or underpressurized cabins (72,141,142). A wide variety of physiological training protocols are (or have been) in use (143). However, the majority of these exposures might be characterized as involving short oxygen prebreathes, short exposures to modest altitudes, with little or no exercise performed at altitude. Definitive diagnostic and therapeutic resources are usually immediately available. This situation can be contrasted with current EVA operations in space where activities involve relatively long oxygen prebreathes (up to 4 h), long exposures (4 to 6 h), high altitudes (9100 m), moderate exercise, and remote treatment resources. Certain microgravity-induced cardiovascular changes likely provide additional differences in the two types of exposures (144). By some measures, these two types of activities are as different as they can be. Therefore, comparison of the two clinical experiences may have certain limitations. However, since no cases of DCS have been formally reported in the course of EVA conducted by any country, the outcomes of physiological training activities will be discussed to characterize the clinical course of altitude DCS.

In the experience of U.S. Air Force physiological training, the onset of signs and symptoms of DCS occurred as early as during the altitude exposure itself to as long as 36 h after return to ground level (76,142). The median time of onset was 2 h after return to ground level. From a pathophysiological viewpoint, it is interesting to note that most cases of hypobaric DCS appeared hours after termination of the hypobaric exposure. Type I cases accounted for 89% of the total, and 11% were Type II. About three-fourths of the patients received hyperbaric therapy (see below), and all appeared to have had complete resolution of clinical disease.

The U.S. Navy observed symptom onset at altitude in 46% of the cases and at ground level in the remaining 54% (145). Of those cases that began at ground level, the median time of onset was about 1 h after the flight, and 4% of the cases began more than 20 h after the exposure. The case mix was evenly divided between Type I and Type II disease (see Table 2 for specifics). Hyperbaric therapy was administered to 84% of these patients, and about 4% continued to have mild symptoms following treatment.

The above indicates that altitude DCS arising from physiological training profiles is usually fairly mild, and it is generally responsive to prompt therapy. Exceptions to this benign picture of altitude DCS have been reported, including death (74). Exercise at altitude has been known since the 1940s to be a contributing factor to hypobaric DCS (146,147). Henry (147) further reported that the magnitude of physical work performed at altitude, expressed in terms of foot-pounds of work, was more of a determinate of the severity of DCS than brief, intermittent isometric musculoskeletal strain or unloaded movement. In recent studies (148), subjects performed, while at altitude, either dynamic or isometric exercise that induced the same whole-body oxygen consumption; these two forms of exercise seemed equally provocative of DCS. These authors agreed with Henry that the type of exercise is not as important as the magnitude of exercise expressed in terms of a factor such as foot-pounds of work or, perhaps, tissue metabolic rate.

Do Inflight Conditions Modify the Probability of Decompression Sickness?

As of July, 2000, 99 individual, inflight exposures to EVA conditions have taken place during the Shuttle era (48 EVA ''sorties'' involving two crew in each sortie have been conducted, along with one three-person sortie, totaling 99 individual exposures). To date, no incident of DCS of any type has been reported in these operations. In contrast, investigations of EVA-like pressure profiles in terrestrial hypobaric chambers found a 4.6% incidence of Type I symptoms that were severe enough to interfere with the performance of required tasks (149). The following factors might contribute to this discordance between the terrestrial and inflight databases.

1.   Statistical chance: If the expected probability of DCS is 0.046, the probability of accomplishing simply by chance 99 exposures without any ''hits'' is 0.009 (per the binomial distribution). Therefore, mere luck is not a likely complete explanation for the lack of inflight disease.

2.   More extensive denitrogenation prior to hypobaric exposures inflight than in ground-based laboratory experiments: All but four EVAs performed to date have used prebreathing protocols that require a reduction in cabin pressure to 70.3 kPa (10.2 psi) while breathing approximately 27% oxygen for a period of either 12 or 24 h prior to a short oxygen prebreathe and final depressurization to suit pressure. Terrestrial laboratory studies have faithfully reproduced these timelines. In reality, prior to all but four actual EVAs using these staged-decompression methods, far more time was spent at this intermediate pressure than is called for in the protocol. The average duration of this stage has been about 40 h, with a range extending to over 100 h (150). Such extensions in the staged decompressions have not been evaluated in terrestrial studies, and, therefore, their effect on the probability of DCS is unknown. However, these extensions probably decrease the risk of DCS. Consequently, inflight experience may not coincide with laboratory data because substantially different denitrogenation routines are used in these two groups.

3.   Masking of Type I symptoms by suit-induced strains and injuries: Working in the EMU exposes the subject to unusual and severe musculoskeletal strains as well as suit-induced trauma from abrasion and pressure points. The symptoms of these injuries may closely mimic musculoskeletal DCS. Similarly, compression of peripheral nerves by the suit, or unusual, repetitive patterns of

exercise may cause paresthesia that imitate neurological DCS. Consequently, mild or moderate intensities of DCS occurring inflight may have been mistaken for suit-induced phenomena. In contrast, subjects in terrestrial laboratory experiments have worn only light clothing, not pressure suits, and may, therefore, have been more likely to recognize DCS symptoms.

4. Underreporting inflight: In laboratory experiments, subjects perform repetitive, boring tasks for many hours, and they have little to do but maintain vigilance for the development of DCS symptoms. They are also questioned as to their overall well-being on a regular basis and, therefore, are reminded of DCS concerns. In contrast, crew during EVAs are fully occupied with the task at hand, and DCS symptoms may not attract their attention until the phenomena are severe. Also, these individuals are highly motivated to complete their missions, and, since DCS might threaten mission goals, crew may have a tendency to deny or minimize sensory phenomena that might be related to DCS. By analogy, reports of DCS are very rare during the active careers of high-altitude U-2 pilots, but, following their retirement, 76% of surveyed pilots stated that they had experienced one or more episodes of DCS symptoms during their flying days, and 9.6% described an incident that was severe enough to force a descent or abort a mission (141). Similarly, Michael Collins, a participant in Gemini and Apollo missions, including Apollo 11, revealed in a layman's book written after his retirement from NASA, two incidents of knee pain that are consistent with Type I DCS (in these spacecraft, a significant decompression stress occurred during launch as the cabin atmosphere was permitted to vent down to about 34 kPa [5.0 psi], and his book describes the development of knee pain shortly after two launches) (151). His pain responded to aspirin and resolved over a period of a few hours. This treatment of mild, Type I DCS was probably not inappropriate (assuming that disease progression does not occur) since no long-term sequelae of such disease have been reported (141).

5. Analgesic use in the peri-EVA period inflight: Current flight rules require oral administration of 650 mg of aspirin within 2 h of final depressurization to suit pressure, with an additional 325 mg following EVA. A review of inflight medical records reveals that additional doses of aspirin, large doses of ibuprofen, and intramuscular injections of promethazine have been taken within 24 h of EVA (probably within a few hours of EVA, but the information in the records is, in most cases, insufficient to precisely determine the time of dosing) (152). In contrast, the protocols for laboratory studies have specifically excluded analgesics for some period before and after altitude exposure. Consequently, the inflight and laboratory databases may be dissimilar because analgesics are *required* before and after EVA whereas they are *prohibited* in laboratory experiments. Analgesic use in the peri-EVA period may simply mask mild musculoskeletal symptoms while having no direct effect on the probability of serious DCS. If analgesics do only mask Type I symptoms, then the dearth of inflight reports of mild disease

might not provide much support for the assumption that the risk of serious DCS is also low.

6. Reduction in gas-phase formation under space flight conditions: Previously in this chapter, the concept was introduced that musculoskeletal strain and activity participate in the process of bubble formation during hypobaric exposures. Since ambulation involves musculoskeletal activity, it follows that walking about in an altitude chamber may potentiate DCS. In contrast, crew do not ambulate during EVA, so their risk of DCS might be less. Also, the suit is rather inflexible, functioning as a sort of whole-body air splint that drastically limits the range of motion of the musculoskeletal system. Additionally, the distribution and magnitude of blood flow in microgravity may, on balance, favor inert gas elimination and decrease the probability of DCS. A number of studies have investigated how these aspects of microgravity conditions might modify the risk of DCS, including one series that demonstrated a decrease in VGE scores in subjects who experienced strict bed rest for 3 days before and throughout an EVA-like hypobaric exposure, in comparison with the same subjects when they performed a separate, ambulating experiment (149). Microgravity conditions seem to confer some protection against DCS, but the magnitude of these effects has not yet been defined. Additionally, how readily these effects might be negated by other factors (e.g., exercise prior to EVA) is also unknown.

The above review of inflight EVA experience has made no mention of Soviet and Russian operations. To date, approximately 183 individual exposures have been conducted in these programs. No incident of any form of DCS has been reported, an important finding since, in theory, the decompression stress in these operations was greater than that of Shuttle-based EVAs (the pressure in the Russian Orlan suit is higher than that of the American EMU, but the Russian oxygen prebreathing procedure is substantially shorter, yielding an overall decompression stress greater than that of Shuttle operations, assuming that actual operations take place as specified; see below). However, this database is difficult to evaluate. It appears that the cabin pressure in Soviet/Russian spacecraft was often less than its nominal, sea-level specification, the oxygen content of the cabin atmosphere was often higher than 21%, and the oxygen prebreathe prior to EVA was sometimes longer than required. Additionally, many of these EVAs were relatively short: 21% were less than 3 h in duration, whereas 2% of Shuttle-era EVAs were this brief (laboratory studies typically expose subjects to hypobaric conditions for 4 to 6 h). Taken together, these factors make comparisons of this inflight experience with terrestrial databases very difficult.

Given the above considerations, a discordance between the inflight and terrestrial databases regarding the incidence of DCS from EVA-like pressure exposures is not surprising. Inflight conditions are simply very different from those in a terrestrial chamber. How these differences affect the risk of serious DCS is a central, unanswered question.

III. Prevention of Decompression Sickness

A. In Space

Oxygen Prebreathing

Historically, prevention of DCS in space has been achieved by implementation of either of two protocols that reduce tissue nitrogen prior to an EVA. The first involves an oxygen prebreathe at ambient pressure for 4 h and the second a staged decompression prior to EVA. Oxygen prebreathing prior to decompression was proposed in 1940 (153), and its protective effect was shown to be dependent upon the duration of the prebreathe period (31,154). Complete freedom from all symptoms was reported with extended prebreathe periods of 8 and 12 h (35). Shorter prebreathe periods were reported to prevent symptoms by Kazakova et al. (155) working with the Russian space program, and Waligora et al. (156) observed neither symptoms of DCS nor VGE with an 8-h prebreathe. Oxygen prebreathing by tight-fitting face mask for 4 h seems long enough to wash nitrogen out of tissues with the least amount of blood flow (66). The rate of tissue denitrogenation varies with the solubility and diffusivity of the gas in the particular tissue and, perhaps most importantly, on the magnitude of blood perfusion of the tissue.

Staged Decompression

The second procedure involves an initial 1-h oxygen prebreathe, then a staged decompression from the ISS/Shuttle cabin pressure of 101 kPa (14.7 psia) to an intermediate pressure of 70.3 kPa (10.2 psia), ''camp-out'' for at least 12 h while breathing 26.5% oxygen, and finally an abbreviated oxygen prebreathe of 40 to 75 min prior to an EVA (66). This staged decompression results in a tissue ratio (see earlier section for definition) of 1.70 vs. 1.68 for the protocol involving a 4-h oxygen prebreathe at 101 kPa (14.7 psia) (157). Staged decompression is currently the procedure preferred by NASA, and to date no DCS in space has been reported. With assembly and maintenance of the ISS, the ''camp-out'' procedure poses a number of limitations, including an inability to perform full-duration 6-h EVAs within the allocated crew scheduling times, an excessive utilization of the limited high pressure oxygen stores (these stores are a limiting factor in the number of EVAs that can be performed on a given mission), and various operational and crew comfort factors (158).

Exercising Prebreathing

Some 484 EVAs will be conducted over the course of ISS construction. NASA has sought alternative protocols to develop a 2-h prebreathe procedure for EVA from the 101 kPa (14.7) psia ambient cabin pressure of the Shuttle/ISS. The basis for a 2-h protocol resulted from recent studies indicating that adynamic individuals subjected to decompression develop fewer VGE and symptoms of DCS than ambulating subjects (149,159,160). Essentially all astronauts in space are adynamic, thus Waligora et al. (161) postulated this factor to explain why no DCS has been reported in space. Additionally, exercise during the prebreathe process has been demonstrated to reduce VGE and Type I DCS symptoms (157,162). However, to date, no body of experimental evidence directly establishes an effect of exercise during prebreathing on the risk of *Type II* altitude DCS in EVA-like pressure profiles. Also, the ''dose–response curve'' for the risk of DCS as a function of exercise during prebreathing has not been clearly determined, an important consideration since exercise during prebreathing may produce physiological changes with conflicting influences on the risk of DCS, e.g., risk reduction as a result of increased tissue perfusion but risk enhancement as a consequence of bubble nuclei generation (in theory, at least).

Selection of a 2-h prebreathe protocol was based largely on the operational constraints of donning the EMU and on the results of human trials employing various combinations of exercise and adynamia during the prebreathe period. A 2-h protocol is planned for use by NASA that consists of 10 min of upper and lower body exercise on a dual cycle ergometer at 75% of maximum oxygen consumption, followed with the light exercise involved in EVA equipment preparation and donning (158,163,164).

B. In Immersion Facilities

Members of the crew on every Shuttle mission are trained to perform EVA tasks. This training takes place in a number of environments, each with its particular strengths and limitations. For example, parabolic flight provides a very faithful reproduction of the microgravity conditions that are encountered in flight, but the experience is limited to 25-s parcels of time. In contrast, the entire timeline of a 6-h EVA can be practiced during water immersion, although strong sensory clues regarding ''which way is up'' remain. Currently, both methods are used extensively.

To accommodate the sprawling mockups of components of ISS, NASA has placed into operation a new water immersion training facility, the Neutral Buoyancy Laboratory (NBL) (Fig. 3). The pool at the heart of the NBL is 31 $\times$ 62 m in lateral dimensions, and 12.2 m deep. Persons undergoing training, evaluating equipment, and developing procedures dive in this pool while wearing space suits that are weighted such that the tendency of the relaxed diver to both change depth and rotate to a particular orientation (e.g., head-up or head-down) is minimized. To reproduce the inflexibility of a pressurized suit in space, breathing gas flow from the suit is adjusted to maintain the absolute pressure of the gas within the suit 29.6 kPa greater than the absolute pressure in the surrounding water column

Figure 3 The Neutral Buoyancy Laboratory, Johnson Space Center, National Aeronautics and Space Administration, Houston, Texas. This facility is used to develop equipment and techniques for extravehicular activities and to train crew members for specific missions. (From NASA.)

(for an EMU; 38.6 kPa for an Orlan suit). Because ambient pressure increases about 10.1 kPa per meter of depth, suited divers are exposed to pressures equal to those 2.92 m deeper in the water column than their current location. Consequently, a suited diver at the bottom of the NBL is performing a dive, from a physiological point of view, to 2.92 m deeper than the depth of the pool, or a total "physiological depth" of 15.1 m.

Nitrox

As mentioned above, a major advantage of immersion training is that the entire timeline of an EVA can be rehearsed in one "sitting." Such dives typically last 6 h. However, the Standard Air Decompression Table of the *U.S. Navy Diving Manual* limits dives to 15.1 m that do not include staged decompression stops (so-called no-decompression dives) to a total duration of 100 min (165). Even if one were to accept the limitations and hazards associated with decompression diving, the decompression penalty associated with a dive to 15.1 m for 6 h while breathing air would be on the order of several hours. A solution to this operational problem is to breathe something other than air.

Because the risk of DCS is a function of both the partial pressure of inert gas in the breathing medium (nitrogen, in the case of air dives) and the duration of the dive, one way to decrease the risk of DCS is to decrease the nitrogen content of the divers' breathing gas. The logical extension of this concept is to eliminate all nitrogen and simply breathe 100% oxygen. Indeed, oxygen would seem to be the ideal breathing gas for diving since it is metabolically consumed by tissues, and its metabolism is coupled to the production of carbon dioxide, a compound that is quite soluble and is efficiently eliminated from the body. Unfortunately, breathing a partial pressure of oxygen (P_{O_2}) in excess of about 1.4 ata is associated with the development of neurological oxygen toxicity, the most dramatic manifestation of which is a grand mal seizure (166). So a balance must be struck between the risk of DCS (caused by a high partial pressure of nitrogen) and neurological oxygen toxicity (provoked by a high P_{O_2}). In the case of the NBL, this compromise has resulted in the use of a breathing gas containing 46% oxygen and 54% nitrogen. Such a gas is known as oxygen-enriched air or, more simply, "nitrox." A suited diver at the bottom of the NBL is exposed to a P_{O_2} of about 1.1 ata, an acceptable dose. Under these conditions, the partial pressure of inspired nitrogen is equivalent to that which would be encountered if air were breathed at a depth of about 7.3 m, a dive with a small risk of DCS, even if it lasted days (10). One more risk remains to be managed, however: the pulmonary effects of breathing a high P_{O_2} for many hours.

Pulmonary Oxygen Toxicity

Because provision of a high inspired P_{O_2} is of value in many clinical situations as well as commercial diving operations, much is known about the pathophysiology of pulmonary oxygen toxicity (POT). A detailed discussion of this topic is well beyond the scope of this chapter. Several reviews have recently been published regarding POT in divers (167,168). The following discussion focuses on methods of managing the risk of this disorder during dives at immersion training facilities.

In normal individuals, symptoms of a "chemical burn" of pulmonary tissue begin to develop when a gas with a P_{O_2} above about 0.5 ata is breathed for prolonged periods of time. These symptoms initially consist of burning retrosternal pain on inspiration and a dry, persistent cough, and they are associated with the development of many abnormalities in pulmonary function tests, most notably declines in vital capacity. Detailed studies with human volunteers breathing pure oxygen without interruption in a hyperbaric chamber (167) indicate that significant, but reversible, toxicity (defined as a 2% reduction in vital capacity) begins to develop after about 615 min of oxygen breathing at 1 ata, and that this time limit decreases as chamber pressure increases. A method has been developed to normalize periods of oxygen breathing at a variety of partial pressures to familiar units. In this scheme, one unit of pulmonary toxic dose (UPTD) induces the same amount of pulmonary stress as breathing pure oxygen for 1 min at 1 ata. As mentioned above, significant toxicity begins to appear after about 615 min of oxygen breathing at sea level, so, in the terms of the UPTD method, a "dose" of 615 UPTDs [or, in terms preferred by some authors, cumulative pulmonary toxicity dose (CPTD)] is the maximum that can be administered without inducing clinically significant disease. Specifically (169):

$$\text{UPTD (or CPTD)} = t_x * ((0.5/(P_x - 0.5))^{\wedge -0.833})$$

where

$\qquad t_X$ = duration of oxygen breathing in minutes
$\qquad P_X$ = partial pressure of oxygen in ata

Per this scheme, 615 UPTDs are reached after breathing pure oxygen for 615 min at sea level. In the case of the NBL where 46% oxygen is breathed at a maximum physiological depth (depth of the suit plus the depth-equivalent imposed by overpressurization of the suit) of 15 m, the inspired P_{O_2} is 1.1 ata. Per the above equation, the limit of 615 UPTDs would be reached after diving to the bottom of the pool for about 510 min, substantially more than the 360-min dive duration that is needed. At the NBL, a real-time reading of absolute pressure within each suit is available, and UPTDs are continuously calculated throughout each dive. Since most of the typical dive takes place in the middle of the water column rather than on the bottom of the pool, UPTDs rarely exceed 300.

Theoretically, an individual crew member might wish to perform long dives at the NBL on consecutive days. In reality, "back-to-back" dives are exceedingly rare, but these operations do raise the question of the possibility of accumulation of POT over the course of several days of intense diving. Methods have been devised to handle this type of situation (170), and adaptation of these approaches to the NBL would seem possible.

IV. Treatment of Decompression Sickness

This discussion of the treatment of DCS elaborates on the fact that four therapeutic interventions are known to be effective in the treatment of altitude DCS:

1. Increased ambient pressure
2. High partial pressure of inspired oxygen
3. Fluids to maintain normovolemic status
4. Supportive care, such as airway management, cardiac life support, urinary bladder catheterization, etc.

Although other therapies have been suggested, the above is the cornerstone of treatment. With this overall perspective in mind, therapeutic interventions are discussed in order of escalating "invasiveness."

A. Treatment Options

Do Nothing

Even if the subject is still in a hypobaric environment, very mild symptoms do not necessarily demand immediate action. To date, over 500 subjects have been exposed to pressure profiles similar to those used during Shuttle operations (140,149,171,172). A variety of DCS symptoms have been observed ranging from mild musculoskeletal pain to, rarely, cardiovascular instability and severe central nervous system dysfunction. In many of these experiments, mild Type I DCS was not considered a reason to terminate an altitude exposure, and subjects who developed such disease completed their flights without untoward sequelae.

Return to "Ground Level" Ambient Pressure

In a variety of hypobaric operations, complete and lasting resolution of symptoms has been observed during the course of pressurization to ground level (the pressure from which the hypobaric excursion originated) in most cases. In one report, 37% of altitude DCS cases with onset at altitude resolved upon descent, although 17% of these case relapsed (145). Although no definitive study has been performed to specifically evaluate this intervention alone (pressurization is usually combined with a period of oxygen breathing as a minimal treatment for established DCS), it seems likely that pressurization alone can diminish or terminate some early, mild cases of DCS (173). In operational environments with limited treatment capabilities, such as space flight, relatively early termination of exposures that are producing symptoms may forestall serious disease.

Ground-Level Oxygen "Postbreathe"

In recent years, ground-level oxygen breathing (GLO) has become established as a treatment modality for altitude DCS (174,175). It seems effective for prevention of recurrence of Type I symptoms that resolve during descent, and, perhaps, treatment of mild Type I symptoms that persist to ground level. Table 4 depicts results from one study of the efficacy of GLO. Overall success rates of 77 to 99% have been reported in treating Type I altitude DCS with GLO alone (176,177). Of subjects with symptoms that resolved completely during descent and who were subsequently treated with GLO as a precaution against recurrence, no further symptoms were observed in 99.2% of subjects in one series (178) and 96.2% in another (177). Ground-level oxygen breathing alone was unsuccessful in treating 40% of cases in which symptoms began at altitude and persisted to ground level

Table 4 Efficacy of Ground-Level Oxygen
Treatment for Altitude Decompression Sickness

Onset of DCS at altitude	
Clinical course	Total cases
Initial onset at altitude	104
Asymptomatic upon return to ground	78
Received HBO	0
Received GLO; symptoms recurred	3
Resolved with HBO	3
Symptomatic upon return to ground	26
Received immediate HBO	6
Received GLO	20
Successful GLO	12
Unsuccessful GLO, successful HBO	8

Onset of DCS at ground level	
Clinical course	Total cases
Initial onset at ground	117
Received immediate HBO	39
Received GLO	78
Successful GLO without recurrence	40
Successful GLO but with recurrence	5
Successful HBO	5
Unsuccessful GLO	33
Successful HBO	33

(177). Treatment of Type I symptoms that begin at ground level with GLO alone
is controversial. Ground-level oxygen breathing is not considered appropriate by
most investigators as the sole therapy for any Type II symptoms or any recurrent
symptoms. A "dose–response curve" has not been established to determine the
optimal length of GLO, although a 2-h period is commonly used (175).

Hyperbaric Oxygen Therapy

Originally developed to treat "the bends" in divers, hyperbaric oxygen has become the mainstay of therapy for altitude DCS. Hyperbaric oxygen therapy
(HBO) involves both the application of increased atmospheric pressure and the
provision of high inspired partial pressures of oxygen. Four salutary effects are
accomplished:

1. Minimizes bubble volume by compression according to Boyle's law, e.g., the volume of a bubble at 2.8 ata, equivalent to a depth of 18 m (the usual depth of a treatment "dive"), is about 36% of that at sea level
2. Maximizes the gradient between the partial pressure of inert gas in bubbles (which is raised as the bubble is compressed) and that of surrounding tissue (which approaches 0 when pure oxygen is breathed), thereby accelerating resorption of gas
3. Enhances delivery of oxygen to tissues rendered ischemic by DCS-induced microvascular disease
4. Brings to bear other therapeutic effects of high partial pressures of oxygen, such as a reduction of tissue edema and intracranial pressure (179,180) and inhibition of platelet aggregation and leukocyte adhesion to damaged capillary endothelium (181,182)

In many cases of altitude DCS, particularly those in which treatment has been delayed or in which clinical improvement continues during the course of repeated treatments over many days, HBO has been observed to be effective long after tissue bubbles would be expected to have resolved. In these cases, items 3 and 4 above may be the factors that are providing the greatest benefit. If, in fact, what is being treated in these cases is not a tissue gas phase itself but rather the aftermath of bubbles that have since disappeared, then the benefits of HBO may be more dependent upon the *absolute* partial pressure of oxygen administered, and the compression of a gas phase by items 1 and 2 above may not be involved. In such cases, oxygen can be thought of as a drug with a "dose" expressed in terms of partial pressure; a chamber is employed simply as a means of delivering oxygen at a "dose" in excess of 1 ata. As will be discussed below, the typical "dose" of oxygen involved in HBO is 2.8 ata.

B. Hyperbaric Treatment Tables

Clinical experience and laboratory data gained over the past few decades have generated a degree of consensus regarding the optimal partial pressure of oxygen for the initial treatment of DCS. Tables constructed around oxygen delivered at 2.8 ata have proved successful in treating both altitude (36) and diving (183) casualties. Laboratory studies of an animal model of hyperbaric DCS indicate that the optimum partial pressure of oxygen in this setting is 2.0 to 2.5 ata (184,185). Most modern clinical treatment tables center on administration of 2.8 ata of oxygen. The table that seems to strike the best, most effective balance between benefit and risk is the U.S. Navy Treatment Table 6 (TT6) (Fig. 4) (186) or the nearly identical version developed by the United States Air Force. Notable features of this table include a maximum depth of 2.8 ata, prolonged periods of oxygen breathing interspersed with short air breaks (to reduce the risk of neuro-

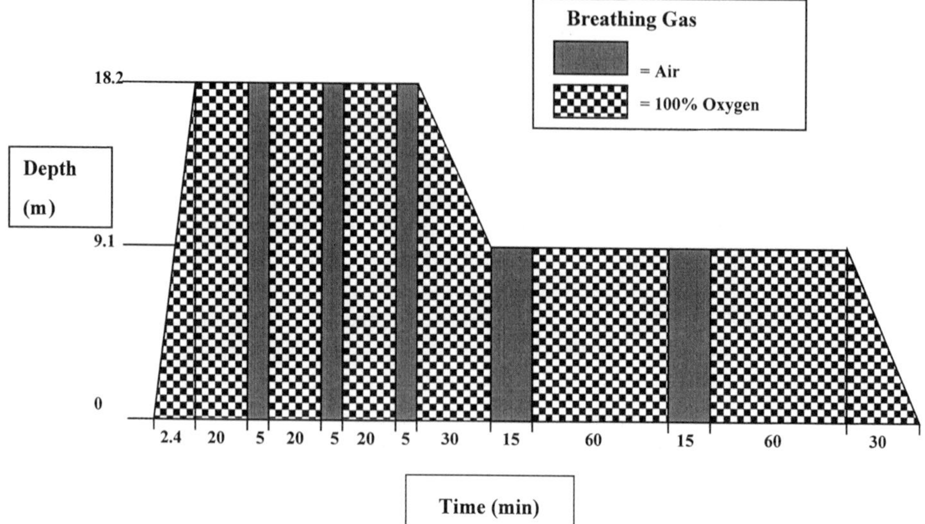

Figure 4   Treatment Table 6, as published by the U.S. Navy. This table (or the nearly identical version developed by the U.S. Air Force) is the most common schedule utilized for treating altitude decompression sickness. (Adapted from Ref. 186.)

logical and pulmonary oxygen toxicity), a gradual depressurization to 1.9 ata, a prolonged "stage" at this pressure, then a gradual decompression to the surface. The total duration of the table is 4 h and 45 min, although extensions at 2.8 ata and/or 1.9 ata can be made as the patient's condition dictates.

C. Adjunctive Therapy

An important part of the treatment of any case of DCS is the provision of fluids to maintain intravascular euvolemia. Hemoconcentration occurs commonly in DCS (42). Fluids are provided per conventional clinical methods, recognizing that catheterization of the urinary bladder is necessary is some cases of Type II DCS. A case can be made for exclusion of glucose from intravenous fluids administered to patients with neurological involvement (187).

Beyond pressure, oxygen, fluids, and supportive care, many adjunctive therapies have been investigated (41,188). A truncated list of these includes lidocaine (189–191), corticosteroids (192,193), perfluorocarbons (194,195), aspirin, and dipyridamole (196,197). None of these adjunctive interventions has yet been widely accepted in routine clinical practice.

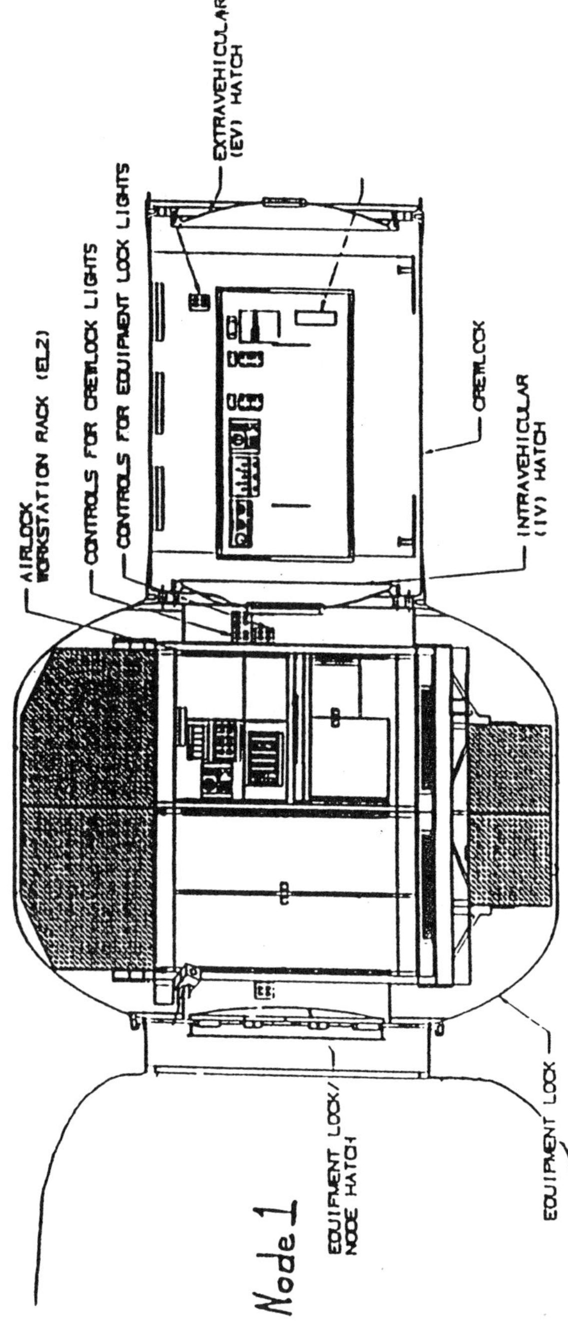

Figure 5 The hyperbaric facility originally planned for incorporation in Space Station Freedom. The Crew Lock portion of the Airlock complex was designed to serve two functions: (1) to provide a "door" to the exterior of the spacecraft and (2) to function as a multiplace hyperbaric chamber capable of performing a Treatment Table 6 (see Fig. 4). Controls for operation of the hyperbaric facility (pressurization and depressurization valves, gas monitoring equipment, etc.) were to be located in the equipment lock. (From Ref. 202.)

D. Treatment in Space Flight

This discussion of the treatment of DCS would not be complete without mention of how cases arising during space flight will be handled. Because the decompression stress involved in Shuttle and ISS operations is large, some DCS seems inevitable, although the magnitude of this risk is unclear. Laboratory investigations indicate that most cases will be mild (140,149,171). Indeed, mild Type I DCS might not even be recognized given the discomfort of the suit, the focus of the crew member on the task at hand, and the fact that analgesic drugs are used before and after EVA (152). However, serious DCS, including cardiovascular instability (149,198) and severe cognitive impairment, has been observed during ground-based investigations. Consequently, a small but real risk of serious DCS seems to exist in current space flight operations.

If hyperbaric treatment of altitude DCS is delayed, progression of symptoms from mild Type I pain to death has been observed (74). Also, delay in treatment decreases the efficacy of hyperbaric therapy (199). To prevent delay, an onboard treatment chamber would provide the greatest protection. Space Station Freedom originally included a structure that was designed to function both as an airlock and as a multiplace, monolock hyperbaric treatment chamber (Fig. 5) (this program was canceled in 1992). Such on-site treatment capability affords many advantages: the progression of disease can be halted at an early stage, complete

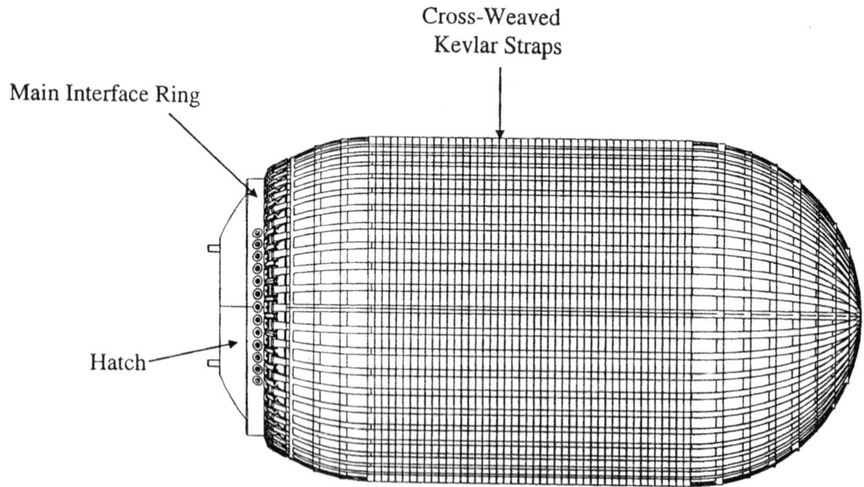

Figure 6 An inflatable, stowable, two-person hyperbaric chamber capable of performing a complete Treatment Table 6. This chamber is currently under development at the Johnson Space Center. The outside dimensions are 122 cm in diameter and 229 cm in length. (Courtesy of C. Hansen, Johnson Space Center, Houston, Texas.)

cure with one treatment is more likely, the costs and mission-impact of evacuation are avoided, oxygen prebreathing requirements can be reduced until inconvenient forms of DCS appear without unduly risking the eruption of catastrophic forms of disease, and the crew can be returned rapidly to full duty, operating under a "treat it and forget it" philosophy.

The International Space Station currently includes no conventional hyperbaric capability. The equivalent of GLO can be provided by keeping patients in their space suits. Additionally, delivery of nearly 100% oxygen at a pressure as great as 1.54 ata is possible through installation of the Bends Treatment Apparatus in the suit (200). As of this writing, an effort is underway at the Johnson Space Center to develop an inflatable, stowable, two-person, hyperbaric chamber that could provide ISS with the capability of performing a complete TT6 (Fig. 6). This chamber includes a unique means of interfacing the flexible pressure restraint layer with the metal hatch ring that seems to solve the problems with structural failure at this point that have plagued some previous designs for inflatable chambers. A prototype of this chamber has been built and is currently undergoing structural testing. The pressure vessel itself (exclusive of subsystems such as oxygen masks and communications equipment) weighs 93 kg, and the flexible restraint layer stows within the inner dimensions of the hatch ring.

V. Conclusions

The lungs play an important role in EVA operations. In addition to their obvious role in oxygen and carbon dioxide exchange, they are critically important filters of gas bubbles present in venous blood as a result of the decompression stress involved in EVA operations. This decompression stress is a consequence of current specifications for the atmosphere contained in spacecraft cabins and space suits. Therefore, the lungs are, in essence, another spacecraft subsystem, one that protects important biological tissues from embolization by bubbles generated as a consequence of the design of other subsystems. However, the efficacy of this filter is not always perfect, and heavy embolization of the lungs is not without consequence. The designers of future spacecraft should regard the crew member as an integral part of the spacecraft and carefully select the composition of cabin and suit atmospheres.

References

1. Golding FC, Griffiths P, Hempleman HV, Paton WDM, Walder DN. Decompression sickness during construction of the Dartford Tunnel. Br J Ind Med 1960; 17: 167–180.

2. United States Navy. U.S. Navy Diving Manual. NAVSEA 0994-LP-9010. Washington, DC: Naval Sea Systems Command, 1993:22–26.

3. Weien RW. Transcribed comments. In: Pilmanis AA, ed. Proceedings of the 1990 Hypobaric Decompression Sickness Workshop. AL-SR-1992-0005. Brooks Air Force Base, TX: Air Force Systems Command, 1992:371.

4. United States Navy. U.S. Navy Diving Manual. NAVSEA 0994-LP-9010. Chapter 8. Washington, DC: Naval Sea Systems Command, 1993:66.

5. Francis TJR, Smith DJ, eds. Describing Decompression Illness. Proceedings of the 42nd Workshop of the Undersea and Hyperbaric Medical Society. Bethesda, MD: Undersea and Hyperbaric Medical Society, 1991.

6. Wirjosemito SA, Touhey JE, Workman WT. Type II altitude decompression sickness (DCS): U.S. Air Force experience with 133 cases. Aviat Space Environ Med 1989; 60:256–262.

7. Dutka AJ. Clinical findings in decompression illness: A proposed terminology. In: Moon RE, Sheffield PJ, eds. Treatment of Decompression Illness. Kensington, MD: Undersea and Hyperbaric Medical Society, 1996:1–9.

8. Zheng Q, Durben DJ, Wolf GH, Angell CA. Liquids at large negative pressures: Water at the homogeneous nucleation limit. Science 1991; 254:829–832.

9. Weathersby PK, Homer LD, Flynn ET. Homogeneous nucleation of gas bubbles in vivo. J Appl Physiol 1982; 53:940–946.

10. Eckenhoff RG, Osborne SF, Parker JW, Bondi KR. Direct ascent from shallow air saturation exposures. Undersea Biomed Res 1986; 13:305–316.

11. Dixon GA, Krutz R. Evaluation of 9.5 psia as a suit pressure for prolonged extravehicular activity. Proceedings of 23rd Annual Survival and Flight Equipment Symposium. Las Vegas, NV, 1985:122–123.

12. Vann RD, Grimstad J, Nielsen CH. Evidence for gas nuclei in decompressed rats. Undersea Biomed Res 1980; 7:107–112.

13. Evans A, Walder DN. Significance of gas micronuclei in the aetiology of decompression sickness. Nature 1969; 222:251–252.

14. McDonough PM, Hemmingsen EA. Bubble formation in crustaceans following decompression from hyperbaric gas exposures. J Appl Physiol 1984; 56:513–519.

15. McDonough PM, Hemmingsen EA. A direct test for the survival of gaseous nuclei in vivo. Aviat Space Environ Med 1985; 56:54–56.

16. Hemmingsen EA. Bubble formation mechanisms. In: Vann RD, ed. The Physiological Basis of Decompression. Bethesda, MD: Undersea and Hyperbaric Medical Society, 1989:153–169.

17. Hayward ATJ. Tribonucleation of bubbles. Br J Appl Physiol 1967; 18:641–644.

18. McDonough PM, Hemmingsen EA. Bubble formation in crabs induced by limb motions after decompression. J Appl Physiol 1984; 57:117–122.

19. McDonough PM, Hemmingsen EA. Swimming movements initiate bubble formation in fish decompressed from elevated gas pressures. Comp Biochem Physiol A 1985; 81:209–212.

20. Okang GI, Vann RD. Bubble formation in blood and urine. In: Vann RD, ed. The physiological basis of decompression. Bethesda, MD: Undersea and Hyperbaric Medical Society, 1989:177–178.

21. Bayne CG, Hunt WS, Johnson DC, Flynn ET, Weathersby PK. Doppler bubble

detection and decompression sickness: A prospective clinical trial. Undersea Biomed Res 1985; 12:327–332.

22. Conkin J, Foster PP, Powell MR, Waligora JM. Relationship of the time course of venous gas bubbles to altitude decompression illness. Undersea Hyperb Med 1996; 23:141–149.

23. Nishi RY. Doppler and ultrasonic bubble detection. In: Bennett PB, Elliot DH, eds. The Physiology and Medicine of Diving. London: Saunders, 1993:433–453.

24. Eatock BC, Nishi RY. Analysis of Doppler ultrasonic data for the evaluation of dive profiles. In: Bove AA, Bachrach AJ, Greenbaum LJ, eds. Proceedings of 9th International Symposium on Underwater and Hyperbaric Physiology. Bethesda, MD: Undersea and Hyperbaric Medicine Society, 1987:183–195.

25. Hemmingsen BB, Steinberg NA, Hemmingsen EA. Intracellular gas supersaturation tolerances of erythrocytes and resealed ghosts. Biophys J 1985; 47:491–496.

26. Hemmingsen EA, Hemmingsen BB. Bubble formation properties of hydrophobic particles in water and cells of Tetrahymena. Undersea Biomed Res 1990; 17:67–78.

27. Hills BA, James PB. Spinal decompression sickness: Mechanical studies and a model. Undersea Biomed Res 1982; 9:185–201.

28. Powell MR, Spencer MP. The Pathophysiology of Decompression Sickness and the Effects of Doppler Detectable Bubbles. Technical Report on ONR Contract N00014-73-C-0094. Seattle, WA: Institute of Applied Physiology and Medicine, 1980.

29. Gersh I, Catchpole HR. Appearance and distribution of gas bubbles in rabbits decompressed to altitude. J Cell Comp Physiol 1946; 28:253–268.

30. Chryssanthou C, Palaia T, Goldstein G, Stenger R. Increase in blood-brain barrier permeability by altitude decompression. Aviat Space Environ Med 1987; 58:1082–1086.

31. Ferris EB, Engel GL. The clinical nature of high altitude decompression sickness. In: Fulton JF, ed. Decompression Sickness. Philadelphia: Saunders, 1951:4–52.

32. Inman VT, Saunders JB. Referred pain from skeletal structures. J Nerv Ment Dis 1944; 99:660–667.

33. Vann RD. Vacuum phenomena: An anotated bibliography. In: Vann RD, ed. The Physiological Basis of Decompression. Bethesda, MD: Undersea and Hyperbaric Medical Society, 1989:179–195.

34. Behnke AR. Decompression sickness following exposure to high pressures. In: Fulton JF, ed. Decompression Sickness. Philadelphia: Saunders, 1951:53–89.

35. Bateman JB. Preoxygenation and nitrogen elimination. In: Fulton JF, ed. Decompression Sickness. Philadelphia: Saunders, 1951:242–277.

36. Bason R. Altitude chamber DCS: USN experience 1981–1988. In: Pilmanis AA, ed. Proceedings of the 1990 Hypobaric Decompression Sickness Workshop. AL-SR-1992-0005. Brooks Air Force Base, TX: Air Force Systems Command, 1992: 395–413.

37. Lee WH, Jr., Hairston P. Structural effects on blood proteins at the gas-blood interface. Fed Proc 1971; 30:1615–1622.

38. Chenoweth DE, Cooper SW, Hugli TE, Stewart RW, Blackstone EH, Kirklin JW. Complement activation during cardiopulmonary bypass: Evidence for generation of C3a and C5a anaphylatoxins. N Engl J Med 1981; 304:497–503.

39. Ward CA, Koheil A, McCullough D, Johnson WR, Fraser WD. Activation of complement at plasma-air or serum-air interface of rabbits. J Appl Physiol 1986; 60: 1651–1658.

40. Philp RB, Inwood MJ, Warren BA. Interactions between gas bubbles and components of the blood: Implications in decompression sickness. Aerosp Med 1972; 43: 946–953.

41. Bove AA. The basis for drug therapy in decompression sickness. Undersea Biomed Res 1982; 9:91–111.

42. Philp RB, Ackles KN, Inwood MJ, et al. Changes in the hemostatic system and in blood and urine chemistry of human subjects following decompression from a hyperbaric environment. Aerosp Med 1972; 43:498–505.

43. Thorsen T, Lie RT, Holmsen H. Induction of platelet aggregation in vitro by microbubbles of nitrogen. Undersea Biomed Res 1989; 16:453–464.

44. Albertine KH, Wiener-Kronish JP, Koike K, Staub NC. Quantification of damage by air emboli to lung microvessels in anesthetized sheep. J Appl Physiol 1984; 57: 1360–1368.

45. Moosavi H, Utell MJ, Hyde RW, et al. Lung ultrastructure in noncardiogenic pulmonary edema induced by air embolization in dogs. Lab Invest 1981; 45:456–464.

46. Staub NC, Schultz EL, Albertine KH. Leucocytes and pulmonary microvascular injury. Ann NY Acad Sci 1982; 384:332–343.

47. Benestad HB, Hersleth IB, Hardersen H, Molvaer OI. Functional capacity of neutrophil granulocytes in deep-sea divers. Scand J Clin Lab Invest 1990; 50:9–18.

48. Rothe G, Kellermann W, Valet G. Flow cytometric parameters of neutrophil function as early indicators of sepsis- or trauma-related pulmonary or cardiovascular organ failure. J Lab Clin Med 1990; 115:52–61.

49. Worthen GS, Schwab Bd, Elson EL, Downey GP. Mechanics of stimulated neutrophils: Cell stiffening induces retention in capillaries. Science 1989; 245:183–186.

50. Doerr TA, Rosolia DL, Peters SP, Gee MH, Albertine KH. PGE1 inhibited PMN attachment to air emboli in vivo during infusion of ZAP without preventing lung injury. J Appl Physiol 1992; 72:340–351.

51. Kindt GC, Gadek JE, Weiland JE. Initial recruitment of neutrophils to alveolar structures in acute lung injury. J Appl Physiol 1991; 70:1575–1585.

52. Wang D, Li MH, Hsu K, Shen CY, Chen HI, Lin YC. Air embolism-induced lung injury in isolated rat lungs. J Appl Physiol 1992; 72:1235–1242.

53. Flick MR, Perel A, Staub NC. Leukocytes are required for increased lung microvascular permeability after microembolization in sheep. Circ Res 1981; 48:344–351.

54. Flick MR, Hoeffel JM, Staub NC. Superoxide dismutase with heparin prevents increased lung vascular permeability during air emboli in sheep. J Appl Physiol 1983; 55:1284–1291.

55. Binder AS, Nakahara K, Ohkuda K, Kageler W, Staub NC. Effect of heparin or fibrinogen depletion on lung fluid balance in sheep after emboli. J Appl Physiol 1979; 47:213–219.

56. Binder AS, Kageler W, Perel A, Flick MR, Staub NC. Effect of platelet depletion on lung vascular permeability after microemboli in sheep. J Appl Physiol 1980; 48:414–420.

57. Warren BA, Philp RB, Inwood MJ. The ultrastructural morphology of air embo-

lism: Platelet adhesion to the interface and endothelial damage. Br J Exp Pathol 1973; 54:163–172.

58. Philp RB. A review of blood changes associated with compression-decompression: Relationship to decompression sickness. Undersea Biomed Res 1974; 1:117–150.

59. Nossum V, Koteng S, Brubakk AO. Endothelial damage by bubbles in the pulmonary artery of the pig. Undersea Hyperb Med 1999; 26:1–8.

60. Haller C, Sercombe R, Verrecchia C, Fritsch H, Seylaz J, Kuschinsky W. Effect of the muscarinic agonist carbachol on pial arteries in vivo after endothelial damage by air embolism. J Cereb Blood Flow Metab 1987; 7:605–611.

61. Anderson DK, Means ED. Iron-induced lipid peroxidation in spinal cord: Protection with mannitol and methylprednisolone. J Free Radic Biol Med 1985; 1:59–64.

62. Hardman JM, Beckman EL. Pathogenesis of central nervous system decompression sickness. Undersea Biomed Res 1990; 17(suppl):95–96.

63. Anderson D, Nagasawa G, Norfleet W, Olszowka A, Lundgren C. O_2 pressures between 0.12 and 2.5 atm abs, circulatory function, and N_2 elimination. Undersea Biomed Res 1991; 18:279–292.

64. Van Liew HD, Schoenfisch WH, Olszowka AJ. Exchanges of N_2 between a gas pocket and tissue in a hyperbaric environment. Respir Physiol 1968; 6:23–28.

65. Vann RD, Thalmann ED. Decompression physiology and practice. In: Bennett PB, Elliot DH, eds. The Physiology and Medicine of Diving. London: Saunders, 1993: 376–432.

66. Waligora JM, Horrigan D, Jr, Conkin J, Hadley AT III. Verification of an Altitude Decompression Sickness Prevention Protocol for Shuttle Operation Utilizing a 10.2 psi Pressure Stage. NASA Technical Memorandum 58259. Houston, TX: National Aeronautics and Space Administration, 1984.

67. Hempleman HV. History of decompression procedures. In: Bennett PB, Elliot DH, eds. The Physiology and Medicine of Diving. London: Saunders, 1993:342–375.

68. Erde A, Edmonds C. Decompression sickness: A clinical series. J Occup Med 1975; 17:324–328.

69. Brew SK, Kenny CT, Webb RK, Gorman DF. The outcome of 125 divers with dysbaric illness treated by recompression at HMNZS PHILOMEL. SPUMS J 1990; 20:226–230.

70. Bennett PB, Coggin R, Roby J. Control of HPNS in humans during rapid compression with trimix to 650 m (2131 ft). Undersea Biomed Res 1981; 8:85–100.

71. Piccard J. Aeroemphysema and the birth of bubbles. Proc Staff Meet Mayo Clin 1941; 16:700–704.

72. Bason R, Yacavone DW. Loss of cabin pressurization in U.S. Naval aircraft: 1969–90. Aviat Space Environ Med 1992; 63:341–345.

73. Neubauer JC, Dixon JP, Herndon CM. Fatal pulmonary decompression sickness: A case report. Aviat Space Environ Med 1988; 59:1181–1184.

74. Dixon JP. Death from altitude-induced decompression sickness: Major pathophysiologic factors. In: Pilmanis AA, ed. Proceedings of the 1990 Hypobaric Decompression Sickness Workshop. AL-SR-1992-0005. Brooks Air Force Base, TX: Air Force Systems Command, 1992:97–105.

75. Fryer DI. Subatmospheric Decompression Sickness in Man. AGARDograph 125. Brussels: North Atlantic Treaty Organization, 1969:141–189.

76.   Baumgartner N, Weien RW. Decompression sickness due to USAF altitude chamber exposure (1985–1987). In: Pilmanis AA, ed. Proceedings of the 1990 Hypobaric Decompression Sickness Workshop. AL-SR-1992-0005. Brooks Air Force Base, TX: Air Force Systems Command, 1992:363–376.

77.   Weien RW. Altitude decompression sickness: The U.S. Army experience. In: Pilmanis AA, ed. Proceedings of the 1990 Hypobaric Decompression Sickness Workshop. AL-SR-1992-0005. Brooks Air Force Base, TX: Air Force Systems Command, 1992:379–383.

78.   Harding RW. DCS experience outside North America. In: Pilmanis AA, ed. Proceedings of the 1990 Hypobaric Decompression Sickness Workshop. AL-SR-1992-0005. Brooks Air Force Base, TX: Air Force Systems Command, 1992:467–471.

79.   Catron PW, Flynn ET, Jr., Yaffe L, et al. Morphological and physiological responses of the lungs of dogs to acute decompression. J Appl Physiol 1984; 57:467–474.

80.   Levin LL, Stewart GJ, Lynch PR, Bove AA. Blood and blood vessel wall changes induced by decompression sickness in dogs. J Appl Physiol 1981; 50:944–949.

81.   Barnard EE, Hanson JM, Rowton-Lee MA, Morgan AG, Polak A, Tidy DR. Postdecompression shock due to extravasation of plasma. Br Med J 1966; 5506:154–155.

82.   Wells CH, Bond TP, Guest MM, Barnhart CC. Rheologic impairment of the microcirculation during decompression sickness. Microvasc Res 1971; 3:162–169.

83.   Davis JC, Elliott DH. Treatment of decompression disorders. In: Bennett PB, Elliott DH, eds. The Physiology and Medicine of Diving. London: Bailliere Tindall, 1982:475–476.

84.   Butler BD, Katz J. Vascular pressures and passage of gas emboli through the pulmonary circulation. Undersea Biomed Res 1988; 15:203–209.

85.   Butler BD, Hills BA. The lung as a filter for microbubbles. J Appl Physiol 1979; 47:537–543.

86.   Butler BD, Hills BA. Transpulmonary passage of venous air emboli. J Appl Physiol 1985; 59:543–547.

87.   Katz J, Leiman BC, Butler BD. Effects of inhalation anaesthetics on filtration of venous gas emboli by the pulmonary vasculature. Br J Anaesth 1988; 61:200–205.

88.   Vik A, Brubakk AO, Hennessy TR, Jenssen BM, Ekker M, Slordahl SA. Venous air embolism in swine: Transport of gas bubbles through the pulmonary circulation. J Appl Physiol 1990; 69:237–244.

89.   Lynch PR, Brigham M, Tuma R, Wiedeman MP. Origin and time course of gas bubbles following rapid decompression in the hamster. Undersea Biomed Res 1985; 12:105–114.

90.   Marquez J, Sladen A, Gendell H, Boehnke M, Mendelow H. Paradoxical cerebral air embolism without an intracardiac septal defect. Case report. J Neurosurg 1981; 55:997–1000.

91.   Bedell EA, Berge KH, Losasso TJ. Paradoxic air embolism during venous air embolism: Transesophageal echocardiographic evidence of transpulmonary air passage [see comments]. Anesthesiology 1994; 80:947–950.

92.   Tommasino C, Rizzardi R, Beretta L, Venturino M, Piccoli S. Cerebral ischemia

after venous air embolism in the absence of intracardiac defects. J Neurosurg Anesthesiol 1996; 8:30–34.

93. Gottdiener JS, Papademetriou V, Notargiacomo A, Park WY, Cutler DJ. Incidence and cardiac effects of systemic venous air embolism. Echocardiographic evidence of arterial embolization via noncardiac shunt. Arch Intern Med 1988; 148:795–800.

94. Black M, Calvin J, Chan KL, Walley VM. Paradoxic air embolism in the absence of an intracardiac defect. Chest 1991; 99:754–755.

95. Pilmanis AA, Meissner FW, Olson RM. Left ventricular gas emboli in six cases of altitude-induced decompression sickness. Aviat Space Environ Med 1996; 67: 1092–1096.

96. Yahagi N, Furuya H. The effects of halothane and pentobarbital on the threshold of transpulmonary passage of venous air emboli in dogs. Anesthesiology 1987; 67: 905–909.

97. Butler BD, Leiman BC, Katz J. Arterial air embolism of venous origin in dogs: Effect of nitrous oxide in combination with halothane and pentobarbitone. Can J Anaesth 1987; 34:570–575.

98. Vik A, Jenssen BM, Brubakk AO. Effect of aminophylline on transpulmonary passage of venous air emboli in pigs. J Appl Physiol 1991; 71:1780–1786.

99. Butler BD, Conkin J, Luehr S. Pulmonary hemodynamics, extravascular lung water and residual gas bubbles following low dose venous gas embolism in dogs. Aviat Space Environ Med 1989; 60:1178–1182.

100. Butler BD, Luehr S, Katz J. Venous gas embolism: Time course of residual pulmonary intravascular bubbles. Undersea Biomed Res 1989; 16:21–29.

101. Mandelbaum I, King H. Pulmonary air embolism. Surg Forum 1963; 14:236–238.

102. Meltzer RS, Tickner EG, Popp RL. Why do the lungs clear ultrasonic contrast? Ultrasound Med Biol 1980; 6:263–269.

103. Verstappen FT, Bernards JA, Kreuzer F. Effects of pulmonary gas embolism on circulation and respiration in the dog. I. Effects on circulation. Pflugers Arch 1977; 368:89–96.

104. Verstappen FT, Bernards JA, Kreuzer F. Effects of pulmonary gas embolism on circulation and respiration in the dog. III. Excretion of venous gas bubbles by the lung. Pflugers Arch 1977; 370:67–70.

105. Sergysels R, Jasper N, Delaunois L, Chang HK, Martin RR. Effect of ventilation with different gas mixtures on experimental lung air embolism. Respir Physiol 1978; 34:329–343.

106. Hlastala MP, Robertson HT, Ross BK. Gas exchange abnormalities produced by venous gas emboli. Respir Physiol 1979; 36:1–17.

107. Ohkuda K, Nakahara K, Binder A, Staub NC. Venous air emboli in sheep: Reversible increase in lung microvascular permeability. J Appl Physiol 1981; 51:887–894.

108. Butler BD. Pulmonary effects of decompression stress in the rat. Undersea Biomed Res 1991; 18(suppl):74.

109. Spencer MP, Oyama Y. Pulmonary capacity for dissipation of venous gas emboli. Aerosp Med 1971; 42:822–827.

110. Levy SE, Stein M, Totten RS, Bruderman I, Wessler S, Robin ED. Ventilation-

perfusion abnormalities in experimental pulmonary embolism. J Clin Invest 1965; 44:1699–1707.

111. Soloff LA, Rodman T. Acute pulmonary embolism. 1. Review. Am Heart J 1967; 74:710–724.

112. Neuman TS, Spragg RG, Wagner PD, Moser KM. Cardiopulmonary consequences of decompression stress. Respir Physiol 1980; 41:143–153.

113. Perkett EA, Brigham KL, Meyrick B. Continuous air embolization into sheep causes sustained pulmonary hypertension and increased pulmonary vasoreactivity. Am J Pathol 1988; 132:444–454.

114. Perkett EA, Brigham KL, Meyrick B. Granulocyte depletion attenuates sustained pulmonary hypertension and increased pulmonary vasoreactivity caused by continuous air embolization in sheep. Am Rev Respir Dis 1990; 141:456–465.

115. Bernard GR, Korley V, Chee P, Swindell B, Ford-Hutchinson AW, Tagari P. Persistent generation of peptido leukotrienes in patients with the adult respiratory distress syndrome. Am Rev Respir Dis 1991; 144:263–267.

116. Bonsignore MR, Rice TR, Dodek PM, Staub NC. Thromboxane and prostacyclin in acute lung injury caused by venous air emboli in unanesthetized sheep. Microcirc Endothel Lymphat 1986; 3:187–212.

117. Fukushima M, Kobayashi T. Effects of thromboxane synthase inhibition on air emboli lung injury in sheep. J Appl Physiol 1986; 60:1828–1833.

118. Butler BD, Powell MR, Little T. Dose response levels of 11-dehydrothromboxane B2 (TxB2) and leukotriene E4 (LTE4) with venous air embolism (abstr). Undersea Biomed Res 1995; 22(suppl):40.

119. Butler BD, Little T. Effects of venous air embolism and decompression on thromboxane and leukotriene levels in dogs (abstr). Undersea Hyperb Med 1994; 21(suppl):21–22.

120. Butler BD, Little T, Doursout MF. Experimental decompression sickness: role of inflammatory mediators (abstr). FASEB J 1999; 13:A407.

121. Butler BD, Doursout MF, Little T. Plasma nitric oxide (NO) levels with venous air embolism and decompression (abstr). Undersea Hyperb Med 1999; 26(suppl): 58.

122. Butler BD, Little T, Smolensky M. Circadian rhythm influences on decompression outcome in rats. In: Tovitov Y, ed. Biological Clocks Mechanisms and Application. Amsterdam: Elsevier, 1998:541–544.

123. Adornato DC, Gildenberg PL, Ferrario CM, Smart J, Frost EA. Pathophysiology of intravenous air embolism in dogs. Anesthesiology 1978; 49:120–127.

124. Vik A, Jenssen BM, Eftedal O, Brubakk AO. Relationship between venous bubbles and hemodynamic responses after decompression in pigs. Undersea Hyperb Med 1993; 20:233–248.

125. Brubakk AO, Flook V, Vik A. Gas bubbles and the lungs. In: Lundgren C, Miller JN, eds. The Lung at Depth. New York: Marcel Dekker, 1999:237–294.

126. Holt EP, Jr., Webb WR, Cook WA, Unal MO. Air embolism. Hemodynamics and therapy. Ann Thorac Surg 1966; 2:551–560.

127. Geissler HJ, Allen SJ, Mehlhorn U, Davis KL, Morris WP, Butler BD. Effect of body repositioning after venous air embolism. An echocardiographic study. Anesthesiology 1997; 86:710–717.

128. Bove AA, Hallenbeck JM, Elliott DH. Circulatory responses to venous air embolism and decompression sickness in dogs. Undersea Biomed Res 1974; 1: 207–220.

129. Powell MR, Spencer MP, von Ramm O. Ultrasonic surveilance of decompression. In: Bennett B, Elliot DH, eds. The Physiology and Medicine of Diving. London: Bailliere Tindall, 1982:404–434.

130. Gronert GA, Messick JM, Jr., Cucchiara RF, Michenfelder JD. Paradoxical air embolism from a patent foramen ovale. Anesthesiology 1979; 50:548–549.

131. Butler BD, Hills BA. Effect of excessive oxygen upon the capability of the lungs to filter gas emboli. In: Bachrach AJ, Matzen MM, eds. Underwater Physiology VII. Bethesda, MD: Undersea Medical Society, 1981:95–102.

132. Brubakk AO, Peterson R, Grip A, et al. Gas bubbles in the circulation of divers after ascending excursions from 300 to 250 msw. J Appl Physiol 1986; 60:45–51.

133. Stone DA, Godard J, Corretti MC, et al. Patent foramen ovale: Association between the degree of shunt by contrast transesophageal echocardiography and the risk of future ischemic neurologic events. Am Heart J 1996; 131:158–161.

134. Di Tullio M, Sacco RL, Venketasubramanian N, Sherman D, Mohr JP, Homma S. Comparison of diagnostic techniques for the detection of a patent foramen ovale in stroke patients. Stroke 1993; 24:1020–1024.

135. Job FP, Ringelstein EB, Grafen Y, et al. Comparison of transcranial contrast Doppler sonography and transesophageal contrast echocardiography for the detection of patent foramen ovale in young stroke patients. Am J Cardiol 1994; 74:381–384.

136. Germonpre P, Dendale P, Unger P, Balestra C. Patent foramen ovale and decompression sickness in sports divers. J Appl Physiol 1998; 84:1622–1626.

137. Knauth M, Ries S, Pohimann S, et al. Cohort study of multiple brain lesions in sport divers: Role of a patent foramen ovale. Br Med J 1997; 314:701–705.

138. Clark JB, Hayes GB. Patent foramen ovale and type II altitude decompression sickness (abstr). Aviat Space Environ Med 1991; 62:445.

139. Gallagher KL, Hopkins EW, Clark JB, Hawley TA. U.S. Navy experience with type II decompression sickness and the association with patent foramen ovale (abstr). Aviat Space Environ Med 1996; 67:712.

140. Conkin J, Powell MR, Foster PP, Waligora JM. Information about venous gas emboli improves prediction of hypobaric decompression sickness. Aviat Space Environ Med 1998; 69:8–16.

141. Bendrick GA, Ainscough MJ, Pilmanis AA, Bisson RU. Prevalence of decompression sickness among U-2 pilots. Aviat Space Environ Med 1996; 67:199–206.

142. Weien RW, Baumgartner N. Altitude decompression sickness: Hyperbaric therapy results in 528 cases. Aviat Space Environ Med 1990; 61:833–836.

143. Garrett JL, Bradshaw P. The USAF chamber training flight profiles. In: Pilmanis AA, ed. Proceedings of the 1990 Hypobaric Decompression Sickness Workshop. AL-SR-1992-0005. Brooks Air Force Base, TX: Air Force Systems Command, 1992:346–359.

144. Robinson RR, Doursout MF, Chelly JE, Powell MR, Little TM, Butler BD. Cardiovascular deconditioning and venous air embolism in simulated microgravity in the rat. Aviat Space Environ Med 1996; 67:835–840.

145. Bason R, Yacavone D. Decompression sickness: U.S. Navy altitude chamber expe-

rience 1 October 1981 to 30 September 1988. Aviat Space Environ Med 1991; 62: 1180–1184.

146. Gray JS. The Effect of Exercise at Altitude on Aeroembolism in Cadets. Reports 156-1 and 169. Washington, DC: U.S. NRCC Comm Aviat Med, 1943.

147. Henry FM. The role of exercise in altitude pain. Am J Physiol 1946; 145:279–284.

148. Pilmanis AA, Olson RM, Fischer MD, Wiegman JF, Webb JT. Exercise-induced altitude decompression sickness. Aviat Space Environ Med 1999; 70:22–29.

149. Powell MR, Waligora JM, Norfleet W, Kumar KV. Project Argo—Gas Phase Formation in Simulated Microgravity. NASA technical memorandum 104762. Houston, TX: National Aeronautics and Space Administration, 1993.

150. Waligora JM, Pepper LJ. Physiological experience during Shuttle EVA. 25th International Conference on Environmental Systems. SAE Technical Paper Series 951592. Warrendale, PA: Society of Automotive Engineers, 1995:1–6.

151. Collins M. Carrying the Fire. New York: Ballantine, 1974:383.

152. Norfleet W. Analgesic use by astronauts during the peri-EVA period (abstr). Aviat Space Environ Med 1993; 64:423.

153. Boothby WM, Lovalace JR, Benson OO. High altitude and its effect on the human body. J Aerosp Soc Am 1940; 7:1.

154. Allen TH, Maio DA, Bancroft RW. Body fat, denitrogenation and decompression sickness in men exercising after abrupt exposure to altitude. Aerosp Med 1971; 42:518–524.

155. Kazakova RT, Poleschchuk IP. Checking effectiveness of prevention of high altitude decompression disorders by means of prebreathing in an oxygen environment (in Russian). Kosm Biol Med 1978; 2:28–31.

156. Waligora JM, Horrigan DJ, Jr., Conkin J. The effect of extended O_2 prebreathing on altitude decompression sickness and venous gas bubbles. Aviat Space Environ Med 1987; 58:A110–A112.

157. Webb JT, Fischer MD, Heaps CL, Pilmanis AA. Exercise-enhanced preoxygenation increases protection from decompression sickness. Aviat Space Environ Med 1996; 67:618–624.

158. Gernhardt ML, Conkin J, Foster PP, et al. Design of a 2-hour prebreathe protocol for space walks from the International Space Station (abstr). Aviat Space Environ Med 2000; 71:277.

159. Powell MR, Waligora JM, Kumar KV, Loftin KC. Hypobaric decompression in simulated null gravity: a model using chair-rest adynamia (abstr). Undersea Hyperb Med 1995; 22(suppl):67.

160. Conkin J, Edwards BF, Waligora JM, Stanford J, Jr, Gilbert JH, III, Horrigan D Jr. Updating Empirical Models that Predict the Incidence of Altitude Decompression Sickness and Venous Gas Emboli for Shuttle and Space Station Extravehicular Operations. NASA Technical Memorandum 100456 Update. Houston, TX: National Aeronautic and Space Administration, 1990.

161. Waligora JM, Powell MR, Norfleet W. The abaroferic hypothesis: A mechanism for reduction in risk of decompression sickness in microgravity (abstr). Aviat Space Environ Med 1993; 64:421.

162. Vann RD, Gerth WA, Leatherman NE. Exercise and decompression sickness. In:

Vann RD, ed. The Physiological Basis of Decompression. Bethesda, MD: Undersea and Hyperbaric Medical Society, 1989: 119–137.

163. Waligora JM, Conkin J, Foster PP, et al. Operational implementation of a 2-hour prebreathe protocol for International Space Station (abstr). Aviat Space Environ Med 2000; 71:278.

164. Butler BD, Vann RD, Nishi RY, et al. Human trials of a 2-hour prebreathe protocol for extravehicular activity (abstr). Aviat Space Environ Med 2000; 71:278.

165. United States Navy. U.S. Navy Diving Manual. NAVSEA 0994-LP-9010. Washington, DC: Naval Sea Systems Command, 1993.

166. Butler FK, Jr., Knafelc ME. Screening for oxygen intolerance in U.S. Navy divers. Undersea Biomed Res 1986; 13:91–98.

167. Clark JM. Oxygen toxicity. In: Bennett PB, Elliot DH, eds. The Physiology and Medicine of Diving. London: Saunders, 1993:121–169.

168. Parker JC, Taylor AE. Oxygen toxicity. In: Lundgren C, Miller JN, eds. The Lung at Depth. New York: Marcel Dekker, 1999:165–209.

169. Flynn ET, Bayne CG. Diving Medical Officer Student Guide. Course A-6A-0010. Washington, DC: U.S. Government Printing Office, 1977:321–326.

170. Ornhagen H, Hamilton RW. Oxygen Enriched Air—''Nitrox''—in Surface Oriented Diving. Stockholm: National Defence Research Establishment, 1989.

171. Kumar VK, Billica RD, Waligora JM. Utility of Doppler-detectable microbubbles in the diagnosis and treatment of decompression sickness. Aviat Space Environ Med 1997; 68:151–158.

172. Kannan N, Raychaudhuri A, Pilmanis AA. A loglogistic model for altitude decompression sickness. Aviat Space Environ Med 1998; 69:965–970.

173. Ryles MT, Pilmanis AA. The initial signs and symptoms of altitude decompression sickness. Aviat Space Environ Med 1996; 67:983–989.

174. Heimbach RD, Sheffield PJ. Decompression sickness and pulmonary overpressure in accidents. In: DeHart RL, ed. Fundamentals of Aerospace Medicine. Baltimore: Williams & Wilkins, 1996:131–161.

175. Kimbrell PN. Treatment of altitude decompression sickness. In: Moon RE, Sheffield PJ, eds. Treatment of Decompression Illness. Kensington, MD: Undersea and Hyperbaric Medical Society, 1996:43–51.

176. Krause KM, Pilmanis AA. The effectiveness of ground level oxygen treatment for altitude decompression sickness in human research subjects. Aviat Space Environ Med 2000; 71:115–118.

177. Rudge FW. The role of ground level oxygen in the treatment of altitude chamber decompression sickness. Aviat Space Environ Med 1992; 63:1102–1105.

178. Demboski JT, Pilmanis AA. Effectiveness of ground level oxygen as therapy for pain-only altitude decompression sickness (abstr). Aviat Space Environ Med 1994; 65:454.

179. Sukoff MH, Ragatz RE. Hyperbaric oxygenation for the treatment of acute cerebral edema. Neurosurgery 1982; 10:29–38.

180. Miller JD, Ledingham IM, Jennett WB. Effects of hyperbaric oxygen on intracranial pressure and cerebral blood flow in experimental cerebral oedema. J Neurol Neurosurg Psychiat 1970; 33:745–755.

181. Zamboni WA, Roth AC, Russell RC, Kucan J. The effect of hyperbaric oxygen

treatment on the microcirculation of ischemic skeletal muscle. Undersea Biomed Res 1990; 17(suppl):26.

182. Zamboni WA, Roth AC, Russell RC, Nemiroff PM, Casas L, Smoot EC. The effect of acute hyperbaric oxygen therapy on axial pattern skin flap survival when administered during and after total ischemia. J Reconstr Microsurg 1989; 5:343–347; discussion 349–350.

183. Workman RD. Treatment of bends with oxygen at high pressure. Aerosp Med 1968; 39:1076–1083.

184. Leitch DR, Hallenbeck JM. Oxygen in the treatment of spinal cord decompression sickness. Undersea Biomed Res 1985; 12:269–289.

185. Sykes JJ, Hallenbeck JM, Leitch DR. Spinal cord decompression sickness: A comparison of recompression therapies in an animal model. Aviat Space Environ Med 1986; 57:561–568.

186. United States Navy. U.S. Navy Diving Manual. NAVSEA 0994-LP-9010. Washington, DC: Naval Sea Systems Command, 1993:45.

187. Drummond JC, Moore SS. The influence of dextrose administration on neurologic outcome after temporary spinal cord ischemia in the rabbit. Anesthesiology 1989; 70:64–70.

188. Moon RE, Sheffield PJ. Consensus statement. In: Moon RE, Sheffield PJ, eds. Treatment of Decompression Illness. Kensington, MD: Undersea and Hyperbaric Medical Society, 1996:417–426.

189. Cogar WB. Intravenous lidocaine as adjunctive therapy in the treatment of decompression illness. Ann Emerg Med 1997; 29:284–286.

190. Drewry A, Gorman DF. Lidocaine as an adjunct to hyperbaric therapy in decompression illness: A case report. Undersea Biomed Res 1992; 19:187–190.

191. Broome JR, Dick EJ, Jr. Neurological decompression illness in swine. Aviat Space Environ Med 1996; 67:207–213.

192. Kizer KW. Corticosteroids in treatment of serious decompression sickness. Ann Emerg Med 1981; 10:485–488.

193. Francis TJR, Dutka AJ, Clark JB. An evaluation of dexamethasone in the treatment of acute experimental spinal decompression sickness. In: Bove AA, Bachrach AJ, Greenbaum LJ, eds. Proceedings of 9th International Symposium on Underwater and Hyperbaric Physiology. Bethesda, MD: Undersea and Hyperbaric Medical Society, 1987:999–1013.

194. Lynch PR, Krasner LJ, Vinciquerra T, Shaffer TH. Effects of intravenous perfluorocarbon and oxygen breathing on acute decompression sickness in the hamster. Undersea Biomed Res 1989; 16:275–281.

195. Spiess BD, McCarthy RJ, Tuman KJ, Woronowicz AW, Tool KA, Ivankovich AD. Treatment of decompression sickness with a perfluorocarbon emulsion (FC-43). Undersea Biomed Res 1988; 15:31–37.

196. Catron PW, Flynn ET, Jr. Adjuvant drug therapy for decompression sickness: A review. Undersea Biomed Res 1982; 9:161–174.

197. Philip RB, Bennett PB, Andersen JC, et al. Effects of aspirin and dipyridamole on platelet function, hematology, and blood chemistry of saturation divers. Undersea Biomed Res 1979; 6:127–146.

198. Powell MR, Norfleet WT, Kumar KV, Butler BD. Patent foramen ovale and hypobaric decompression. Aviat Space Environ Med 1995; 66:273–275.
199. Rudge FW, Shafer MR. The effect of delay on treatment outcome in altitude-induced decompression sickness. Aviat Space Environ Med 1991; 62:687–690.
200. National Aeronautics and Space Administration. SSP Flight Data File. JSC 48092. Houston, TX: National Aeronautics and Space Administration, 1997:2–13.
201. Butler BD, Kurusz M. Embolic events. In: Gravlee GL, Davis RF, Kurusz M, Utley JR, eds. Cardiopulmonary Bypass: Principles and Practice. Philadelphia: Lippincott, Williams & Wilkins, 2000:320–341.
202. Norfleet W. Options and plans for treatment of decompression sickness aboard Space Station Freedom. In: Pilmanis AA, ed. Proceedings of the 1990 hypobaric decompression sickness workshop. AL-SR-1992-005. Brooks Air Force Base, TX: Air Force Systems Command, 1992: 315–326.
203. Hills BA. Decompression Sickness. Vol. 1. Chichester: Wiley, 1977.

14

Overall Summary

G. KIM PRISK and JOHN B. WEST

University of California, San Diego
La Jolla, California

MANUEL PAIVA

Université Libre de Bruxelles
Brussels, Belgium

I. Lessons Learned

The chapters in this book may be loosely grouped into four major categories, with some chapters crossing the boundaries of these groupings, and some fitting less well. In this chapter, we attempt to draw together the lessons that are spread through the various chapters based on these groupings.

A. Historical Perspective

The first two chapters, by John West and David Glaister, provide the historical perspective. While the chapter on the high-G environment is not strictly historical, almost all of the major work in this field was performed quite some time ago. As a result, these two chapters provide the reader with some basis for the state of the art prior to the studies performed in microgravity. In addition, the studies performed at 1 G and at high G provided the basis for many of the predictions of what would be expected in microgravity (μG).

The historical introduction (Chapter 1) points out that as far back as 1880, Orth predicted low pulmonary perfusion in the apices of the human upright lung

"

(15). Remarkably, approximately 120 years later, there is still debate on how important gravity is in the distribution of pulmonary blood flow in some animals (compare Chapters 6 and 7), although it is clear that gravity plays a major role in humans. However, it was not until the 1950s that radioisotope imaging techniques allowed quantitative evaluation of this.

The Russian visionary Tsiolkovsky began to imagine the effects of the absence of gravity late in the nineteenth century, well before any human had direct experience of μG (11). However, measurements in microgravity are relatively recent, with most of the work being done on various Spacelab missions. The U.S. and Soviet/Russian programs have diverged to some extent. The U.S. program has concentrated on extensive sophisticated measurements exploiting the technical resources of Spacelab. The Soviet/Russian emphasis has been on long-term microgravity, first in the Salyut and then in the Mir space station, which has allowed relatively few measurements on board but has the advantage of looking at much more long-term exposure to μG.

In comparing the chapter on the studies performed in high G (Chapter 2) with chapters later in the book, it is striking that much of what has so far been done in μG is based on very indirect techniques. The innovative studies with inhalations or infusions of radioactive tracers provide considerably more direct information on the distribution of ventilation and blood flow in the lung than do the indirect techniques that have been used in μG to date. Unfortunately, it is likely to be some considerable time before such techniques can be applied in sustained μG.

The rapid development of the airplane as a consequence of World War I led to the study of humans in an environment of altered G. For obvious reasons, most of these studies concentrated on the effects of increased G levels. These studies suggest that the lung essentially behaves as an isotropic elastic material, and that as a consequence, gravity has a major effect on both ventilation and perfusion. However, extrapolation from the high-G studies to μG is fraught with uncertainties. For example, extrapolation of the data shown in Figure 10, Chapter 2, would lead one to believe that pulmonary arterial pressure would be quite high in μG. While no one has actually measured pulmonary artery pressure in μG, direct measurements of central venous pressure (see Chapter 10) suggest that, if anything, pulmonary vascular pressures are actually lower in μG than on the ground. Similarly, based on data collected in high G, it was suggested that an increase in pulmonary capillary pressure might lead to the formation of pulmonary edema in μG (17). However, the evidence clearly suggests that there is no formation of pulmonary edema in normal individuals in μG (see Chapters 6 and 11). Throughout the book, many of the more interesting results recently obtained in sustained μG are those where such extrapolation has proven to be incorrect.

B. Gas Behavior in the Lung

Perhaps the most extensive knowledge base on the lung in μG is that surrounding the behavior of gases in the lung. This is probably a consequence of the experimental ease with which inhaled gases can be manipulated, in contrast to, say, studies of perfusion. With gases, sophisticated experiments are, relatively speaking, easy to perform and rapid measurements of exhaled gas composition and volume are readily available.

Static lung volumes are modestly changed by the removal of gravity (Chapter 3). From the data collected in μG, it is clear that gravity increases functional residual capacity (FRC) in the upright human. While the reduction in FRC seen in μG is explainable through the removal of the weight of the abdominal contents (1), there was, prior to experiments in μG, disagreement by some even in which direction the lung volume would change (29). That something as seemingly straightforward as the determinants of FRC remained in question highlights just how little was really known about the effects of gravity on the lung. It took measurements in sustained μG to provide the definitive answer (6). Unloading of the abdomen also increases abdominal compliance and markedly increases the contribution of the abdominal compartment to tidal volume (16). These changes are qualitatively similar to those induced by a change from the upright to the supine posture. Weightlessness, however, also produces changes that do not occur during the transition from upright to supine, and therefore could not be predicted from ground-based experiments. For example, weightlessness alters the pattern of breathing by decreasing tidal volume and increasing breathing frequency, and it decreases residual volume (RV), likely by eliminating the gravitational stretch on the upper regions of the upright lung.

The studies performed in μG have been instrumental in determining how much of the ventilatory inhomogeneity normally seen in the lung is intrinsic, and how much is a result of the effects of gravity. Prior to these studies, it was generally thought that the lung was largely homogeneous and much of the unevenness of ventilation resulted directly from the effects of gravity. Somewhat surprisingly, at least during tidal breathing, ventilation remains quite inhomogeneous (19,21) and is little different in μG than on the ground. Studies in μG are also showing that many of the markers of inhomogeneity in the lung are related to the effects of airways closure (in other words, to events in the lung at very low lung volume) (5).

A lesson emerging from the studies of ventilation distribution is that there are large differences in the gravitational effects on large-scale (interregional) differences in ventilatory inhomogeneity, and small-scale (intraregional) differences (Chapter 4). The studies using gases of different diffusivity (He and SF_6) allow a differential probe of these effects. The results seen at a large scale are similar between these two gases as would be expected since convective inhomogeneity dominates. However, examination of the exhaled gas concentration profiles thought to be dominated by effects at the acinar level show considerable differences between 1 G and μG. In μG, the acinar-level gas mixing of He and SF_6 is markedly different from that in 1 G (23); however, the cause remains unclear. Similar changes are not seen in the transient μG of parabolic flight (13). That similar differences to those observed between 1 G and μG are seen in lung transplant subjects undergoing acute rejection episodes (24) may provide a potential clue in the interpretation of these results, but at present the exact answer remains to be determined.

While not conclusive, the data also suggest that the scale of the residual nongravitational convective inhomogeneity is relatively small (of the order of that occurring between a few acini) (19). These results help in the interpretation of traditional tests of ventilation distribution performed on the ground, such as the multiple-breath nitrogen washout. This test has shown only small changes in μG, when performed at identical lung volumes (19), demonstrating that inhomogeneities in lung mechanical properties, rather than gravitational distortion, are dominant in determining the inhomogeneities of ventilation. This gave support to the proposed interpretation of new indices that separate conductive from acinar-zone contributions to ventilation inhomogeneities. This new interpretation is finding application in the study of different pathologies (25,26).

The deposition of inhaled particulate matter (PM) in the lung is coming to be recognized as a significant matter of public health. Particulate matter exposure in the form of environmental pollution is linked to both short-term (days) and long-term (years) increases in mortality. Interestingly, the risk increases as the particle size gets smaller, with substantial improvements in the correlation between pollutant load and risk as one moves from considering total suspended particles to PM below 10 μm in size to PM below 2.5 μm in size. This suggests that the particles with the higher risk of damage are those able to be transported into the periphery of the lung. Because the primary mechanism of deposition of particles in the 0.5- to 2.0-μm size range is gravitational sedimentation, the studies discussed in Chapter 5 are of particular note. In altered gravity, the effects of sedimentation on the deposition and dispersion of aerosol can be directly manipulated. The total deposition studies showed the expected increase in deposition as a result of gravitational sedimentation. However, for particles near 1 μm in size, total deposition was unexpectedly high in microgravity, likely because of previously ignored transport mechanisms (additional mixing due to the nonreversibility of the flow) (4). While the overall change in total deposition caused by this process in 1 G might be small, the effect might be disproportionately large if the deposition occurs in the sensitive alveolar region of the lung. These results may provide a link to understanding the unexpectedly strong effect of environmental pollution on the generation of pulmonary diseases.

The studies using inhaled boluses of aerosol, while to date only performed in the transient μG available in parabolic flight, provide a good adjunct to these studies of ventilatory inhomogeneity using gases. In the absence of gravitational sedimentation, small particles (~1 μm) behave, to a large extent, as a nondiffusing gas. They are thus well suited to studying convective gas transport in the lung. The results (Chapter 5) confirm the observations made with gases that there is considerable convective inhomogeneity in the ventilation of the lung, even in the absence of gravity. Thus, the picture emerging from the studies in μG is one of a lung that is far less intrinsically homogeneous than originally believed to be the case.

C. Blood Flow, Hydrostatic Effects, and Distributions of Pressures

There are two chapters devoted to the distribution of pulmonary perfusion in μG in this book. These two chapters present the extremes that currently exist in the field. On one hand, the studies in humans in sustained microgravity described in Chapter 6, while providing important and useful information on the continuing presence of inhomogeneity of pulmonary perfusion in μG, utilize very indirect and inferential techniques (22). At the other extreme, Chapter 7 describes the highly detailed and direct measurements possible using animal models and invasive techniques utilizing microspheres (7). What is clearly lacking is some middle ground where imaging techniques can be used in humans to provide direct spatial information. Such studies using radioactive tracers have provided most of our early understanding of the effects of gravity on the distribution of pulmonary perfusion (Chapters 1 and 2).

The direct studies of animal lungs using destructive techniques (Chapter 7) show that the single most important factor determining perfusion heterogeneity in the normal lung is the scale-dependent heterogeneity of flow that arises as a consequence of the progressive branching of the pulmonary vasculature. It should, however, be pointed out that the technique used in these studies (injection of microspheres at a normal lung volume followed by inflation and drying of the lungs at TLC) likely serves to minimize the signal from any gravitational gradient in blood flow that might be present. Recent high-resolution studies of perfusion heterogeneity in animal lungs during transient microgravity have demonstrated that overall perfusion heterogeneity is nearly unchanged by the transition from 1 G to microgravity, although significant macroscopic alterations can be demonstrated at higher G forces (8). While both temporal changes and exercise-associated changes in pulmonary perfusion are demonstrable, high-flow regions remain high, and low-flow regions remain low in many different experimental interventions. This leads the authors of Chapter 7 to the conclusion that pulmonary blood flow heterogeneity in animals is determined primarily by anatomical features of the pulmonary vasculature, with a smaller influence from both gravity and vasomotion. However, in humans, it is clearly the case that a great deal of the heterogeneity in pulmonary perfusion is gravitational in origin (Chapters 2 and 6).

While there are marked differences between these two chapters, the message that emerges is remarkably similar. Namely, that pulmonary perfusion, while markedly affected by gravity, has significant and important nongravitational components. In both cases, the inference that can be drawn from the data is that there is considerable small-scale inhomogeneity of pulmonary perfusion intrinsic to the lung, while gravity imposes a vertical gradient in perfusion over and above this.

There are two chapters that deal with the distribution of pressures in the body or the consequences of changes in those distributions. Prior to any measurements being made, there were predictions that the capillary pressure within the lung would rise in μG, and as a consequence, the lung might become edematous (17). However, the direct measurements of central venous pressure (CVP) made by Buckey et al. (2), and by others show convincingly that contrary to expectations, CVP (referenced to the outside of the body) falls in μG (Chapter 10). These observations came as a surprise to many. How CVP falls in the face of a decrease in lung volume (and hence an increase in intrathoracic pressure) (see Chapter 3) yet with increases in cardiac output and stroke volume (see Chapter 6), has yet to be adequately explained. The following questions remain unanswered: Does intrathoracic pressure decrease in weightlessness compared with the standing or supine position on Earth? Does the removal of tissue compression result in a decrease in extracardiac pressure and is there a way to model this effect? Does mean circulatory filling pressure decrease in weightlessness? The answers to these questions are important not just for understanding the effects of weightlessness, but may also have relevance for the care of obese patients, where the weight of tissues can have important effects.

The lack of understanding of the findings from μG suggests that insufficient attention had been paid to understanding the distribution of pressures within the thorax and how those pressures interact to affect cardiovascular dynamics. The data show that more attention to reference pressures is required as it is clear that vessel and cardiac transmural pressures are important, as opposed to vascular pressures per se.

The chapter on the fluid balance of the pulmonary interstitium (Chapter 11) provides insight into the gravitational effects on the lymphatic system. Notably, however, none of these studies were performed in μG, and so the inferences drawn must await direct experimental confirmation. Despite this, the suggestion that the safety factor in preventing pulmonary edema may be reduced by the removal of gravity because of impaired lymphatic flow is intriguing (14). While microgravity causes a marked reduction in mechanical stresses in the weight-bearing organs (e.g., muscles and bones) there is an increase in mechanical stresses in the pulmonary interstitial matrix (which is by nature quite delicate) due to the sustained condition of vascular engorgement causing an increase in microvascular filtration. Thus, in contrast with other more robust tissue structures, remodeling of interstitial pulmonary matrix might be expected as a result of long-term microgravity exposure.

D. Gas Exchange

Regional differences in pulmonary gas exchange occur in the normal upright human lung because of the regional differences of ventilation and blood flow. These have some clinical implications. For example, the predilection of adult pulmonary tuberculosis for the apex of the lung can be explained by the relatively high P_{O_2} there. The effects of normal gravity on overall gas exchange are minor because the normal distribution of ventilation–perfusion ratios is relatively narrow. Increased acceleration causes major impairment of overall pulmonary gas exchange with most of the blood flow going to the base of the lung, and most of the inspired ventilation going to the apex. In μG, overall pulmonary gas exchange at rest is largely unaltered compared with the situation in 1 G (Chapter 8). There are modest changes in the pattern of ventilation that result in slight changes in total ventilation appropriate for the changes seen in physiological dead space. There is some evidence of slight changes in alveolar gas concentrations but these are, for the most part, small.

Perhaps the most intriguing result is the indication that despite improvements in the inhomogeneity of both ventilation and perfusion (9,22), there is little change in the range of ventilation–perfusion ratios seen in the lung in μG (18). Closer examination of the data suggests that the cause of this seemingly paradoxical result is that gravity imposes gradients in both ventilation and perfusion such that areas of high ventilation are in the same location as areas of high perfusion. The effect is to narrow the spread of ventilation–perfusion ratios seen. However, the data from cardiogenic oscillations in μG suggest that in the absence of gravity such spatial matching is not present, and thus the distribution of ventilation–perfusion ratios is wider than would have otherwise been the case (12). These studies are hampered by the very indirect nature of the measurements. Even such seemingly routine measurements as arterial blood gases have not been performed in space flight, and techniques such as the multiple inert gas elimination technique (27) are too time consuming and complex for use at this time.

The changes are somewhat more pronounced when the system is stressed by exercise (Chapter 9). Our knowledge of the normal effects of gravity on the ground has not been such that we could accurately predict how the exercising human body would perform during exercise without gravity. Once again, the data are somewhat limited by the noninvasive nature of the studies performed in μG; however, the message appears reasonably clear. Increasing the G level results in a marked decrease in efficiency of the lungs for gas exchange (see also Chapter 2). However, there are offsetting factors at work in μG. While the O_2 needs seem to be at a minimum for performing work in μG, there are confounding losses (principally in the circulatory component). Thus, the overall system seems close to being optimized for performing work in the 1 G environment, with little, if any, overall gain in efficiency in μG, in contrast to what would be expected by extrapolations from experiments performed in hypergravity.

Chapter 12 deals with changes in the control of ventilation. The results show a halving of the ventilatory response to hypoxia in μG, likely related to the increase in systemic blood pressure at the carotid bodies in the neck (20). It has previously been shown that decreasing blood pressure at the carotid bodies increases the hypoxic response and vice versa. Thus, in μG, the hypoxic response is approximately the same as that measured supine in 1 G, when there is no hydrostatic pressure gradient between the heart and the neck. These results serve to highlight an important point, namely that as we change from the upright to supine and back every day, we change the response of many of our body systems. How much effect the reduction in the hypoxic response in the supine position has on subjects with sleep-disordered breathing is at present unknown, but it is clear that such effects are often not considered.

Finally, the problems of gas exchange in the hypobaric environment of a space suit present a number of special considerations (Chapter 13). In this circumstance, the issue with gas exchange is not related to O_2 and CO_2, but to the transport of inert (dissolved) gas (mostly N_2) to and from the various tissues of the body, and the resultant possibility of decompression sickness in the low operating pressure of a space suit. Once again, the area is constrained by a lack of direct measurement, especially in μG, where blood flow to various tissues might be markedly different from that in 1 G. This would have the effect of altering tissue washout times, perhaps raising the risk of decompression sickness. However, at this time, we are faced with the paradox that while studies performed in hypobaric chambers on the ground suggest that decompression sickness would be expected to be quite a common occurrence during extravehicular activity (EVA) (28), there are no solid reports of this ever occurring in space flight. Is this simply under-reporting, or is there something fundamentally different about hypobaric exposure in μG?

The lungs play an important role in EVA operations. In addition to their obvious role in oxygen and carbon dioxide exchange, they are critically important filters of gas bubbles present in venous blood as a result of the decompression stress involved in EVA operations. Therefore, the lungs are, in essence, another spacecraft subsystem, one that protects important biological tissues from embolization by bubbles generated as a consequence of the design of other subsystems. However, the efficacy of this filter is not always perfect, and heavy embolization of the lungs is not without consequence.

II. Future Research in Microgravity

With only a few exceptions, virtually all of the data actually collected in µG come from either transient µG (that obtained during parabolic flight) or relatively short exposures (up to ~2 weeks in shuttle flights). Thus, the effects of the long-term removal of gravity on the lung remain largely unknown.

It is now clear that unlike some organ systems such as bone and muscle, there is little to suggest that the lung undergoes any changes that are likely to be detrimental to the health of the space traveler while in µG or upon eventual return to 1 G. However, this does not necessarily mean that studies of pulmonary function are unnecessary.

The closed environment of a spacecraft brings with it the possibility of a contaminated atmosphere both in the form of gases evolved from materials within the vehicle and even exhaled CO_2. It is possible that a buildup of toxic contaminants might affect pulmonary function during long-term space flight because of the closed environment, and it is particularly important to monitor pulmonary function in a new facility such as the International Space Station (ISS). Outgassing of materials may occur. Indeed, during a recent visit (in May/June 1999) to the ISS, the crew of the mission STS-96 reported various symptoms that seemed to be related to poor ventilation and possibly outgassing of materials, primarily in the Zarya portion of the ISS. While these symptoms (headaches, burning or itching eyes, flushed face, and nausea and vomiting) were transient in nature, they emphasize the potentially hostile nature of a closed environment. The lung can be particularly sensitive to such insults because it presents a large surface area to the environment.

Long-term space flight also presents a situation in which aerosol deposition may be an important health consideration. In the spacecraft environment, the potential for significant airborne particle loads is high since the environment is closed and no sedimentation occurs. Some of these particles are probably contaminated and include hair, food, paint chips, synthetic fibers, etc. While routine spacecraft particle loads may be low for much of the time a crew spends in the vehicle, there is no doubt that unexpected events cause very large particle loads indeed. For example, the fire aboard the Mir space station in 1997 produced so much particulate matter that the air was described as so dense that "Korzun . . . cannot see his own hands" (3). This took in excess of 6 hr to clear and since no measurements were done on air quality, it is unknown how much PM may have remained in the Mir atmosphere once the air was considered "clear." Certainly the crew continued to wear masks for some nights to sleep following the fire. Similarly, microgravity provides for potentially high particle concentrations in the airways, since particles that normally sediment will not be removed from the airways, leaving them potentially available for transport to the alveolar regions. Because of the increased alveolar deposition resulting from the enhanced diffusion, there may well be long-term health implications to astronauts exposed to these potentially high concentrations of particulates.

Another important issue in which inhaled PM may be important is on the surface of Mars. In the $^1/_3$ G environment of the Martian surface, sedimentation will be much less than on Earth, which may increase deposition in the alveolar region. In addition, the Mars surface dust is likely to stick to boots, etc., and be tracked into habitats. This dust is thought to have free oxygen radicals on its surface because of the UV environment on the Martian surface and this may prove toxic when brought into contact with the lining of the human respiratory system (10).

The other area of concern that may well arise in the age of the International Space Station is the risk of barotrauma, or decompression sickness associated with extravehicular activity (spacewalks). The assembly of the ISS will require more EVAs than have been performed in the history of U.S. space flight. Thus, the means of assessing the health status of a crew member exposed to some dysbaric insult becomes important in order to adequately assess the need for a possible return to earth earlier than originally planned. Of course, in the process of providing for such contingencies, this will also provide for routine health monitoring.

However, perhaps the most important aspect of continuing pulmonary research in space is simply that it allows the investigation of the ever-present (on Earth) gravitational effects on the lung in a manner impossible in the terrestrial laboratory. As the chapters in this book show, such investigations have provided considerable insight into normal pulmonary physiology on the ground. As a part of the Human Research Facility planned for the International Space Station, the National Aeronautics and Space Administration (NASA) and the European Space Agency (ESA) are jointly developing the Pulmonary Function System. This facility will allow sophisticated tests of pulmonary function to be performed aboard the ISS providing for a continuation of the fruitful and productive research of the effects of gravity on the lung.

References

1. Agostoni E, Mead J. Statics of the respiratory system. In: Fenn WO, Rahn H, eds. Handbook of Physiology, Section 3: Respiration. Vol. 1. Washington, DC: American Physiological Society, 1964:387–428.
2. Buckey JC Jr, Gaffney FA, Lane LD, Levine BD, Watenpaugh DE, Wright SJ, Yancy CW Jr, Meyer DM, Blomqvist CG. Central venous pressure in space. J Appl Physiol 1996; 81:19–25.
3. Burrough B. Dragonfly, NASA and the Crisis Aboard Mir. New York: Harper-Collins, 1998.

4. Darquenne C, Paiva M, West JB, Prisk GK. Effect of microgravity and hypergravity on deposition of 0.5- to 3-μm-diameter aerosol in the human lung. J Appl Physiol 1997; 83:2029–2036.

5. Dutrieue B, Lauzon A-M, Verbanck S, Elliott AR, West JB, Paiva M, Prisk GK. Helium and sulfur hexafluoride bolus washin in short-term microgravity. J Appl Physiol 1999; 86:1594–1602.

6. Elliott AR, Prisk GK, Guy HJB, West JB. Lung volumes during sustained microgravity on spacelab SLS-1. J Appl Physiol 1994; 77:2005–2014.

7. Glenny RW, Lamm WJE, Albert RK, Robertson HT. Gravity is a minor determinant of pulmonary blood flow distribution. J Appl Physiol 1991; 72:620–629.

8. Glenny RW, Lamm WJE, Bernard SL, Pool SL, Chornuk M, Wagner WW Jr, Hlastala MP, Robertson HT. Redristribution of pulmonary perfusion during weightlessness and increased gravity. J Appl Physiol 2000; 89:1239–1248.

9. Guy HJB, Prisk GK, Elliott AR, Deutschman RAI, West JB, Inhomogeneity of pulmonary ventilation during sustained microgravity as determined by single-breath washouts. J Appl Physiol 1994; 76:1719–1729.

10. Klein HP. The Viking biological investigations: Review and status. Orig Life 1978; 9:157–160.

11. Kosmodemiansky A. Konstantin Tsiolkovsky, 1857–1935. Moscow: General Editorial Board for Foreign Publications. Moscow: Nauka Publishers, 1985.

12. Lauzon A-M, Elliott AR, Paiva M, West JB, Prisk GK. Cardiogenic oscillation phase relationships during single-breath tests performed in microgravity. J Appl Physiol 1998; 84:661–668.

13. Lauzon A-M, Prisk GK, Elliott AR, Verbanck S, Paiva M, West JB. Paradoxical helium and sulfur hexafluoride single-breath washouts in short-term vs. sustained microgravity. J Appl Physiol 1997; 82:859–865.

14. Miserocchi G, Negrini D. Pleural lymphatics as regulators of pleural fluid dynamics. News Physiol Sci 1991; 6:153–158.

15. Orth J. Atiologisches und anatomisches uber Lungenschwindsucht. Berlin: August Hirschwald, 1887.

16. Paiva M, Estenne M, Engel LA. Lung volumes, chest wall configuration, and pattern of breathing in microgravity. J Appl Physiol 1989; 67:1542–1550.

17. Permutt S. Pulmonary circulation and the distribution of blood and gas in the lungs. In: Physiology in the Space Environment. Washington, DC: NAS NRC 1485B, 1967: 38–56.

18. Prisk GK, Elliott AR, Guy HJB, Kosonen JM, West JB. Pulmonary gas exchange and its determinants during sustained microgravity on Spacelabs SLS-1 and SLS-2. J Appl Physiol 1995; 79:1290–1298.

19. Prisk GK, Elliott AR, Guy HJB, Verbanck S, Paiva M, West JB. Multiple-breath washin of helium and sulfur hexafluoride in sustained microgravity. J Appl Physiol 1998; 84:244–252.

20. Prisk GK, Elliott AR, West JB. Sustained microgravity reduces the human ventilatory response to hypoxia but not hypercapnia. J Appl Physiol 2000; 88:1421–1430.

21. Prisk GK, Guy HJB, Elliott AR, Paiva M, West JB. Ventilatory inhomogeneity determined from multiple-breath washouts during sustained microgravity on Spacelab SLS-1. J Appl Physiol 1995; 78:597–607.

22. Prisk GK, Guy HJB, Elliott AR, West JB. Inhomogeneity of pulmonary perfusion during sustained microgravity on SLS-1. J Appl Physiol 1994; 76:1730–1738.

23. Prisk GK, Lauzon A-M, Verbanck S, Elliott AR, Guy HJB, Paiva M, West JB. Anomalous behavior of helium and sulfur hexafluoride during single-breath tests in sustained microgravity. J Appl Physiol 1996; 80:1126–1132.

24. Van Muylem A, DeVuyst P, Yernault J-C, Paiva M. Inert gas single-breath washout and structural alteration of respiratory bronchioles. Am Rev Respir Dis 1992; 146: 1167–1172.

25. Verbanck S, Schuermans D, Noppen M, Van Muylem A, Paiva M, Vincken W. Evidence of acinar airway improvement in asthma. Am J Respir Crit Care Med 1999; 159:1545–1550.

26. Verbanck S, Schuermans D, Van Muylem A, Melot C, Noppen M, Vincken W, Paiva M. Conductive and acinar lung-zone contributions to ventilation inhomogeneity in COPD. Am J Respir Crit Care Med 1998; 157:1573–1577.

27. Wagner PD, Saltzman HA, West JB. Measurement of continuous distributions of ventilation-perfusion ratios: theory. J Appl Physiol 1974; 36:588–599.

28. Waligora JM, Horrigan D Jr, Conkin J, Hadley AT III. Verification of an altitude decompression sickness prevention protocol for shuttle operations utilizing a 10.2-psi pressure stage. In: NASA Technical Memorandum 58259. Houston, TX: NASA, 1984:1–44.

29. Wilson TA, Liu S. Effect of acceleration on the chest wall. J Appl Physiol 1994; 76:1242–1246.

SUBJECT INDEX

E

F